From Maverick to Mainstream
A History of No-Till Farming

Founding *No-Till Farmer* Editor Frank Lessiter's
personal stories of the people, innovations
and influences that saved the soil —
and the family farm

By Frank Lessiter

With contributions from 54 writers and editors who served
as part of the *No-Till Farmer* team during the past 5 decades

Book design and layout by Jayne Laste and Jeff Lazewski

*This book is dedicated to the no-till pioneers that sat
alone in the back of coffee shops to avoid being laughed
at by the die-hard moldboard plowing fanatics who
ridiculed no-tillage's 'ugly' fields and 'lazy' farmers.*

Lessiter Media, Brookfield, Wis.

From Maverick to Mainstream
A History of No-Till Farming

By Frank Lessiter

International Standard Book Number: ISBN-13: 978-0-944079-21-8

Published by
Lessiter Media
P.O. Box 624, Brookfield, WI 53008-0624
For additional copies or information on other books or publications
offered by No-Till Farmer and Lessiter Media, write to the above address.
Telephone (866) 838-8455 (U.S. and Canada only) or (262) 432-0388.
Fax: (262) 786-5564
Email: info@no-tillfarmer.com
Website: www.No-TillFarmer.com

Manufactured in the United States of America

"Even after 50 years,
no-till is still a journey — not a destination —
since it's being perfected every year."
— *Glover Triplett, retired Ohio State University*
and Mississippi State University, agronomist, Starkville, Miss.

Part of No-Till
From the Very Beginning

Famous French author Victor Hugo once wrote, "All the forces in the world are not so powerful as an idea whose time has come."

In the mid 20th century, especially for agriculture, no-tillage was such an idea. Now, well into the 21st century, it remains the best way ever found to conserve our greatest, God-given, natural resource — the life sustaining topsoils of the world.

Huge Changes Ahead

I was there at the beginning of modern no-tillage crop production. In the spring of 1962, just south of Herndon, Ky., my father, Harry M. Young, Jr., was the first American farmer to try this revolutionary idea. I was 11 years old then, and didn't recognize the huge changes that were about to take place in farming.

My father, however, did have the vision. A mental light had turned on in 1961 when he saw a small University of Illinois test plot in Dixon Springs, Ill. George McKibben, a man well ahead of his time, was the one behind that pivotal plot.

Since my father had been a University of Kentucky statewide extension specialist during the 1940s and 50s, he understood the importance of test plots. He decided to try McKibben's idea on his own farm.

That winter, he adapted an old mule-

WHERE NO-TILL GOT ITS START. John Young (left) and his son, Alex, represent the fifth and sixth generations to run the Young's 4,200-acre farm near Herndon, Ky. A historical marker sits at the site where John's father, Harry Young Jr., planted the first commercial no-till plot on 7/10ths of an acre in 1962.

drawn two-row planter and mounted it underneath a Farmall 140 tractor. (See page 51.) That became the first no-till planter ever used by a farmer.

No-tillage worked as promised, and continues to work. (As I write this, we are in the process of harvesting no-tilled corn for the 56th consecutive year on our family farm.)

During the next several years, along with University of Kentucky agronomist Shirley Phillips, my father put his visionary insight to work, both in writing his books and speaking at many meetings across the grain belt. He was always willing to speak to a group of farmers, even when many in the crowd were skeptical, or even cynical.

In those meetings, he loved to get tough questions about no-tillage. He had already asked, and answered, those same tough questions while making no-tillage work in his own farming operation.

Another Pioneer

Another man who saw the vision in those early days was Frank Lessiter, the founder of Lessiter Media, the long-time editor of *No-Till Farmer* and the author of this book. My father, who died in 1988, always spoke highly of Frank

Back in the early days, Frank flew down to our western Kentucky farm on more than one occasion to visit with my father. He was a keen enthusiast for no-till from the beginning, and remains so today.

No-Till Farmer has served as the unofficial extension service for the world. Everything from innovative machinery and weed control, to variety selection and cover crops

WORTH THE RISK. After no-tilling corn in 1962, Harry Young (shown here in both photos) worked with University of Kentucky agronomist Shirley Phillips to perfect his no-till system. At the time, Young didn't anticipate how no-till would grow. "His training as an ag economist, though, told him it would save labor, fuel, machinery and soil, and it was well worth trying," says his son, John Young.

have been studied, scrutinized and presented by the editors to farmers and agribusinessmen. They have been the preeminent leaders in spreading the word about major agricultural improvements that have come along through the years.

Frank knows that even the best idea remains a dream until it is put into actual practice. He has been both coach and cheerleader for countless farmers and innovators who want to find a better way of growing crops.

This book, which is all about the history of no-tillage, will be a welcome addition to the library of anyone who likes to eat. Since our food comes from the land, we should all care about conserving our soil and water. After all, the next generation will want to eat as well.

No-tillage is the best tool we have as we search for truly sustainable agriculture throughout the world. That should be reason enough to see just what has

been done during the past 56 years in the life of modern no-tillage.

Look to the Future

One final word is in order for the readers of this historical book. Don't ever assume that what has already been done is all that needs to be done. While many no-till pioneers are featured here, I believe they would all acknowledge that their discoveries and research were only scratching the surface in the search for the perfect way to feed a hungry world.

Thanks to the efforts of Frank Lessiter and many other contributors, this book will show you that no-tillage is much different today than it was in 1962. I firmly believe it will be much different in another 50 years than it is today.

We should never give up the effort to improve on what we have. At the same time, we should never forget where we have been.

— *John S. Young, Herndon, Ky.*

"For nearly 5 decades, No-Till Farmer has served as the unofficial extension service for the world. Everything from innovative machinery and weed control, to variety selection and cover crops have been studied, scrutinized and presented by the editors to farmers and agribusinessmen."
— *John S. Young*

My 'Ringside Seat' to 5 Decades
of No-Till's Evolution

Back home on the family farm in Michigan one afternoon in late July of 1972, I'm sure Dad questioned whether I'd lost my sanity in giving up a secure job at a livestock magazine to move to Wisconsin (from Illinois) to help launch *No-Till Farmer.*

Even after I explained what no-till was and its potential, Dad still wasn't convinced I was making a sensible choice for my wife and three kids (ages 8, 6 and 3 — with a fourth that would arrive soon after) by rolling the dice on a magazine startup and on an unproven method of farming, no less. I handed him a few brochures on no-till and the contact herbicide paraquat and drove home to Illinois to get ready for the family's move.

A Scary Decision

After being the first family member to leave our farm in six generations, this was another new road for me. A little scary too, but I guess at age 33 I was excited to get in on the ground floor with a new idea that someday might lead to major changes in food production around the world.

Two weeks later, I was back at the farm again and Dad greeted me at the door. "Frank, where can I get a few gallons of paraquat?" he asked. After reading the leave-behind brochures, he'd figured out that the contact herbicide would be a good way to control some problem weeds around the barns and in a few fence rows.

Our 6-generation family dairy farm was 35 miles north of Detroit. We'd tried wheel-track planting (see page

NO-TILL PARTNERS. Pam handled the business side of the operation virtually all by herself during our first 30 years. Without her faith, advice, wisdom and work ethic, we couldn't have built a meaningful, independent family-owned business while helping farmers navigate a new way of farming.

46) in the early 1950s when corn was planted directly into plowed ground without any additional tillage. Even as a kid I still recall how those bumpy trips across the fields were tough on the back and kidneys — difficult enough that we abandoned the wheel-track system after a couple of years.

But it was a start in reducing unnecessary and costly tillage trips.

As Dad slowed down in the 1980s, he rented part of our ground to a neighbor who no-tilled corn that first season. And together we'd see

how that no-tilled corn flourished during a very dry summer.

Same Job for 45 Years

Writing, editing and producing *No-Till Farmer* has been part of my job description since 1972 when I signed on as its first editor at Reiman Publications (the original publisher). And I've continued to hold the same editor title all these years after my wife, Pam, and I bought *No-Till Farmer* and launched the family publishing business in 1981.

My partner in this risky endeavor was Pam.

She had the faith and belief in my dream that we could build a meaningful, independent business while helping farmers navigate a new way of farming. She handled the business side of the operation virtually all by herself for our first 30 years. Without her contributions, advice and wisdom and both finance and cash-flow expertise, we may not have made it.

Timing is everything. Soon after going out on our own in the early '80s, we saw huge crop surpluses, free-falling corn and soybean prices due to the government's push for fence row-to-fence row planting and interest rates on farm loans that reached as high as 23%. Farm loan payments were tough to make due to huge decreases in farm income and skyrocketing interest rates. Unfortunately, many growers ended up with forced exits from farming.

Yet many innovative farmers embraced no-till as a way to trim cropping costs, and that's how a lot of them got through those dark times. As a result, the no-till acres more than doubled during the 1980s.

Still Learning About No-Till

I've had the pleasure of reporting on many changes in no-till since those early days in 1972, not to mention strip-till, zone-till, ridge-till, interseeding, relay intercropping, vertical tillage, cover crops and other important aspects of the zero-tillage movement. My best days in this business have always been getting out and visiting with no-tillers across the country.

While there are "desk jockeys" that move up in the trade media business, I've held the same "editor" title for nearly 50 years. But I wouldn't trade my no-till experiences

Frank Lessiter

NO-TILL INNOVATORS. One of my great days was walking down no-till's memory lane in 2011 while sitting in the living room of the late Eugene Keeton (left) along with Howard Martin. Both Kentucky farmers-turned-inventors were key to the "ground zero" efforts of no-till farming that would catch on nationally and globally. Farming only a few miles apart, the two formed a close relationship as they developed planter accessories that made no-till work better. For more on their stories, see pages 37 and 255.

for anything. It's kept me close to my farm roots as I tried to help growers find new innovative ways to stay on the land, protect the environment, be sustainable, earn good profits and help feed the world.

No-Till Had Arrived

After more than 55 years in the farm publishing business, some comments tend to stick with you. One came from Charlotte Sine in the mid-1980s. The long-time editor of *Farm Chemicals* magazine said, "Frank, your time has come with no-till."

She was congratulating us on what we were doing in those days with a very different type of publication that was *No-Till Farmer.* But more importantly, I appreciated her recognition of the impact no-till was starting to have on agriculture around the world.

Thank you for reading this collection of my life's work. We felt it was important to tell these stories of no-tillage as agriculture undergoes changes, and share for all generations how positive, lasting change can

happen through the personal perseverance, ingenuity and grit that these farmers showed. They sought to find ways to make something different work, instead of finding reasons it could not.

I can't think of anything else a writer and editor could ever hope for than helping to make an impact on their audience's lives.

Thanks for inviting me and *No-Till Farmer* into your homes since 1972, and for spending your time with these pages about a reduced tillage practice that is still revolutionizing farming around the world.

—Frank Lessiter, Editor, No-Till Farmer, August 2018

P.S. Please note that like *No-Till Farmer*, this is a "different" kind of book. We purposely included many different points of entry for you. You'll find lots of short digestible items, over 650 photos, illustrations and charts, and more than 200 items from our archives that showed the no-till thinking at that point in time. ✻

Let Me Tell You a No-Till Story...

Throughout this book, you'll find numerous short items like the one shown below. These "from the archives" pieces cite the month and year they appeared in *No-Till Farmer* issues.

NO-TILL LOOSE ENDS...

HURRICANE AGNES STRENGTHENED NO-TILL'S BENEFITS

November 1972

In late June of 1972, conventionally-tilled corn in Pennsylvania felt the full force of Hurricane Agnes, which dropped 18 inches of rain in just a few hours in some areas of Pennsylvania. Many corn plants were washed away, and the eventual harvesting of what was left was tough in fields that suffered from serious erosion.

By comparison, most of the area's no-till corn came through in excellent shape.

These two photos demonstrate the enormous erosion control differences between no-tillage and conventional tillage at the start of Hurricane Agnes. The photos were shot out of the opposite front seat windows of a car parked along a country lane near York, Pa.

George Dell grew the no-till corn (right) while the conventionally-tilled corn across the lane (left) was in a neighbor's field. Dell credited the mulch cover and a good root system with keeping his 270 acres of no-tilled corn from washing away.

"There was a lot of erosion eventually in my neighbor's conventionally-tilled corn," says Dell. "He will likely have to bulldoze some of those gullies before he can harvest. Where water ran across his fields, some gullies were 3-4 feet deep and 4 feet wide. Plus, a lot of corn simply washed away." ✳

MAJOR EROSION DIFFERENCES WITH TILLAGE SYSTEMS. These photos of conventional-tilled (left) and no-tilled (right) corn fields were taken in 1972 on opposite sides of a Pennsylvania country lane just 5 minutes after Hurricane Agnes started to dump as much as 18 inches of rain on area fields.

HOLDING DOWN THE SAME JOB FOR NEARLY 5 DECADES

Over the years, a number of journalists have held editorial titles on *No-Till Farmer*. While I'm the only one that's written articles for every issue since the first edition in November of 1972, other writers made major contributions to the publication over the years.

Most of the editors listed here have earned promotions, switched jobs and changed careers as they moved on to new and better things, and I'm proud of how we helped train many of them in an actionable way of service journalism that helped them pursue their goals.

Since I've been holding down the same job since 1972, I can relate to many farmers who stay in the same job for most of their lives. I wouldn't trade my staying put for the world, as I've stayed close to the land and helped farmers find new ways to stay on their land, protect the environment, be sustainable, earn good profits and feed the world.

Having been in at no-till's virtual beginning and having seen no-till grow and break through to mainstream agriculture gave me a long-term perspective most other editors never realize. While there are other ag editors that have worked more years than I have, I'm not aware of anyone else working on the same "beat"— no-till in my case — for as many years as I have. And I'm proud of that accomplishment.

Where No-Till Got Its Start

A week after I came aboard as editor of *No-Till Farmer* in 1972, an early August trip served as my introduction to no-till, Others on the Reiman Publications staff had been following no-till for a few years. In fact, Gaylin Morgan had made the first no-till trip in June of 1967 and had 11 trips to the Hopkinsville, Ky., area to his credit by the summer of 1972. All told, eight members of the staff had made 20 trips to the southwest Kentucky area by 1972.

At right is our roster of editors and editorial support staffers since the beginning in 1972. Some were out on no-till farms reporting on what was and was not working, while others were doing the blocking and tackling in a support role for me and the other editors. But all contributed to spreading the no-till word. ✳

VISITING NO-TILL'S BIRTHPLACES. In August of 1972, five members of the original *No-Till Farmer* staff (at that time owned by Reiman Publications) flew to Kentucky and Illinois for a firsthand look at no-till. During a 3-day trip that served as my first look at no-till, we visited with no-till pioneers George McKibben, Harry Young and other no-tillers. Shown here at the Dixon Springs Agricultural Center in southern Illinois are Editor Frank Lessiter, Managing Editor Andy Holum, Advertising Manager Chuck Hagen, Editorial Director Gaylin Morgan and Publisher Roy Reiman.

> "Having been in at the beginning and having seen no-till grow to where it is today offers a **unique long-term perspective most other editors don't have...**"

NO-TILLING THROUGH THE YEARS

Since 1972, the *No-Till Farmer* editors have turned out thousands of articles and other educational items through print, digital, audio, video and in-person events to help growers farm much more effectively and profitably.

Over the years, *No-Till Farmer* has served as the authority for in-depth knowledge on no-till crop production. Not only has the community given no-tillers the tools to farm more efficiently, but it's also helped growers learn how to improve soil quality, curb erosion, become more sustainable and protect the environment for future generations.

Like farming, folks in the information business must be ready — and eager — to change with the dynamics around us. Long-time readers of *No-Till Farmer* have witnessed a number of magazine and newsletter developments and innovations, some of which are shown above.

Today's subscribers gain access to numerous digital offerings and recordings, eight monthly newsletters and four *No-Till Farmer* magazines that deliver timely, unbiased no-till information 12 times a year to paid subscribers.

Niche-Specific Markets

In recent years, we've expanded our publishing formula by launching niche-specific projects such as *Strip-Till Farmer* and *Dryland No-Tiller*, a digital newsletter serving the specific needs of growers in the Great Plains area that stretches from Texas to the provinces of western Canada.

Along the way, we've developed no-till and strip-till websites, an extensive archive of previously published no-till articles, daily email newsletters, special reports, white papers, eguides, conference proceedings, podcasts, webinars, videos and books to help no-tillers stay abreast of the latest developments.

In addition, the National No-Tillage Conference has been a truly unique mid-winter learning experience for growers since the first one launched in 1993. Following a similar format, we expanded our one-on-one learning opportunities with the introduction of the mid-summer National Strip-Tillage Conference in 2014.

No-Tillers Taught Us Well

Learning so much from no-tillers has led to other niche-market publishing opportunities at Lessiter Media. One example is our *Precision Farming Dealer* magazine for the dealership side of the ag business that came about because of what we learned about precision agriculture from no-tillers. This print publication, numerous digital offerings and a precision ag meeting have become an important part of our business that serves the folks from whom you buy farm equipment.

We're mighty proud of the educational role we've played over the years in expanding the no-till movement here in the states and around the world. ✳

Be Careful ... Until You
Know the Language

> "It's easy to get the impression that **the minute a farmer stops plowing** that he or she is practicing conservation tillage..." — *Rolf Derpsch*

Over the years, there's been plenty of confusion as to what various "tillage" terms actually mean, with different interpretations found not only in North America, but also around the world. So here's our shot at explaining some of the confusion behind a few terms that have played a key role in the growth of no-till.

February 1990

Major Tillage Systems Defined

Here's how the major tillage practices were described in the early 1990s.

No-Tillage: Except for nutrient injection, the soil is left undisturbed from harvest to planting. Planting is completed in a narrow seedbed or slot created by coulters, row cleaners, disc openers, in-row chisels or other row attachments. Weed control is accomplished primarily with herbicides. Cultivation may be used for emergency weed control.

Minimum-Tillage: This includes mulch-till or ridge-tillage, which leaves at least 30% of the soil surface covered by residue after planting. Where soil erosion due to wind is the primary concern, there must be at least 1,000 pounds of residue left on the surface during the critical erosion period.

In mulch-tillage, chisel plows, field cultivators, discs, sweeps or blades are used. Weed control is accomplished with herbicides and/or cultivation.

Ridge-Tillage: Like no-till, ridge-till leaves the soil undisturbed from harvest to planting except for nutrient injection. Planting is completed in a seedbed prepared on ridges built with sweeps, disc openers, coulters or row cleaners. Residue is left on the surface between the ridges. Weed control is accomplished with herbicides and/or cultivation. Ridges are rebuilt with one, two or even three cultivation trips.

Conventional-Tillage: This includes tillage systems that do not leave at least 30% of the soil surface covered by residue after planting.

This includes moldboard plowing, which often leaves less than 15% of residue after planting, along with other types of tillage that result in anywhere from 5%-30% residue after planting.

How Do You Spell No-Till?

Folks in the farm equipment industry today spell no-till with two "Ls," along with a hyphen between the two words. But that wasn't always the case.

An early-day Allis-Chalmers trademark referred to the concept as No-Til, with both words capitalized, only one "L" and a hyphen. Chevron Chemical Co. (the original U.S. firm who marketed paraquat that later became part of Syngenta Crop Protection and ChemChina in a 2017 merger), had trademarked "Ortho-Til," also with only one "L" in the '60s. At that time, paraquat was manufactured in Great Britain by ICI.

In the early years of *No-Till Farmer*, we removed the hyphen for a year or two. This was done

battles over their trademarks with other equipment manufacturers.

We were delighted to add the hyphen back in no-till in March of 1976, since it always belonged there anyway.

There's also another hyphen story worth sharing. In the late 1960s, I spent 3 1/2 years editing the Massey-Ferguson *Farm Profit* magazine that was mailed six times-a-year to over 400,000 U.S. farmers. At that time, there was a hyphen between Massey and Ferguson, but today's brand no longer includes a hyphen. And the folks at AGCO, who now own the Massey Ferguson brand, have no idea when the hyphen got dropped.

at the insistence of the Allis-Chalmers attorneys who argued we were infringing on their trademark by using the hyphen. Rather than argue with them and gear up for an expensive courtroom battle and because Reiman Publications, the parent publishing company (before the Lessiter family purchase), was producing Allis-Chalmers promotional materials, the hyphen was dropped in April of 1974.

Several years later, the Allis-Chalmers attorneys asked us to replace the hyphen in no-till. They explained that returning to the original style would help them in legal

November 1972

No-Till Terms Very Confusing

When a no-tiller describes his operation as *"rotated double-cropping with barley-beans, following corn in a winter rye cover,"* a tillage-happy grower might respond: *"How are you going to harvest a mess like that?"*

Some terms used by no-till farmers are confusing, even to other no-tillers. It's especially confusing when a term doesn't mean what it says, such as the word *no-till.*

No-till or no-tillage does not mean a total absence of tillage. Rather, it refers to a variety of crop production systems employing a reduced or limited amount of tillage. How "limited" depends on who you are and where you farm.

For the sake of communication, *no-tillage* and *no-till*, as used in the name and content of this publication, refers to any number of crop production systems that eliminate unnecessary tillage operations.

No Consistency

The terminology for several no-tilled crops and rotations has also created a number of confusing terms.

Double-cropped soybeans, for example, does not mean two crops of soybeans, but refers to soybeans planted after the harvest of wheat, barley, peas, forages or another crop during the same growing season.

NO-TILL-AGE®

February 1990

What Counts in Making Tillage Decisions?

Practice	% Farmers
Soil erosion	42%
Weed control	28%
Machinery/fuel costs	26%
Labor requirements	25%
Government rules, programs	8%

— Garst Seed Co.

The two crops in a double-cropping sequence are often linked together. For example: *barley-beans*, means soybeans grown after barley, while *hay-corn* means planting corn following the first cutting of hay.

The date of planting and crop maturity also often becomes confusing. When a no-till farmer refers to *full-season corn*, he generally means it was planted early. *Late corn* doesn't mean a full-season maturity rating, but rather a late planting date.

Early beans refers to the planting date and not to early crop maturity. Double-cropped soybeans are often referred to as *late beans*, which again has nothing to do with the bean variety or maturity.

In addition, a number of other terms have been used to describe no-till systems, some of which are common to various areas of North America. These include terms such as *zero tillage*, *zone tillage*, *sod planting*, *direct planting*, *direct seeding* and probably some other terms we don't remember.

Soybean Tillage Systems Compared

Tillage System	Bean residue after planting
No-till	36%
Blade plow, plant	15%
Field cultivate, plant	20%
Blade plow, till-plant	15%
Till-plant	8%
Disc, plant	6%
Disc, field cultivate, plant	16%
Chisel plow, disc, plant	13%

— University of Nebraska

What's Conventional Tillage?

While conventional tillage can mean many things, we use it to describe the generally accepted method of seedbed preparation for any given crop or area, whether that method is a plow-disc-harrow system, a subsoil-disc-bed procedure or whatever.

Defined Differently Around the World

Rolf Derpsch maintains a major problem associated with conservation tillage is the fact that there are different definitions of what conservation, tillage and no-till mean in different areas of the world.

November 2002

"Given the impact of globalization, it would be extremely helpful if the tillage terminology could be globalized," says the no-till consultant and agronomist from Asuncion, Paraguay. "It's very difficult to achieve a mutual understanding when everybody speaks a different language."

Derpsch says it's easy for uninformed folks to get the impression, especially in Europe, that the minute a farmer stops moldboard plowing that he or she is practicing conservation tillage.

Tillage Definitions Not Being Met

The definition of conservation tillage, according to Soil Conservation Service officials, states that at least 30% of the soil must be covered by residue after planting.

September 1987

Conservation tillage usually includes no-till, which leaves much more residue cover than ridge-tillage, chisel plow, strip-till, disc tillage or other reduced tillage systems.

Only No-Till Qualifies

However, a study by University of Nebraska ag engineers indicates that among eight tillage and planting systems used with soybeans, only no-till left more than 30% residue cover. Compared with both narrow- and wide spaced rows, residue coverage was measured after fall tillage, after spring tillage and again at planting time.

With soybeans, the field cultivate/plant, blade plow/till-plant, blade plow/plant and disc/field cultivate/plant systems retained 15%-20% residue. The till-plant (ridge-till) system and systems that included a discing or chisel plowing operation retained less than 15% residue.

Will Conservation Tillage be Redefined?

Based on these results, only two systems are capable of achieving acceptable soil erosion control with soybeans. It means you can't have an acceptable residue cover if the commonly used soil and residue disturbing implements associated with conservation tillage are used with soybeans.

If you are serious about keeping your soil at home and using it and valuable moisture to boost your yields, no-till is probably the best way to go. ❁

BE SURE TO ASK THE
RIGHT QUESTIONS

For many years, Jim Johnson and his brother Larry were die-hard ridge tillers with a sizeable acreage of corn and soybeans. In their area near Chaska, Minn., they were among the only ridge-tillers, as most growers were using minimum tillage or conventional tillage.

As the two brothers got older and busier with off-farm activities, the day came when they decided to retire and rent their land.

As Jim explained, they didn't demand that other farmers ridge-till the ground, but they definitely wanted to rent to someone who took conservation tillage seriously. After interviewing several prospects, they settled on a grower who indicated he would only disc ahead of planting.

When the planting season rolled around, the brothers were shocked at how little crop residue was left and how black the dirt looked.

I remember Jim telling me the hard lesson they learned: While using minimum tillage with discing had seemed OK, they neglected to ask how many times the farmer planned to disc the fields. In some fields, three discings had been done, which led to very little crop residue left on the soil surface to fight soil erosion.

In fact, the final result that year wasn't any better than if the fields had been moldboard plowed...A reduced tillage lesson learned the hard way. ❁

Notching the No-Till Milestones
Through the Years

"Technology, research and sheer determination **transformed no-till into a profitable practice** that now spans more than 100 million U.S. acres..."

As the success of no-till farming closes in on 60 years, it's important to remember how its acceptance and growth came about.

Had it not been for the vision and courage of numerous researchers, engineers, scientists and pioneering farmers, no-till might only represent a small segment of agriculture, rather than encompassing more than 100 million acres as it does today.

Concept Started 75 Years Ago

The ideas behind no-till started in 1943 when Edward Faulkner's book, *Plowman's Folly,* dared to challenge the plow's supremacy in agricultural production. Faulkner asserted that

moldboard plows were the most undesirable tool for crop production.

The no-till revolution among farmers began in Herndon, Ky., in 1962, when Harry Young Jr. modified a former mule-driven planter to establish the first commercial no-till production plot in the U.S. A historical marker is erected in this field, which has been continuously no-tilled since 1962.

From there, curious researchers and farmers started experimenting with no-till crops as they sought to answer challenging questions about planting, fertilization and weed control.

Early farm meetings produced many concerns about weeds in no-till systems. Herbicides like atrazine, 2,4-D, dicamba (then known as Banvel) and paraquat were introduced that offered broad-spectrum protection. In addition, several chemical companies continued to roll out new products to address weed resistance concerns.

As successes began to mount, manufacturers like Allis-Chalmers, John Deere, The Tye Co. and Great Plains Mfg. began churning out planters and drills equipped with coulters that did a better job of handling residue and closing the seed slot.

Essential planter attachments like row cleaners helped clear residue from the path of double-disc openers, while closing wheels and the

Keeton seed firmer helped no-tillers get better results in challenging soil conditions.

Genetically modified crops like Roundup Ready corn, soybeans and cotton, stacked insecticide traits, fungicides and inoculants eventually made it easier for no-tillers to mange their operations.

Rapid Expansion in the 1990s

During the 1990s, the advent of precision technology — including auto-steer, RTK guidance, auto-row shutoff and spray-boom height control — helped no-tillers cut fertilizer and seed costs, improve crop stands and reduce their environmental footprint.

Cover crops have also returned to popularity as new research recognizes their potential in no-till systems to reduce compaction, trim erosion problems and even fix soil-nitrogen levels.

In recent years, soil amendments like gypsum show promise in helping rehabilitate weathered, low-nutrient soils, while urease inhibitors help growers improve nitrogen-use efficiency.

What follows is a timeline of key developments in the growth of no-till. This timeline was compiled from nearly five decades of *No-Till Farmer* publications and a variety of other sources. This timeline was displayed in a first-ever museum-like experience on wooden display boards for attendees at the 15th annual National No-Tillage Conference in 2017 in St. Louis. ✾

A LOOK BACK IN TIME. There was great interest in the no-till timeline that was displayed in a museum-like experience for farmers and suppliers attending the 25th annual National No-Tillage Conference in 2017 in St. Louis.

NO-TILL FARMING: A TIMELINE OF EVENTS

1830

■ About 250 to 300 labor hours were required to produce 5 acres of wheat with a walking plow, harrow, hand broadcast of seed, sickle and flail.

1837

■ John Deere and Leonard Andrus begin manufacturing steel plows.

1862-75

■ Change from hand work to horses characterizes the first American agricultural revolution.

1935

■ The National Soil Dynamics Laboratory is established at Auburn, Ala., as part of the USDA's Agricultural Research Service. Conceived by Mark I. Nichols and John W. Randolph, it was a full-size laboratory for the controlled study of the relationship between tillage tools, traction equipment and soil types.

1943

■ University of Illinois agronomist George McKibben establishes plots in Dixon Springs, Ill., to begin no-till research.

1962

■ Harry Young no-tills corn on 0.7 of 1 acre in Herndon, Ky. — the first commercial onfarm no-till acres in the U.S.

1961

1967

■ No-till soybeans are double-cropped after wheat.

■ Chevron Chemical Co. releases paraquat for use as a burndown herbicide and to eliminate weedy plants in wheat fields.

1969

■ Ohio State researchers Glover Triplett (above) and David Van Doren find mulch-covered no-till soils reduce surface evaporation, maintain soil moisture and create a favorable environment for root development.

1966 ■ Allis-Chalmers introduces the first fluted-coulter no-till planter.

■ *Plowman's Folly*, authored by Edward H. Faulkner, is released. "No one has ever advanced a scientific reason for plowing," the book asserts.

✺ NO-TILL FARMING: A TIMELINE OF EVENTS

1972

■ Frank Lessiter edits and produces the first issue of *No-Till Farmer* magazine while at Reiman Publications.

■ Planting of U.S. no-till crops is estimated at 3.3 million acres, according to the first survey of its kind published by *No-Till Farmer*.

1973

■ Kentucky no-till pioneers Harry Young Jr. and S.H. Phillips honored at the first national no-till conference held in Honolulu, Hawaii.

■ Ohio State University researcher Glover Triplett suggests a classification system for tillage, after recognizing differences in crop response to no-till on different soils.

■ The 224-page book titled *No-Tillage Farming*, authored by Shirley Phillips and Harry Young, is published by *No-Till Farmer*.

■ *No-Till Farmer* Editor Frank Lessiter testifies before the U.S. House Agriculture Committee on the importance of no-till to agriculture.

1974

■ Ben Overstreet, winner of the 1973 Georgia corn-growing contest, shares how he no-tilled his way to a yield of 204 bushels per acre.

■ John Deere introduces its MaxEmerge plateless planter with finger-pickup seed metering, Tru-Vee openers and angled closing wheels.

■ 5.4 million acres of no-till crops are planted in the U.S.

■ A study indicates over 10.5 million acres of Ohio land is suitable for no-till corn production.

■ Farmers, researchers evaluate a new John Deere rig with power-driven circular blades as a possible no-till corn or soybean planter.

1975

■ The third annual *No-Till Farmer* tillage survey shows an 18% increase in no-till acres.

■ USDA researchers estimate 95% of U.S. farmland will include reduced tillage by 2010.

■ Introduction of Basagran herbicide (BASF) expands no-till soybean potential.

1976

■ EPA approves Roundup for use in no-till and conventional-tillage systems.

■ Roy Applequist's starts Great Plains Mfg. The company's first product, a 30-foot folding grain drill, gets its first field test under dryland Kansas conditions.

■ Melroe's no-till drill features a triple-disc system that makes direct seeding possible.

1977

■ No-till U.S. wheat production jumps to 614,255 acres, compared to 222,833 acres in 1973.

❄ NO-TILL FARMING: A TIMELINE OF EVENTS

1979

■ *No-Till Farmer* publishes an 8-year tillage comparison showing no-till acres increased from 3.3 million acres in 1972 to 7.1 million acres in 1978.

1980

■ U.S. no-till acres are estimated at 16 million.

■ 1980-85 — Many selective and non-selective herbicides compatible with no-till are produced.

1981

■ South American no-till acres reach the same level in 3 years that it took the U.S. 30 years to reach.

■ Great Plains Mfg. builds its first no-till drill prototypes.

■ Frank and Pam Lessiter acquire *No-Till Farmer* and launch new publishing company in their name.

1982

■ For the first time, U.S. farmers use reduced-tillage practices on more than 100 million acres of land, compared to only 29.6 million acres in 1972.

1986

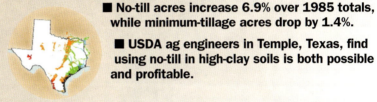

■ 20 years of research shows well-drained soils are the key to producing competitive no-till yields.

■ No-till acres increase 6.9% over 1985 totals, while minimum-tillage acres drop by 1.4%.

■ USDA ag engineers in Temple, Texas, find using no-till in high-clay soils is both possible and profitable.

1987

■ A *No-Till Farmer* "Tillage Practices Survey" shows an estimated 2.1 million acres were farmed with ridge-till in 1987 — a 4.2% nationwide increase over the previous year. This is basically no-till planted on a ridge that was formed the previous year with 1-2 cultivations.

1988

■ No-till pioneer Harry Young Jr. passes away.

■ University of Illinois agronomist and no-till pioneer George McKibben dies.

1989

■ The 18th annual *No-Till Farmer* "Tillage Practices Survey" shows 17,694 acres of no-tilled cotton.

■ About 1.9 million acres of no-till, double-cropped soybeans are planted — up from 1.8 million acres in 1988.

1985

■ Introduction of the John Deere 750 no-till drill helped producers switch to no-till.

■ Burndown herbicide options experience market expansion.

■ The Food Security Act of 1985 requires erosion to be reduced on the nation's most erodible lands.

■ Conservation tillage is the method of choice for a vast majority of grain farmers.

❋ NO-TILL FARMING: A TIMELINE OF EVENTS

1991

■ No-till acreage grows 24%, to 17.6 million acres, over 2 years.

■ CrustBuster 3800 no-till drill lets farmers control seed depth independently of rig's coulters.

■ USDA concerns about residue lead to discussion of possible erosion laws and regulations.

1992

■ No-till reaches 22.5 million acres.

■ 2,4-D is approved for no-till soybeans.

1993

■ *No-Till Farmer* hosts its first-ever National No-Tillage Conference in Indianapolis with 814 attendees and 45 speakers slated over 3 days.

■ Illinois farmer Roger Denhart plans to no-till 50,000 acres of soybeans near Odessa in the southern Ukraine in the spring of 1993.

■ 32 million acres of no-till cropland is reported, a 60% increase from 1991.

1994

■ Russia reports success in using no-till to bring back 11 million acres of farmland destroyed by Chernobyl nuclear power-plant explosion.

1996

■ U.S. no-till acres reach 42.5 million.

■ Monsanto introduces Roundup Ready soybeans.

1997

■ Monsanto introduces biotech corn, cotton and canola.

■ Studies dealing with the impact of row spacing on cereals and pulse crops over 30 years demonstrate benefits for no-till in western Canada.

1998

■ Southern Illinois University agronomist George Kapusta says Roundup Ready is the "new kid on the block" for no-till soybeans.

■ Research indicates ammonium sulfate can be an effective source of nitrogen, particularly with no-till corn.

1995

■ Purdue University researchers say adoption of precision technology is the next big step toward efficient no-till farming.

1999

■ No-tillers gear up to sell carbon credits after studies show tillage releases 500 to 8,000 pounds of carbon dioxide per acre into the atmosphere each year.

2000

■ U.S. no-till acres are estimated at 50 million.

■ Four Washington no-tillers form an LLC to share a no-till single drill and alleviate the financial burden of each buying bigger equipment.

■ The U.S. EPA places new restrictions on raising *Bt* Corn, requiring a 20% refuge area with non-*Bt* hybrids.

2005

■ The 13th National No-Tillage Conference showcases interest in annual ryegrass as a no-till cover crop.

■ Environmental, consumer groups question safety of Roundup Ready crops, saying they create "super weeds," among other problems.

NO-TILL FARMING: A TIMELINE OF EVENTS

2006

■ Research results from the U.S. Department of Energy's Oak Ridge National Laboratory describe the great potential for biomass — specifically gathered from no-till farms — to fuel cars and trucks in the U.S.

■ Ohio State University soil scientist Warren Dick says gypsum applications on no-till fields can decrease surface crusting and improve aeration and water infiltration, aiding crop emergence.

2007

■ Glyphosate-resistant horseweed, or marestail, is found in 16 states.

■ 3.6 million acres of corn is strip-tilled ... 19% of no-till corn acres. It's basically no-till on 6-8 inch tall berms that were built the previous fall or in the spring.

2008

■ The percentage of acres containing either Roundup Ready or LibertyLink corn grew from 30% in 2005 to 60% in 2007.

■ Oilseed radishes hold great promise as a cover crop.

2009

■ Rolf Derpsch, a no-till consultant from Paraguay, determines that there are 232.8 million acres of no-till worldwide, including 88.2 million acres of no-till in North America.

2010

■ A USDA report says 35% of U.S. cropland (88 million acres) has no-tillage operations in 2009. This includes barley, corn, cotton, oats, rice, sorghum, soybeans and wheat acres.

■ USDA says no-till could play a role in U.S. efforts to reduce and control greenhouse gas emissions.

■ The President's Cancer Panel expresses concern about use of atrazine, nitrogen and phosphate due to possible health impacts.

■ David Hula who farms at Charles City, Va., wins National Corn Growers Association national yield contest at 368.4 bushels per acre in the no-till irrigated category.

2011

■ The Chicago Climate Exchange, which ran the only national carbon-trading market in the U.S., scales back its operations.

■ The 19th annual National No-Tillage Conference attracts 867 growers, crop consultants, and ag educators, a record crowd for the Cincinnati location.

■ No-tillers find that planting winter wheat with oilseed radishes boosts no-till wheat yields by as much as 18 bushels per acre.

2012

■ 96 million acres no-tilled — 35% of U.S. total crop acres.

2013

■ No-tillage and cover crops receive major play in U.S. newspapers and magazines.

2014

■ The Buffet Foundation and USDA hold national cover crops workshop.

2015

■ 73% of no-tillers seed cover crops.

2016

■ No-Till on the Plains holds 20th anniversary winter conference.

■ 9% of total U.S. crop acres seeded to cover crops, up from 4% in 2005 and 1% in 2006.

■ Growers surveyed for 8th annual *No-Till Farmer* Benchmark Study average 1,146 acres.

2017

25TH ANNIVERSARY
NATIONAL NO-TILLAGE CONFERENCE
January 10-13, 2017 • St. Louis, MO
Quarter Century of No-Till Learning

■ No-Till Farmer holds 25th National No-Tillage Conference.

■ Illegal off-label use of dicamba herbicide damages 3.6 million acres of soybeans.

Jeff Lazewski

How No-Till Farmer
Repurposed Plowshares

"When the parts man would ask **what make and model we needed the plowshares for,** he was always skeptical when we told him it didn't matter..."

In the early days of promoting no-till, each year we honored several outstanding no-till growers and educators who were researching and promoting reduced tillage with our special "Plowshare Award."

For youngsters who have never plowed, a hardened steel plowshare represents the cutting or leading edge on the bottom of the moldboard that slices through the ground. The sliced dirt is then turned over by the moldboard.

Something Different

Mounted on a wooden base, the chrome-plated plowshares were chosen for our awards to symbolize the end of the plowing era. From 1973 to 1981, 25 outstanding no-till farmers, educators and pioneers were presented with these plowshare awards.

With no-till eliminating the need for plowing, we felt it was appropriate to use a plowshare as the centerpiece for these awards.

Plus, the use of plowshares in these awards offered a new use for an item that wasn't going to enjoy as much demand in the future as

GLOVER TRIPLETT

BOB STEWART   ERNIE BEHN

GAIL WICKS   PAUL SCHAFFERT

JOE NEWCOMER   NEIL SPRINGER

MEET THE 25 PLOWSHARE AWARD WINNERS

With chrome-plated plowshares symbolizing the end of the plowing era in American agriculture, 25 outstanding farmers, educators and pioneers were honored with these no-till awards each year from 1973 to 1981.

KEN REHM

1973
- ◆ **HARRY YOUNG, JR.**
  Herndon, Ky.
- ◆ **SHIRLEY PHILLIPS**
  University of Kentucky,
  Lexington, Ky.

1974
- ◆ **RALPH RASNIC**
  Riner, Va.
- ◆ **GEORGE MCKIBBEN**
  Dixon Springs
  Agricultural Center,
  Simpson, Ill.
- ◆ **W.W. MOSCHLER &**
  G. MYRON SHEAR
  Virginia Tech,
  Blacksburg, Va.

1975
- ◆ **ERNIE BEHN**
  Boone, Iowa,
- ◆ **GLOVER TRIPLETT, JR.**
  Ohio State University,
  Wooster, Ohio
- ◆ **KEITH BARRONS**
  Dow Chemical Co.,
  Midland, Mich.

1976
- ◆ **BOB STEWART**
  Jewett, Ohio
- ◆ **JOE NEWCOMER**
  University of Maryland,
  College Park, Md.

- ◆ **MILTON SPRAGUE**
  Rutgers University,
  New Brunswick, N.J.

1977
- ◆ **PAUL SCHAFFERT**
  Indianola, Neb.
- ◆ **BILL LEWIS**
  North Carolina State
  University,
  Raleigh, N.C.
- ◆ **WALT BUESCHER**
  Allis-Chalmers,
  West Allis, Wis.

1978
- ◆ **JERRELL & LEO**
  HARDEN
  Banks, Ala.

- ◆ **ALLEN WIESE**
  Texas A&M Agricultural
  Research and
  Education Center,
  Amarillo, Texas
- ◆ **C.M. WOODRUFF**
  University of Missouri,
  Columbia, Mo.

1979
- ◆ **NEIL SPRINGER**
  Mount Vernon, Ohio
- ◆ **GAIL WICKS**
  University of Nebraska,
  North Platte, Neb.
- ◆ **HAROLD LLOYD**
  USDA North
  Appalachian

Experimental
Watershed Station,
Coshocton, Ohio

1980
- ◆ **KEN REHM**
  Kirkland, Ill.
- ◆ **RAYMOND GALLAGHER**
  University of Florida,
  Gainesville, Fla.
- ◆ **J.N. JONES**
  Virginia Polytechnic
  Institute,
  Blacksburg, Va.

1981
- ◆ **JIM MCCUTCHEON**
  Homewood, Manitoba

SLICING DIRT. A hardened steel plowshare serves as the cutting edge along the bottom of the moldboard. The fresh sliced soil is then turned over by the moldboard as the plow travels through the field.

the number of plowed acres became smaller and smaller.

When one of our staff would go to a farm equipment dealership to buy the plowshares, the parts counter man would always ask what make and model plow we needed the plowshares for. He was always skeptical when we told him it didn't matter.

After explaining what we had in mind and seeing the look of surprise on the parts guy's face, we'd take the plowshares to a Milwaukee area chrome plater and have them add the silver color.

A Conversation Starter

The honored educators and growers usually found a place for these unusual plowshare awards somewhere on their desks. And I'm sure the special plaques sparked many a conversation as to what the awards represented in an era where the number of plowed acres were being dramatically reduced. ❁

"These chrome-plated plowshares **symbolized the end of the moldboard plowing era...**"

NO-TILL-AGE®

November 2007

Plowing vs. No-Tilling in Great Britain

Tillage System	Cost Per Acre	Minutes Per Acre	Fuel Per Acre
No-till	$18.04	10.2	0.97 gal.
Plow	$72.16	40.5	6.10

— *Irish Farmers Weekly*

No-Till Movement Reaps Major Benefits from These Innovators

A number of innovators in the no-till field have been honored each year since 1996 in promoting the advancement of no-till, encouraging better soil health and protecting the environment.

Sponsored by Syngenta and *No-Till Farmer*, the No-Till Innovator Awards program honors farmers, researchers, educators, suppliers and organizations who have identified ways to no-till more efficiently and economically while having a significant impact on the environment.

The commitment and willlingness of these innovators to share what they have learned with others throughout the no-till world has resulted in a leap-frogging of technology transfer and knowledge that has helped give many farmers a head start when it comes to moving into no-till.

These innovators have played a leading role in both the acceptance and the rapid expansion of no-tilled acres across the country.

CROP PRODUCTION

1996 — **John McLarty,**
London, Ontario
1997 — **Ron Jaques,**
Hutchinson, Kan.
1998 — **Steve Groff,**
Holtwood, Pa.
1999 — **Doug Harford,**
Mazon, Ill.
2000 — **Mike Ellis,**
Worth & Dee Ellis Farm,
Eminence, Ky.
2001 — **Russ Zenner,**
Genesee, Idaho
2002 — **Mike, Ted, Paul, Tom and Nick Guetterman,**
Bucyrus, Kan.
2003 — **Tom, Jeff and Doug Martin,**
Mt. Pulaski, Ill.
2004 — **Randy Schwartz,**
Great Bend, Kan.
2005 — **John Aeschliman,**
Colfax, Wash.
2006 — **Joe Breker,**
Havana, N.D.
2007 — **Bill Richards,**
Circleville, Ohio
2008 — **Allen Berry,**
Nauvoo, Ill.
2009 — **Rodney, Ken and Roy Rulon,**
Rulon Enterprises,
Arcadia, Ind.
2010 — **Ray McCormick,**
Vincennes, Ind.
2011 — **Terry Taylor,**
Geff, Ill.
2012 — **Dan DeSutter,**
Attica, Ind.
2013 — **Jack Maloney,**
Brownsburg, Ind.
2014 — **Mike Beer,**
Keldron, S.D.
2015 — **David Brandt,**
Carroll, Ohio
2016 — **Steve Berger,**
Wellman, Iowa
2017 — **Annie Dee,**
Aliceville, Ala.

EDUCATION & RESEARCH

1996 — **John Bradley,**
University of Tennessee,
Milan, Tenn.
1997 — **George Kapusta,**
Southern Illinois University,
Carbondale, Ill.
1998 — **Paul Jasa,**
University of Nebraska,
Lincoln, Neb.
1999 — **Dwayne Beck,**
Dakota Lakes Research Farm,
Pierre, S.D.
2000 — **John Walker,**
Ricks College,
Rexburg, Idaho
2001 — **Wayne Reeves,**
National Soils Lab, Agricultural Research Service, U.S. Department of Agriculture,
Auburn, Ala.
2002 — **Glover Triplett,**
Mississippi State University,
Starkville, Miss.
2003 — **Norman Widman,**
Natural Resources Conservation Service, Columbus, Ohio
2004 — **Dan Towery,**
Conservation Technology Information Center, West Lafayette, Ind.
2005 — **Jim Cook,**
Washington State University,
Pullman, Wash.
2006 — **Jim Leverich,**
University of Wisconsin Cooperative Extension Service,
Sparta, Wis.
2007 — **Bud Davis,**
Natural Resources Conservation Service, Salina, Kan.
2008 — **Randy Raper,**
National Soil Dynamics Lab, Agricultural Research Service, U.S. Department of Agriculture, Auburn, Ala.
2009 — **Barry Fisher,**
National Resources Conservation Service,
Indianapolis, Ind.
2010 — **Harold Reetz,**
Reetz Agronomic Services,
Monticello, Ill.
2011 — **Randall Reeder,**
Ohio State University,
Columbus, Ohio.

2012 — **Randy Pryor,**
University of Nebraska.
Wilber, Neb.
2013 — **Jill Clapperton,**
Rhizoterra, Inc.,
Florence, Mont.
2014 — **Joel Gruver,** Western Illinois University,
Macomb, Ill.
2015 — **Lloyd Murdock,**
University of Kentucky,
Princeton, Ky.
2016 — **Dave Franzen,**
North Dakota State University.
Fargo, N.D.
2017 — **Hans Kok,**
Conservation consultant,
Indianapolis, Ind.

BUSINESS & SERVICE

1996 — **David Swaim,**
Swaim Crop Consulting,
Crawfordsville, Ind.
1997 — **Philip Anderson,**
Southern States Cooperative,
Owensboro, Ky.
1998 — **Joe Nester,**
Nester Ag Management,
Antwerp, Ohio
1999 — **Ed Winkle,**
Hymark Consulting,
Blanchester, Ohio
2000 — **David Savage,**
Resourceful Agronomics,
Farley, Iowa
2001 — **David Cole,**
ITAC, Inc.,Prairie du Sac, Wis.
2002 — **Brad Mathson,**
Whitehall Agricultural Services, Whitehall, Wis.
2003 — **Dave Moeller,**
Moeller Ag Service,
Keokuk, Iowa
2004 — **Guy Swanson,**
Exactrix Inc.,
Spokane, Wash.
2005 — **Paul Schaffert,**
Schaffert Mfg. Co.,
Indianola, Neb.
2006 — **Karl Kroeck,**
Kroeck Crop Consulting,
Knoxville, Pa.
2007 — **Don Hoover,**
Binkley & Hurst, Lititz, Pa.
2008 — **Rich Follmer,**
Progressive Farm Products,
Hudson, Ill.

2009 — **Dave Nelson,**
Brokaw Supply Co.,
Fort Dodge, Iowa

2010 — **Roy Applequist,**
Great Plains Mfg., Salina, Kan.

2011 — **Jamie Scott,**
J.A. Scott Farms Inc.,
Pierceton, Ind.

2012 — **Daryl Starr,**
Advanced Ag Solutions.
Lafayette, Ind.

2013 — **Bill Lehmkuhl,**
Precision Agri Services,
Minster, Ohio

2014 — **Terry Metzger,**
Ag Spectrum, DeWitt, Iowa

2015 — **Phil Needham,**
Needham Ag Technologies,
Calhoun, Ky.

2016 — **Betsy Bower,**
Ceres Solutions,
Lafayette, Ind.

2017 — **John Baker,**
Cross Slot USA,
Pullman, Wash.

NO-TILL ORGANIZATIONS

1997 — **Manitoba-North Dakota Zero Tillage Farmer's Association,**
Brandon, Manitoba

1998 — **Clark County Farm Bureau,**
Clark County, Ind.

1999 — **Conservation Action Project,**
Defiance, Ohio

2000 — **Ohio No-Till Council,**
Columbus, Ohio

2001 — **Coffee County Conservation Tillage Alliance,**
Coffee County, Ga.

2002 — **West Tennessee No-Till Farmers Association,**
Milan, Tenn.

2003 — **Pacific Northwest Direct Seed Association,**
Pullman, Wash.

2004 — **Lower Elkhorn Natural Resource District,**
Norfolk, Neb.

2005 — **Innovative Farmers Assn.,**
Port Stanley, Ontario

2006 — **Innovative Cropping Systems Team,**
Quinton, Va.

2007 — **Clearwater Direct Seeders,**
Clearwater, Idaho

2008 — **Embarras River Management Association,** Toledo, Ill.

2009 — **No-Till on the Plains,**
Salina, Kan.

2010 — **Delta Conservation Demonstration Center,**
Metcalfe, Miss.

2011 — **Pennsylvania No-Till Alliance,**
Harrisburg, Pa.

2012 — **Midwest Cover Crops Council,**
West Lafayette, Ind.

2013 — **Virginia No-Till Alliance,**
Harrisonburg, Va.

2014 — **Conservation Tillage Conference,**
Ada, Ohio

2015 — **Conservation Cropping Systems Initiative**

2016 — **Champlain Valley Farmer Coalition,** Middlebury, Vt.

2017 — **Conservation Agriculture Systems Innovation,**
Five Points, Calif.

EQUIPMENT DESIGN

1996 — **Howard Martin,**
Martin Industries, Elkton, Ky.

1997 — **Ray Rawson,**
Farwell, Mich.

1998 — **Eugene Keeton,**
Clarksville, Tenn.

1999 — **Jon Kinzenbaw,**
Kinze Manufacturing Co.,
Williamsburg, Iowa

2000 — **Terry Schneider,**
Shirley, Ill.

2001 — **The Reed Family,**
Washington, Iowa

2002 — **Herb Stamm,**
Hi-Pro Manufacturing Co.,
Watseka, Ill.

AG SUPPLIER

1997 — **Jim and Tom Boyd,
B & B Farm Service,**
Fredericktown, Ohio

1998 — **Eric Laux,**
Laux Farm Service,
New Madison, Ohio

1999 — **Jeremiah Jones,**
Crop Production Services,
Morganfield, Ky.

2000 — **Miles Farm Supply and Opti-Crop,**
Owensboro, Ky.

2001 — **Roger Strand,**
Strand of Milan,
Milan, Minn.

2002 — **Roger Stitzlein,**
Loudonville Farmers Equity,
Loudonville, Ohio

OTHER

ENVIRONMENTAL STEWARDSHIP

1996 — **Jim LeCureux,**
Michigan State University,
Bad Axe, Mich.

PUBLIC IMAGE OF AGRICULTURE

1996 — **Marion Calmer,**
Calmer Farms, Alpha, Ill.

EDUCATION

1996 — **Mike Plumer,**
University of Illinois,
Marion, Ill.

NO-TILL INNOVATOR AWARD PAINTINGS

To honor the No-Till Innovators, Syngenta for many years commissioned a leading water color artist to paint an original rural scene that includes details from each award winner's work. Over the years, the scenes shown in these pages were painted by Dean Brickler of Ames, Iowa, and Gayle Strider of Hartselle, Ala.

"Mother nature **never created a soil that required tillage...**"

— *Jerry Crew, Webb, Iowa*

"50 years ago, **many farmers thought no-till was ludicrous and lazy,** but we got believers when they saw the results..."

— *Bill Haddad, Danville, Ohio*

"Tillage tool manufacturers got kicked out of the University of Illinois test plots **because we weren't contributing as many dollars as the chemical companies were for no-till studies...**"

— *Bill Schmidtgall, Morton, Ill.*

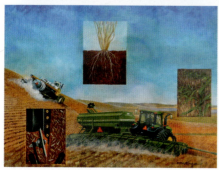

"No-till became popular in the Corn Belt not because of the moisture savings, but **because farmers could farm more land with less equipment, less fuel and less labor...**"

— *Rich Follmer, Hudson, Ill.*

"No-till is not only a different cropping technique, **but a matter of survival...**"
— Franke Dijkstra, Carambei, Brazil

HONORING NO-TILL LEADERS. Lessiter Media president Mike Lessiter congratulates Guy Swanson as one of the "No-Till Legends" during the 25th anniversary National No-Tillage Conference held in St. Louis in 2017.

Jeff Lazewski

Meet the
No-Till Legends

"These highly talented growers, educators and suppliers have played a key role in the growth of no-till..."

As part of the 40th anniversary of *No-Till Farmer* in 2011 and again for the 25th anniversary of the National No-Tillage Conference in 2017, we selected 43 individuals who have made tremendous contributions to the growth, popularity and acceptance of no-till.

This roster of "No-Till Legends" includes outstanding growers, educa-tors and suppliers who have played a key role in the growth of no-till from only 3.2 million acres in the U.S. in 1972 to an expected 145 million acres by 2030.

As we took inventory of the many legends, it became apparent that a number of folks outside North America also played a critical role in no-till's acceptance around the world. As a result, we've added a list of 11 worldwide no-till legends involved in the no-till movement.

Hailing from 17 states and 2 Canadian provinces, the 43 North American "No-Till Legends" follow in alphabetical order. Next, you'll find the names of 11 no-till influenc-ers from seven other countries around the world that have played a major role in the adoption and growth of no-till.

JOHN AESCHLIMAN

Colfax, Wash.

John was among the no-till pioneers in the Palouse area of eastern Washington and a founding member of the Pacific Northwest Direct Seed Association in 2000.

No-tilling some of the steepest hills found anywhere in the country. Aeschliman and his son run a 4,000-acre operation that's been 100% no-tilled since the mid-1970s. They no-till winter wheat, spring wheat and spring barley in 7 1/2-inch paired rows and also no-till corn, peas, sunflowers, garbanzo beans and cover crops in a 100% continuous no-till system.

Aeschliman is a major proponent of the need for no-till to protect topsoil from erosion, having distributed photos of Palouse-area roads covered with topsoil washed from conventionally-tilled fields to prove his point. He's also discouraged the burning of residue in no-till systems.

On visits to his farm, I've walked up no-tilled slopes so steep you think you're about to fall and tumble head first down the hill.

ROY APPLEQUIST

Salina, Kan.

Establishing Great Plains Mfg. Co. in 1976, Applequist was among several shortline farm equipment manufacturers in the early 1980s that saw the need for designing equipment specifically for the no-till market. The company was an active participant in Tennessee's Milan No-Till Field Days, where

several manufacturers introduced narrow-row prototype drills designed specifically for no-till conditions.

Since those early days, Great Plains has become a leader in developing specialized seeding equipment for the no-till market. The started-from-scratch company that sold in 2016 to Kubota for $430 million was the result of a brand name made famous through the design and manufacture of specialized equipment in the no-till and conservation tillage world.

KEITH BARRONS

Midland, Mich.

Keith was the manager of ag chemical development at Dow Chemical Co. and was among several Dow scientists who worked on the development of no-till techniques in 1951. In some of their early herbicide work in no-till, these researchers needed as much as 6 pounds of atrazine per acre to combat weeds.

Over the years at Michigan State University and later at Dow, Barrons authored more than 100 technical papers and books on ag developments, including research results dealing with effective herbicide use in no-till systems.

DWAYNE BECK

Pierre, S.D.

The South Dakota State University agronomist and manager of the farmer-owned Dakota Lakes Research Farm has produced some of the most practical no-till research

found anywhere in the world. Big on introducing additional crops into long-term no-till rotations, Beck is a strong advocate for using cover crops with no-till.

He has shown the importance of seeding a diversified assortment of cover crops and has been a strong advocate for keeping something green and growing on the soil surface every month of the year. Beck is an unusual researcher that understands the need for providing no-tillers with highly practical and profitable ideas they can put to immediate use.

A speaker at the first National No-Tillage Conference in Indianapolis in 1993, I still remember Duane saying: "You folks in the eastern Corn Belt no-till to get rid of the water. In South Dakota, we no-till to capture every raindrop we can."

ERNIE BEHN

Boone, Iowa

Among the early adapters of ridge-till, Boone frequently spoke to numerous farm groups on the benefits of the practice. He wrote a practical guide to ridge-till success entitled, *More Profit with Less Tillage.*

JOHN BRADLEY

Lutts, Tenn.

A long-time University of Tennessee agronomist, Bradley served for 14 years as head of the University of Tennessee's Milan Ag Experiment Station. He directed the annual mid-summer Milan No-Till Field Day that attracted tens of thousands of growers to see the latest innovations in no-till research.

Among the pioneers working with no-till cotton, he has shown thousands of growers how to earn large payoffs with a wide variety of no-tilled crops. He later worked as a no-till educator with Monsanto.

DAVID BRANDT

Carroll, Ohio

 Brandt began farming in 1971 and has seeded numerous cover-crop mixtures on his 1,150-acre corn, soybean and wheat operation since 1978. He's a huge booster of the tremendous value of cover crops in improving soil health.

Brandt is a frequent speaker at many no-till events, including several National No-Tillage Conferences, where he always shares a wealth of practical and low-input cost ideas used on his own farm to increase no-till profitability.

GABE BROWN

Bismarck, N.D.

 The Brown family operates a highly diversified 5,000-acre operation that focuses on holistically integrating livestock grazing and no-till systems. This includes cash crops, multi-species cover crops and the production of all-natural grass-fed beef, poultry and lamb.

Relying extensively on no-tilling, Brown's system has regenerated the natural resources on the ranch without extensive use of synthetic fertilizers, herbicides, insecticides or fungicides.

MARION CALMER

Alpha, Ill.

 The owner of the Calmer Ag Research Center has conducted hundreds of onfarm practical research projects covering all aspects of crop production for nearly three decades.

A firm believer in field-sized replicated research, his plots run the entire length of fields in widths that match his no-till planter, drill and combine. His mid-summer tours and winter meetings are jam-packed with practical results that always include plenty of down-home advice to help growers no-till more effectively while pushing up profits. Much of his research has centered on ways to make the most effective use of fertilizers, reduce seeding rates, planter modifications, combine adjustments and narrow-row corn.

A popular speaker at many no-till events, Calmer led the charge toward 15-inch rows in no-till corn, engineering and marketing corn heads for narrower rows. He's also developed crushing and chopping stalk rollers that aid in the decomposition of *Bt* corn-stalk residue in no-tilled corn.

Over the years, I've been suckered into being the "villain" in more than a dozen of his skits at the National No-Tillage Conference. Always humorous, somehow the on-stage skits always seem to stress a point about no-tilling in a fresh entertaining way.

JILL CLAPPERTON

Spokane, Wash.

 Showing growers how to benefit from the amazing millions of critters living beneath the soil surface, Clapperton says no-till has evolved to where it not only improves soil structure and stops erosion but also is having a far-reaching impact on consumer food preferences and human health.

An internationally known soil health specialist and soil ecologist, Clapperton analyzes the relationships between organisms and their environments to more effectively manage and benefit from the long-term biological fertility of our soils.

Among only a handful of soil ecology scientists in the world, she is known as a rhizosphere ecologist and is a popular speaker at no-till events, including the National No-Tillage Conference. Previously working with Agriculture and Agri-Food Canada at Lethbridge, Alberta, Clapperton now operates Earthspirit Land Resource Consulting.

JIM COOK

Pullman, Wash.

Cook earned an international reputation as a leading plant pathologist at Washington State University when he demonstrated to Pacific Northwest no-tillers why expanding crop rotations could effectively fight many troublesome disease concerns.

Cook discovered how soil borne pathogens were attacking no-till wheat roots and how naturally occurring microorganisms could battle disease.

He started promoting no-till wheat to growers in the Palouse corner of Washington, Oregon and Idaho in the early 1990s as a way to save time, fuel and soil while capturing more water.

Cook pioneered the research that led to the "green bridge" breakthrough that was holding back the growth of no-till in many areas of the world.

ELBERT DICKEY

Lincoln, Neb.

This University of Nebraska educator pioneered many of the early day water-saving techniques used with no-till by western Corn Belt growers. He was known in the 1980s for his extensive knowledge on how no-till planter adjustments and modifications could lead to improved crop stands and seeding efficiencies

DAVID DUKES

Bedford, Iowa

 A long-time no-tiller in southwest Iowa, Dukes demonstrated the benefits of no-tilling when bringing land out of the Conservation Reserve Program. He is an active promoter of no-till and soil stewardship in livestock operations.

EDWARD FAULKNER

Elyria, Ohio

This county ag agent became an author in 1943 when he published *Plowman's Folly*, a revelation based on understanding and appreciating the value of reducing soil disturbance. His famous quotation is as true today as it was in 1943, *"No one has ever advanced a scientific reason for plowing."*

BARRY FISHER

Indianapolis, Ind.

 State agronomist for the Natural Resource Conservation Service in the Hoosier state before being named one of the agency's regional soil health specialists, Fisher has been a long-time booster of no-till. His efforts helped Indiana's no-till acres grow to the point where it is among the states with the highest percentage of no-till acres in the U.S.

Fisher has devoted much of his government career to encouraging the adoption of no-till and cover crops. Over the years, he's worked with thousands of farmers while conducting numerous field days and has worked with numerous no-till equipment options to demonstrate what makes the most sense when it comes to no-till.

LEONARD FLEISCHER

Columbus, Neb.

 Among the pioneers of ridge-till, Fleischer developed and marketed the leading line of ridge-till equipment under the Fleischer Mfg. brand. The company also sponsored the National Ridge-Till Conference for a number of years, which was the leading educational event for this reduced tillage system.

STEVE GROFF

Holtwood, Pa.

 Among the world's leaders on adapting no-till for vegetable production, Groff has hosted numerous field days that have brought thousands of growers and ag leaders to his highly specialized cropping operation. A cover crop enthusiast, Groff has helped thousands of growers improve the health of their soils with cover crops while promoting the concept that "Soil is meant to be covered."

Having successfully no-tilled tomatoes, sweet corn, pumpkins, corn, alfalfa, soybeans, wheat and cover crops for nearly four decades, Groff is known for the introduction of the tillage radish as a cover crop and has demonstrated its ability to scavenge nutrients, reduce weed pressure, trim compaction and boost soil quality.

JIM HALFORD

Indian Head, Saskatchewan

No-tilling since 1979, Halford developed the Conserva Pak seeding system that was sold in 2007 to John Deere. Instrumental in the organization of the Saskatchewan Soil Conservation Association, Halford has relied on low-disturbance cropping systems to curb erosion concerns.

DAVID HULA

Charles City, Va.

A multi-generational no-tiller, the family operation where no-till has been used since the late 1960s is known for its record-breaking corn yields. These include harvesting 542 bushels per acre from an irrigated no-till field in 2017 that represents the highest yield ever reported in the National Corn Growers Assn. competition.

Many of his ideas have dramatically boosted yields for thousands of no-tillers across the country.

Located along the James River, the family's farm is near the Jamestown settlement, which was the first permanent British settlement in the Americas back in 1607. Along with all the early American history to see in the area, a visit to the Hula operation is not complete without touring the farm's ag museum put together by David's father.

PAUL JASA

Lincoln, Neb.

An ag engineer at the University of Nebraska, Jasa's first experience with no-till occurred when he evaluated the planter usage of 150 growers. After nearly four decades of continuous no-till research, Jasa has helped producers recognize the immense value of following a total no-till system approach.

This has included developing a long-term no-till plan to rebuild the soil structure, no-tilling every acre every year, utilizing a diverse no-till rotation, properly managing residue and placing nutrients where crop roots will find them. He's always been a hands-on educator in demonstrating to growers how to make the necessary adjustments

with no-till planters, drills or air seeders for slicing through residue, getting uniform seed depth and obtaining effective seed-to-soil contact.

Jasa championed the value of no-tilling for erosion control, higher yields, less cost and water-conservation benefits long before many other university educators.

GEORGE KAPUSTA

Carbondale, Ill.

The long-time weed scientist at Southern Illinois University was among the early advocates of no-till. He earned a reputation as an educator who always offered growers practical herbicide answers to handle specific weed concerns with no-till.

EUGENE KEETON

Clarksville, Tenn.

A farmer turned inventor, Keeton came up with innovative planter ideas from his Kentucky farm shop that were later adapted by Kinze and John Deere. He invented the Keeton seed firmer to improve no-till seed-to-soil contact, as well as the finger pickup corn meter and the brush meter for soybeans.

JIM KINSELLA

Lexington, Ill.

A long-time no-till and strip-till enthusiast, Kinsella demonstrated how these two reduced tillage practices save time and money while having a positive impact on the environment. No-tilling soybeans

and strip-tilling corn on 2,200 acres, he and his son have seen a dramatic increase in soil quality and reduction in costs while pushing up yields.

In earlier days, Kinsella worked closely with BASF and several other ag chemical companies to promote no-till. He built an onfarm meeting center to host hundreds of growers at winter meetings and summer field days. Kinsella is recognized among the leaders in the development of strip-till.

ELLERY KNAKE

Urbana, Ill.

The University of Illinois weed scientist saw major no-till benefits before many other agronomists and developed extensive herbicide tankmix recommendations to control weeds in no-tilled corn and soybeans. Much of Knake's early research dealing with weed competition pioneered the establishment of thresholds for several annual grass and broadleaf weed species and provided economic justification for using soil-residual herbicides.

GUY LAFOND

Indian Head, Saskatchewan

The no-till researcher with Agriculture and Agri-Food Canada at Saskatchewan's Indian Head Research Farm developed innovative no-till systems for dryland agriculture. Besides evaluating the environmental benefits of no-till in regard to crop production, soil quality and air quality, Lafond's research focused on nitrogen application guidelines using optical sensors to increase nutrient efficiency.

BILL LEWIS

Raleigh, N.C.

The North Carolina State University weed scientist became the go-to guy in the 1970s for southeastern growers who sought practical advice on all aspects of no-tilling, long before most other educators were on the no-till bandwagon.

HOWARD MARTIN

Elkton, Ky.

Martin started no-tilling in order to earn a decent living from poor-quality land and later developed specialized no-till equipment that led to the formation of Martin Industries. The development of row cleaners, fertilizer openers, closing wheels, gauge wheels and other no-till row-unit attachments has come from a humble one-person farm shop beginning to a highly successful manufacturing company with more than 40 employees that's a leader in the no-till planter and drill accessory market.

JEFF MARTIN

Mount Pulaski, Ill.

No-tilling since 1982, Martin has dramatically expanded the family operation and for nearly two decades has been strip-tilling continuous corn on much of the farm's 5,000-plus acres. Based on extensive soil testing and plant analysis, the family has cut back sharply on nitrogen ap-

plications with corn while producing yields of over 250 bushels per acre.

By paying close attention to details, Martin maintains both no-till and strip-till systems have the potential for higher yields and profits with either rotated or continuous corn than any other tillage system. Martin also played a key role in convincing absentee landowners to allow their ground to be no-tilled.

RAY MCCORMICK

Vincennes, Ind.

A no-tiller for more than 30 years in a 3,200-acre operation, McCormick believes nothing pays bigger dividends than investing in conservation. When he talks diversification, it means raising peaches for the fresh market, no-tilling corn, wheat, both full-season and double-cropped soybeans and cover crops, along with managing 1,000 acres of woodlands and wetlands for waterfowl and whitetail deer hunting.

Recognizing the nutrient and soil protection value, McCormick says he would not sell residue from his no-till fields even if someone offered $100 per ton.

JIM MCCUTCHEON

Homewood, Manitoba

McCutcheon was among the first growers in western Canada to give no-till a try and found low-cost ways to adapt existing equipment for no-till.

He was a founder and past president of the Manitoba-North Dakota Zero Tillage Farmers Association.

GEORGE MCKIBBEN

Simpson, Ill.

The agronomist with the University of Illinois at the Dixon Springs Agricultural Center introduced some of the earliest no-till plots in the states.

McKibben's annual summer field days featured hundreds of amazing tankmix concoctions being evaluated for effective no-till weed control.

His 1961 work on no-tilling corn into fescue sod led to the first field-scale introduction of no-till a year later in southwestern Kentucky.

BOB MCNABB

Minnedosa, Manitoba

No-tilling since 1978 with a wide variety of cereals, pulse crops and forages, McNabb also worked with several groups to introduce no-till to small-acreage farmers in Zimbabwe. McNabb was also among the pioneers who started the Manitoba-North Dakota Zero Tillage Farmers Association and served as president of the group.

JOHN MCNABB

Inkom, Idaho

Having no-tilled as many as 43,000 acres in a single year, McNabb and his family are a shining example of how no-till can maximize profits while successfully no-tilling poor-quality soils on steep slopes that otherwise could not be farmed in southern Idaho and northern Utah.

Following a winter and spring

wheat rotation, no-till has enabled McNabb to get along with half the number of tractors, one-fourth of the labor and enjoy dramatic drops in fuel and herbicide costs compared to conventional tillage. His bottom line indicates no-till over the years has returned an extra $35-$40 per acre.

PHIL NEEDHAM

Calhoun, Ky.

 Born and raised in the United Kingdom, Needham came to the states to show American wheat growers how to dramatically boost wheat yields by adopting the intensive British small grain system. Working on this project, the British-trained agronomist soon learned about no-till and its drawbacks due to a lack of needed equipment modifications.

From that experience, he launched Needham Ag Technologies and has built a career around the need to effectively adjust and modify planters, drills, combines and sprayers for peak performance under no-till's heavy residue situations. Numerous equipment design modifications seen over the past several decades have resulted from Needham's ideas.

JOE NESTER

Bryan, Ohio

 Working as a crop consultant with no-tillers in northwestern Ohio, Nester has demonstrated how no-till can return an extra $40-$50 profit per acre. His work with no-till

started in the late 1980s and over the next few years, Nester bought ten 750 John Deere no-till drills and rented them to growers.

Within a few years, most of his clients switched from 100% conventionally-tilled soybeans to nearly 100% no-tilled soybeans.

Nester has also worked on the algae concerns in Lake Erie due to the over application of phosphorus.

He's been a trendsetter in helping no-tillers reduce costly soil and nutrient runoff while turning out high yields with less fertilizer.

SHIRLEY PHILLIPS

Lexington, Ky.

 The University of Kentucky agronomist and Extension administrator was one of the fathers of the no-till movement. He worked closely with Harry Young to put out the very first onfarm no-tillage corn crop in 1962. In 1973, Phillips and Young co-authored the 224-page *No-Tillage Farming book,* which was the first practical guide to this new reduced tillage concept.

MIKE PLUMER

Creal Springs, Ill.

 The long-time University of Illinois Extension educator in southern Illinois relied on no-till to pump up profits on his own farm. Plumer was a big believer in relying on cover crops to dramatically improve soil quality and no-till income.

After retiring as a university ed-

ucator, he consulted with growers and suppliers on finding new cover crop solutions, making more effective use of no-till and improving soil health.

RAY RAWSON

Farwell, Mich.

 The "father of zone tillage," Rawson and his sons run a large-acreage operation that features no-till corn and soybeans. He spent a number of years promoting the zone-till system at hundreds of meetings across the Corn Belt and working one-on-one with growers.

A number of today's reduced tillage tools got their start in the Rawson farm shop. At one time he designed and built a new no-till planter every winter in his farm shop that featured his latest innovative ideas.

BILL RICHARDS

Circleville, Ohio

 Known as the godfather of no-till and among the early adopters of the practice in southern Ohio, no-till enabled the Richards family to grow to more than 3,000 acres — a move that completely changed his life. Early on, he saw the value of utilizing wider equipment to run tires in permanent wheel tracks to reduce compaction and no-till 31 rows of 20-inch corn in the days when most growers thought narrower rows meant a 30-inch spacing.

While Richard's contributions

NO-TILL-AGE®

Follow the 6-Times Rule

What it costs to invest in tillage equipment, tractors and equipment wear and tear.

Tillage System	Equipment Ownership Operation Cost Per Acre
Conventional tillage	$90
Mulch tillage	$45
No-tillage	$15

This rule of thumb for tillage equipment investment and the machinery operating costs for the various number of passes across the field is based on the number of gallons of diesel fuel required per acre to raise a crop times the per gallon cost of fuel times six. These figures are based on diesel fuel costs of $2.50 per gallon.

— *Guy Swanson, Exactrix Global Systems, Spokane, Wash.*

to the no-till movement started on the home farm, he later had a tremendous impact on the growth of no-tillage while serving as chief of the Soil Conservation Service in Washington, D.C. In this position, he convinced many government officials and farmers that no-till was here to stay and could make the environment better for everyone.

Early on, Richards recognized and promoted the major economic benefits of no-till as a result of reducing costly field trips.

MORT AND GUY SWANSON

Palouse, Wash.

Guy Swanson

The father-son team developed the Yielder drill for extremely dry Pacific Northwest slopes and engineered special-ized openers, spraying equipment and fertilizer applicators in the family's farm shop. The Swansons demonstrated how specialized

equipment could seed numerous no-till crops in continuous no-till on the steep hills of the Palouse while sharply reducing erosion. Their high-ly specialized, extremely heavyweight Yielder drills were used to no-till a variety of grain, pulse and forage crops while dramatically reducing soil erosion.

Guy Swanson has continued to help no-tillers earn higher profits by developing the highly efficient Exactrix Global Systems fertilizer application system and other no-till innovations that has allowed grow-ers to trim nutrient rates.

A true farm shop innovator and well ahead of his time in the no-till world, I met Mort Swanson on the first no-till tour held in Hawaii in 1973.

GRANT THOMAS

Lexington, Ky.

The University of Kentucky agron-omist specialized in soil chemis-try and international agricultural

development. He helped introduce no-till to a number of Argentinean growers.

DAN TOWERY

Lafayette, Ind.

A Natural Resources Conservation Service educator in Illinois, Towery quickly rec-ognized the benefits of no-till. Later as a Conservation Technology Informa-tion Center staffer, he supervised the annual tillage practices survey that was conducted nationwide for years. He has been instrumental in the promotion and recent growth of cover crops, particularly annual ryegrass.

Working as a consultant to government agencies, growers, farm groups and suppliers, Towery has played a critical role in the expansion of cover crops across the nation, especially within the no-till ranks.

GLOVER TRIPLETT

Starkville, Miss.

Triplett did some of the early research on no-till as an Ohio State Uni-versity weed scientist in the early 1960s. He also started the longest on-going no-till research plots in the world at Ohio State's Wooster facility. These extensive plots have pro-duced valuable no-till data for more than 55 years with over 75 scientific papers having been published by educators based on results from these plots.

After retiring from Ohio State, Triplett has enjoyed a part-time career as a plant and soil agronomist with Mississippi State University. Over many years, he showed farmers how to use no-till to convert severely nutrient- and soil-depleted fields back into highly productive cropland.

During a summer field day in 1978, I asked the Mississippi native how it felt to have a fellow Southerner, Jimmy Carter, in the White House. He replied, "Frank, it's the first time in my life that our U.S. President doesn't have an accent."

DAVE VAN DOREN

Medina, Ohio

The Ohio State University soil physicist worked closely with Glover Triplett on introducing no-till to the Eastern Corn Belt. Van Doren carried out pioneering research on determining which soil types were most adaptable for no-tilling.

RAY WARD

Kearney, Neb.

This soil scientist recognized the need for evaluating no-till fertility recommendations in a different way than was the case at most soil testing facilities. He runs Ward Laboratories, a soil-testing lab that treats no-till soil test results in innovative ways, such as measuring microbial life and soil biological health.

HARRY YOUNG JR.

Herndon, Ky.

Young started the on-farm no-till movement in 1962 on 0.7 acres of the family farm now operated by his son, John, and grandson, Alexander. That field has been continuously no-tilled for more than 55 years. Young saw the merits of no-tilling corn plus wheat with double-cropped soybeans in a rotation that allowed the family to harvest three cash crops in 2 years from the same ground.

Along with University of Kentucky agronomist Shirley Phillips, Young co-authored the 224-page *No-Tillage Farming* book in 1973, which was the first practical guide to this reduced-tillage practice.

MEET 11 OF THE WORLD'S TOP NO-TILL INFLUENCERS

Since *No-Till Farmer* got its start in 1972, numerous growers and educators in other countries have also played key roles in the growth and acceptance of no-till in many areas of the world outside North America

Here's our list of 11 "No-Till Legends" from seven other countries around the world.

JOHN BAKER

New Zealand

The long-time no-till-age researcher at the University of Massey has published more than 80 international papers on the science of no-till machinery and its inter-action with the soil. Baker developed the cross-slot technology that's now used in many areas of the world.

Baker is convinced the science and practice of no-till has passed the "point of no return."

He believes no-till is within sight of becoming the most common system on the planet for seeding arable crops.

HERBERT BARTZ

Rolandia, Parana, Brazil

Bartz was among the first growers to pioneer no-till in Latin America and has used it continuously since starting to no-till soybeans in 1972. A past president of the Brazilian Conservation Agricultural Federation, he's been a popular speaker on his no-till experiences over the past five decades.

ADEMIR CALEGARI

Lundrina, Parana, Brazil

With over 35 years of experience with no-till, Calegari maintains Brazil's success with no-till is due to the effective selection of cover crops to suppress diseases and increase profitability. The soil scientist has encouraged and helped growers in numerous countries around the world shift to no-till and sustainable soil management.

NO-TILL-AGE®

Are No-Till Soybean Seeding Rates Too High?
(2004-2005 Average Data)

Soybean Population Per Acre	2004 Yield Per Acre	2005 Yield Per Acre
25,000	------	47 bu.
50,000	64 bu.	49
75,000	67	49
100,000	68	50
125,000	68	51
150,000	68	49
175,000	69	50

— Calmer Agronomic Center, Alpha, Ill.

ROLF DERPSCH
Asuncion, Paraguay

Derpsch has researched no-till in South America since 1971. Derpsch worked for the German Agency for Technical Cooperation for 35 years before becoming a no-till consultant.

He has worked on no-till projects in numerous countries, has been recognized with several international awards for helping develop innovative no-till systems and has spoken at numerous no-till meetings, including several National No-Tillage Conferences.

FRANKE DIJKSTRA
Ponta Grossa, Brazil

Dijkstra started no-tilling in 1976 to control erosion and improve soil quality. Always willing to share his no-till experiences, Dijkstra maintains a rotation of 50% corn and 50% soybeans with summer crops, along with winter crops that include barley and annual ryegrass as a cover crop.

Previous to no-till, he struggled with sandy soils, sloping terrains, excessive tillage and frequent downpours that caused serious wind and water erosion. This became more serious each year from the 1950s to the 1970s.

No-tilling corn and soybeans as early as possible offers 15-40 additional days to no-till a second summer crop of corn or soybeans.

In many fields, Dijkstra grows three crops in a year's time. Between 1997 and 2010, his corn and

BILL CRABTREE
Morawa, Western Australia

Having researched and helped expand no-till growth for more than 30 years, Crabtree is well respected for his no-till work throughout the world. Known as "No-Till Bill" in his native land, he has shown the ability to consistently think outside the box in finding new ways for making no-till work.

Crabtree has devoted much of his career to expanding the no-tilled acres in Australia and has worked with many of the country's no-till pioneers.

Besides consulting around the world on no-till, Crabtree now practices what he preaches with the purchase of 7,500 acres near the edge of the desert in the western area of Australia. Some 95% of his cropping program includes no-tilling continuous wheat, but he also no-tills canola, triticale and lupins.

CARLOS CROVETTO
Concepcion, Chile

Crovetto made no-till work on otherwise unproductive soils on steep slopes and has worked since 1985 to promote no-till in many areas of the world. The founder and president of the Soil Conservation Society of Chile, Crovetto highlighted years of no-till experience in his book, *Stubble Over the Soil*. He was also a popular speaker at several of our National No-Tillage Conferences.

soybean yields doubled or tripled with no-till. The farm also includes a 1,000-cow dairy herd and a large farrow-to-finish hog operation.

BRIAN OLDRIEVE

Westgate, Harare, Zimbabwe

With over 30 years of experience with no-till in Africa, Oldrieve has demonstrated to small-acreage African farmers how to produce substantial yield increases with no-till cereal and legume rotations. He has established a number of highly successful programs to spread the no-till message across Africa.

ALLEN POSTLETHWAITE

Esperance, Western Australia

Having successfully no-tilled a dozen crops, he and his family

 have no-tilled 6,000 acres in a 16-inch rainfall zone since 1984. Postlethwaite has developed a successful no-till system while eliminating summerfallow that has dramatically improved the environment. He has hosted numerous no-till field days and tours to encourage farmers to switch to no-till.

Combining controlled traffic with no-till provided Postlethwaite with a much better way to better manage extremely limited amounts of water.

STEVEN POWLES

Perth, Australia

Powles is considered an international authority on all aspects of

herbicide resistance — from a basic biochemical understanding of how plants evolve with resistance to practical onfarm weed control strategies. Powles is a big believer in no-till while stressing the need for growers to diversify their weed control tactics.

WOLFGANG STURNEY

Berne, Switzerland

His long-term no-till trials were built around a number of diversified cropping rotations, and he successfully developed no-till despite serious soil-fertility worries among many European growers.

Sturney's collection of intensive agronomic and economic data has led to greater acceptance of no-till in many areas of Europe. ✺

NO-TILL LOOSE ENDS...

CAN LEGUMES WORK AS A COVER CROP?

Crimson clover, hairy vetch, Australian winter peas and subterranean clover are four legumes Morris Decker is seeding as cover crops in no-till corn fields.

December 1986

The University of Maryland agronomist lets these legumes grow and fix nitrogen in the soil over the winter and spring before baling, grazing or knocking them down with paraquat prior to no-tilling corn.

In addition to broadcast spraying, which kills the legumes, Decker has tried band spraying a 15-inch wide band where corn is no-tilled in 30-inch rows. Killing the legumes is done as late as possible since this allows the cover crops to fix additional nitrogen.

Most problems with this program have been weather related. If there's a dry spell toward the end of April before the legume is killed, moisture is depleted from the soil and the no-tilled corn has a difficult time getting started.

If weather is dry in the fall when the legumes are planted, there can also be problems. Plus, a wet spring can cause a fungus to run rampant through the legumes and result in up to a 50% loss of clover or other legumes.

However, Decker has never had a failure after seeding legumes with a no-till drill into corn residue. ✺

LEVEL YOUR NO-TILL PLANTER!

This is among the most requested articles by No-Till Farmer subscribers.

Getting your planter properly leveled is critical to your no-till success. In fact, several machinery companies are taking steps to help you properly set up their products for no-till conditions.

"We're concentrating our efforts on how to properly set a planter or drill for no-till," says Pat Whalen of Yetter Manufacturing in Colchester, Ill.

"Our goal is to show farmers specifically what to look for when properly setting up their machines. We're trying to give them a 95% success rate by answering the most critical concerns. They will probably have to find the other 5% on their own."

Bob Harwick, field service representative at Yetter, says a typical problem occurs when a no-tiller adds planter attachments. He may have used the rig for a half-dozen years without any problems, but the planter may stop working properly when new attachments are added.

Another problem is when a farmer tries to run the coulters shallower than the double disc openers. This usually can't be done if the planter isn't properly leveled

The key is to keep the planter level. Maintaining the proper drawbar height, not letting your planter run downhill and avoiding running too deep are essential.

"You may be able to adjust your no-till planter once a year due to a particular hydraulic system," says Harwick. "Or you may have to change it for varying field conditions."

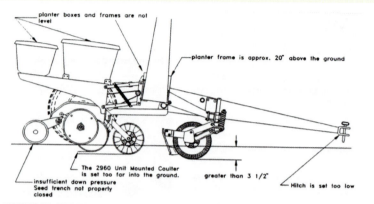

LEVELED VS. NON-LEVELED PLANTER. With this planter set to run in an uneven position, the unit-mounted coulters run below the double-disc openers. The frame-mounted fertilizer coulter is running at a depth greater than 3½ inches, which isn't recommended for no-till conditions. Make sure the frame and seed boxes are level with the ground.

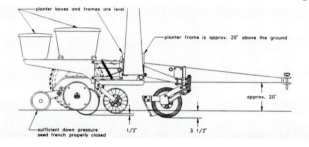

As shown in the illustrations, make sure the planter frames and planter boxes are level with the ground. It's very important for your planter to remain level when no-tilling.

In most cases, a planter will be outfitted with one or more attachments to improve performance under no-till conditions. Remember that most attachments are designed to operate within a specific depth range.

For example, the drawing shows the unit-mounted coulter is running half an inch above the double-disc openers when the frame-mounted coulter runs 3 1/2 inches deep. Yetter officials say these ideal depths are only obtainable if the planter is level.

Non-Level Concerns

Where the planter has not been leveled, this setup may work with conventional tillage, but not with no-till.

With the planter in this position, the unit-mounted coulter is running below the seed double-disc opener and the frame-mounted fertilizer coulter is running deeper than 3 1/2 inches.

Yetter representatives say these depths are not recommended and will greatly affect the performance of each attachment. For example, with the unit-mounted coulter running below the seed double-disc openers, the formation of air pockets below the seed is very likely. ❈

Frank Lessiter

Scientists Toyed
With No-Till in the '50s

"By 1953, successful crops of no-tilled **wheat, oats, flax, soybeans and corn** were being produced by a few ag researchers..."

Some of the earliest no-till research took place in the early 1950s. In those days, a few scientists in different areas of the world were using large amounts of atrazine and a few other herbicides as chemical burndowns to kill grass sod ahead of direct seeding corn and other crops.

The idea behind this early-day work was based on Edward Faulkner's 1943 words in the book *Plowman's Folly,* **"No one has ever advanced a scientific reason for plowing."**

New Approach

In 1951, Keith Barrons, J.H. Davidson and C.D. Fitzgerald at Dow Chemical in Midland, Mich.,

PLOWMAN'S FOLLY. Much of the early-day no-till work was based on Edward Faulkner's 1943 words in the book *Plowman's Folly,* "No one has ever advanced a scientific reason for plowing."

ROUGH RIDES ON THE HOME FARM

While I've tracked no-till developments since its commercial adoption in 1972, much of the early development behind this exciting practice occurred prior to me even hearing the words no-till. The notion of minimum tillage came first.

My first experience with reduced tillage happened on our Michigan Centennial farm where six family generations had lived since 1853. As a kid back in the early 1950s, I remember Dad giving minimum tillage a try with what was known as wheel-track planting.

Among the pioneers in the minimum tillage movement, Ray Cook of the Michigan State University soils department had pioneered wheel-track planting. The concept was never popular with growers, but it served as the fore-runner of today's conservation tillage and no-till systems.

After moldboard plowing, the next trip was made with our 2-row corn planter. Planting took place directly in the tractor wheel tracks. The wheel-track system eliminated at least three moisture-robbing, time-saving trips with discing, harrowing and cultipacking. While the system was effective in controlling both wind and soil erosion, it was a rough, bumpy ride. As a result, we only used the concept for 2 years before going back to a conventional tillage system.

Driving over the rough plowed furrows, the wheel-track system suffered from poor seed placement, was tough on the tractor driver's body and led to excessive machinery wear. Conserving moisture was still a big benefit though.

As a teenager, I figured there had to be a much better system for getting crops in the ground. And I could see the bumpy ride was taking a toll not only on Dad, but also on the planter and tractor. ❀

WHEEL-TRACK PLANTING. This system was used in the early 1950s to plant corn on the Lessiter family farm. After moldboard plowing, corn was planted in the tractor wheel tracks. While it was a rough, bumpy ride that was hard on the kidneys, this Michigan State University reduced tillage concept was the forerunner of later minimum tillage and no-till systems.

Soil Conservation Service

"Farmers were **still not yet ready** to adopt no-tillage..."

reported successful chemical weed control in seedbed preparation. This trail-blazing no-till research demonstrated important soil and moisture conservation benefits leading to a new approach to cropping with little or no soil manipulation.

To chemically control weeds, Barrons used as much as 6 pounds per acre of atrazine in order to burn down the grass sod before no-tilling corn.

Also in the early 1950s, M.A. Sprague at Rutgers University successfully substituted several herbicides as a substitute for mechanical tillage in pasture renovation work.

As early as 1952, several other researchers demonstrated that herbi-

How Many Trips Are Needed?

Tillage System	Trips
Conventional tillage	7 or more
Minimum tillage	4 or 5
No-tillage	2 or 3

— *No-Tillage Farming* textbook

1950 DECADE
AMERICAN AG FACTS

- Chelates are discovered in 1951, which are organic chemicals that better meet the nutritional needs of farm animals.
- Yields increase from use of anhydrous ammonia as a cheap source of nitrogen.
- The first mechanical tomato harvester is introduced in 1959.
- Agricultural products account for 22% of total U.S. exports ($3.53 billion per year), representing the large influence agriculture has had on the economy.
- The Agriculture Trade Development and Assistance Act of 1954, signed by President Eisenhower, lays the basis for permanent expansion of U.S. exports of agricultural products.
- Over 10,000 cooperatives with more than 7 million members lead to the formation of the National Farmers Organization in 1955.
- The Soil Bank Program was authorized in 1956 to pay farmers to stop using marginally productive land for 10 years.
- In 1957, Congress passes the Poultry Products Inspection Act, which required inspection of all slaughtered domesticated birds for human consumption.
- The Humane Slaughter Act of 1957 starts a series of animal slaughter regulations pioneered by a number of animal welfare non-governmental organizations.
- Tractor production peaks in 1951, with over 564,000 built as tractor numbers overtake the horse population.
- At the beginning of World War II there were 1.2 million tractors on about 30 million farms. By 1950, there were 4 million tractors on 23 million farms.
- The 1954 price of Iowa farmland averaged $198.91 per acre.
- The 1955 corn price in Iowa averaged $1.31 per bushel.
- The 1955 soybean price in Iowa averaged $2.24 per bushel.
- The 1955 corn yield in Iowa averaged 48.5 bushels per acre.
- The 1955 soybean yield in Iowa averaged 20 bushels per acre.

WERE FARMERS NO-TILLING IN THE 1930S?

This historical scene from the collection of Indiana photographer J.C. Allen indicates no-till may have been combined with cover crops long before no-till ever got its name.

Allen was a West Lafayette, Ind., photographer who captured Corn Belt ag scenes for many years in a business that eventually included three generations of family photographers. Assembled after Word War I, this collection eventually included more than 80,000 photos depicting many aspects of Midwestern farm life.

The limited information accompanying this photo from the '30s states, "A Farmall tractor pulls the binder to harvest soybeans as four horses pull a grain drill sowing a cover crop."

Likely shot sometime in the 1930s on a sunny fall harvest day on a central Indiana farm, further interpretation of this photo reveals some interesting points.

This photo shows seeding a cover crop directly behind soybeans harvested for hay. The seeding trip with the drill would be classified as no-till today.

What's amazing is how close the cover crop seeding is done directly behind the binder. The four horse-drawn drill had to follow directly behind the binder to avoid hundreds of twin-tied soybean bundles kicked out on the ground during the previous trip around the field.

When the horses stopped to drink or be fed, the tractor driver also had to take a break to keep both soybean hay harvesting and cover crop seeding on the same round.

Here's proof that cover crops have been around for many generations. Based on this photo, some farmers may have no-tilling before they even knew what the system was called. ❋

EARLY-DAY NO-TILL DOUBLE-CROPPING. Cover crops are seeded directly behind a binder cutting soybeans for hay in this 1930s photo.

cides just might make plowing and cultivation for weed control a chore of the past. By 1953, successful crops of no-tilled wheat, oats, flax, soybeans and corn were being produced by a few researchers even though they were not very well publicized.

Back then, all the new cropping know-how, chemicals, equipment and even courage on the part of a few farmers was still not enough to spur the widespread adoption of no-till-age.

No-Till with Grasses

No-till research was also being conducted at New Zealand's Massey University in the 1950s. John Baker, who did his graduate degree work on no-tillage at the school in the mid-1960s, years before forming the Cross-Slot Co., says several papers were published during the 1950s on no-tilling grasses into clover in New Zealand. No-tilling took place after the clovers had been established by broadcasting grass seed into the ashes of burned-over scrub on rough ground.

NO-TILL-AGE®
Does Deep Tillage Pay?
(With Port Byron loess soil at Rochester, Minn., 1997 to 2000)

July 2003

Tillage System	Corn Yield	Residue Cover
Continuous corn...		
Deep till *	163 bu.	54%
Chisel plow	166	26%
Strip-till	162	64%
No-till	155	87%
Corn after soybeans...		
Deep till *	186	41%
Chisel plow	182	23%
Strip-till	183	57%
No-till	182	67%

* Deep tillage at a depth of 15 inches was done each fall with an in-row subsoiler.

— University of Minnesota Southern Research and Outreach Center, Waseca, Minn.

EARLY-DAY NO-TILL RIG BUILT IN 1950S

In 1972, a college student brought his uncle's old no-till planter to Purdue University ag economist Howard Doster It was among 50 planters that had been manufactured in the 1950s based on a prototype created by Purdue agronomist George Scarseth.

The two-row machine featured a huge flat knife-like blade to slice grass sod ahead of each planter unit. But since it was built before herbicides for no-till burndown and weed control such as paraquat and atrazine were introduced, it was never a successful no-till unit. ❀

"This was a new approach with little or no soil manipulation..."

NO-TILL EDUCATOR OF THE YEAR. Keith Barrons (left) of Dow Chemical Co. was honored in 1975 for his early-day no-till research efforts. Shown here with *No-Till Farmer* Editor Frank Lessiter, Barrons was presented with a chrome-plated plowshare symbolizing the end of the moldboard plowing era.

While amitrol and dalapon were used for weed control, a drawback was that these herbicides had long residual effects when mixed in the soil. Seeding could not take place for several weeks after spraying by which time new weeds had often germinated.

In the mid '50s, L.A. Porter of the Levin Horticultural Research Station in New Zealand successfully produced strawberries using no-till techniques in chemically-killed sod.

While they laid the all-important groundwork for no-till, these researchers were well ahead of their time.

But as you will learn in the next chapter, the 1960s would bring major changes to how farmers worked their ground. ✾

WHY SHE HATES NO-TILL

Years ago, my wife, Pam, and I were touring one of the old plantations in southeast Virgina's Jamestown area. In the gift shop, there were a few Civil War bullets, pistols, horseshoes and other relics for sale.

Somehow, the lady behind the counter heard me say something about no-till. And she says, "I hate no-till."

When I asked why, she added, "Moldboard plowing always turns up a few Civil War relics that area farmers bring to us that we can make some money off with tourist sales. We don't get any of those old Civil War items from no-tilled fields." ✾

THE FIRST RIG. The Harry Young family modified this ancient mule-drawn two-row planter for their first no-till planting work.

Harry Young family

No-Tilling
Through the '60s...

> "There's been an increased concentration of organic matter **in the top few inches of soil** after more than five decades of no-tilling..." — *John Young*

Looking back at when no-tillage started to gain momentum, the 1960s quickly come to mind. Not only was there more interest at that time by scientists in no-till, but it also got a start on a few farms around the country.

These early no-till developments have led to where the concept has become widespread on all kinds of land from coast to coast. Not only has no-till helped reduce erosion, but it's reduced the cost of production by decreasing the number of needed trips over many fields. The impact of those fewer trips is compounded by avoiding compaction by no longer making multi-trips with tillage tools over problem soils.

In the 1960s, a number of both university and industry scientists tackled the notion that had represented "conventional wisdom" for farmers since the pioneer days — that clean, deep moldboard plowing, the fine working of soils and mechanical cultivation was the only way to farm.

LOOKING BACK ... ORGANIC MATTER INCREASES WERE DRAMATIC

Harry Young family

SMALL-SCALE NO-TILL PLOTS. Harry Young, shown here doing plot work with a hand-powered no-till unit, and University of Kentucky agronomist Shirley Phillips worked closely on many unanswered no-till concerns. They put out test plots for different crops, herbicides, insecticides, seeding rates, disc openers, closing wheels and much more. At one time, they were applying 5 pounds of atrazine per acre to control weeds in no-tilled corn.

Over the decades, three generations of the Young family have pushed up their no-tilled acreage of corn, wheat, barley and double-cropped beans. Even with waiting to plant until after barley or wheat harvest, a combination of double-cropped beans and small grains often earned more dollars than full-season soybeans.

When John Young came back to the farm after attending the University of Kentucky and Purdue University in the mid-1970, he was skeptical about the value of no-till. Once he got over his skepticism, he learned from his dad how to pay attention to the small details and how one small change could make a huge difference in the results from an on-farm experiment. A good example was the impact that the correct sprayer boom height and the elimination of pesticide and fertilizer overlap could have on the farm's bottom line.

John has seen an increased concentration of organic matter in the top few inches of soil after more than five decades of no-tilling. Further down in the ground, Young believes there hasn't been much soil change since no-till corn and soybean roots don't run very deep.

Some early-day critics of no-till argued compaction would become a problem with continuous no-till, but that has not happened in the Young field that has been continuously no-tilled for more than five decades.

"Nowhere in this field is there the amount of compaction that many people would expect," Young says.

Looking back over more than 55 years of no-till, he says no-till had a huge impact on farming operations:

◆ There is much less erosion, which is a critical concern in many areas.

◆ With no-till, more acres are being farmed today by fewer people.

◆ While chemical usage has gone up over the past five decades, there are fewer fertilizer and pesticide carryover problems with no-till.

◆ Weeds that show up in continuously no-tilled fields are now the same weeds showing up with conventional tillage.

"We find fewer and fewer weeds that are a problem with no-till that aren't also problems with conventional tillage," Young says. "The newer herbicides have really helped, especially over-the-top compounds."

Having seen many benefits, Young and his son, Alex, will definitely continue no-tilling. Plus, he sees many exciting new wrinkles coming with no-till that will make it even more profitable in the years immediately ahead. ❁

NO-TILLAGE PLANTER AMONG THE WONDERS OF AGRICULTURE

In the late 1960s, Allis-Chalmers developed and manufactured the first commercially available no-tillage planter. Instead of turning under 500 to 1,000 tons of topsoil per acre with moldboard plowing, company engineers came up with several unique coulter designs and planter frames that turned no-till into a reality.

Starting with two-row units, the planters were capable of no-tilling corn and soybeans in 20- to 40-inch rows.

In 1993, the Allis-Chalmers no-til planter was recognized by the Equipment Manufacturers Institute as one of the 100 most important developments in agriculture around the world. ❀

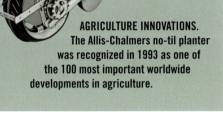

AGRICULTURE INNOVATIONS. The Allis-Chalmers no-til planter was recognized in 1993 as one of the 100 most important worldwide developments in agriculture.

Allis-Chalmers

More No-Till Research Work

Following on the footsteps of a few no-till research projects carried out in the 1950s, the new decade led to no-till experiments in several states.

In the early 1960s, J.E. Moody, G.M. Shear and J.N. Jones Jr., produced excellent corn yields by using herbicides to replace tillage at Virginia Polytechnic Institute.

T.S. Speight, an agronomist with the R.J. Reynolds Tobacco Co., was among the first American educators to successfully double-crop soybeans after small grain harvest. These early tests were conducted at the cigarette company's farm in Merry Hill, N.C., including evaluations with a number of herbicides, spray application methods and other cropping techniques while evaluating no-till opportunities.

George Scarseth of Purdue and Glenn Klingham at North Carolina State University also did extensive basic no-till research work in the early 1960s.

Harry Young family

NO PLOW CORN. Christian County ag agent Reeves Davie points to the planter furrows and some vegetation yet to be killed with atrazine shortly after the first no-till corn was planted in April of 1962 at the Harry Young farm.

At the same time, no-till research was also being done in Ohio and Illinois. A summary of some of the early no-till findings in Ohio is described later in this chapter.

Overseas, no-till research was being carried out in France, Germany, Belgium, Switzerland, the United Kingdom, several Scandinavian countries and New Zealand.

Erodible Land Concerns Drove Studies

George McKibben starting giving no-till a try in some research plots around 1960. The agronomist at the University of Illinois Dixon Springs Experiment Station in Simpson, Ill., believed chemical control of residue and weeds was a possibility. He viewed no-till as a means to allow production of corn and soybeans on hilly, erodible land while conserving valuable moisture.

McKibben soon developed innovative no-till systems that eventually allowed farmers to harvest high yields of corn and soybeans from land that was too steep for conventional tillage. He demonstrated how no-till not only reduced soil losses, but conserved soil moisture, trimmed machinery, fuel and labor costs and offered timelier planting and harvesting.

Later, McKibben saw his early work with no-till lead to widespread adoption on all kinds of farmland. Not only has no-till helped reduce erosion, but it has also dramatically reduced the cost of production by trimming the number of tillage trips over many fields.

ORIGINAL ZERO-TILL PLANTER. This 2-row rig was built and used by McKibben in the early '60s.

Frank Lessiter

NO-TILL LOOKED PROMISING. George McKibben's early no-till work at the Dixon Springs Agricultural Center in southern Illinois convinced Harry Young to give no-till a try.

Harry Young family

BIGGER AND QUICKER. By the late 1960s, Harry Young had switched to larger no-till rigs to no-till double-crop soybeans as soon as possible after barley and wheat harvest.

First Farmer Usage

Harry Young, Jr., had been dreaming for several years about the possibilities of reducing tillage on his southwestern Kentucky farm. Conventional tillage had been practiced on the Herndon, Ky., operation that the family had farmed since the 1830s.

In the summer of 1961, county extension agent Reeves Davie urged Young to join a small group of farmers traveling to view the trail-blazing no-till plots at the Dixon Springs Agricultural Center and visit with McKibben.

Arriving back home after the tour, Young told family members he felt he could make no-till work. His earlier ag economist training convinced

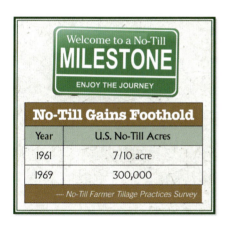

him that no-till would save labor, fuel, machinery and soil.

Young had seen how a few area orchard growers had successfully used paraquat as a contact herbicide to keep weeds away from their trees. He felt this contact herbicide could be used to terminate growing vegetation prior to no-tilling corn and soybeans.

Young decided to try the idea the following year and no-tilled corn on 7/10ths of an acre with a home-modified old mule-drawn two-row planter.

Harry Young family

WHO'S BUYING ALL THAT PARAQUAT?

Sometime in the late 1960s, the top brass in the ag division at Chevron Chemical Co. in San Francisco wondered why ag suppliers were selling so much paraquat in western Kentucky.

Under license from England's Imperial Chemical Inc., (ICI), the herbicide was mainly used in the U.S. to control weeds in and around trees in orchards and to kill weeds in non-cropping situations — such as around giant oil tanks, buildings or along railroad right of ways.

The oil giant's higher-ups didn't think there was much in the way of orchards in western Kentucky. So who was buying all the paraquat?

As the story goes, a Chevron ag marketing manger attending a meeting in New York was told to stop in Kentucky and check it out before heading back to the company's west coast headquarters.

Not happy about extending his trip, he flew into Nashville, Tenn., rented a car and made the 70-mile trip northwest to visit an ag supplier in southwest Kentucky and talk about their purchases of paraquat.

Expecting to be in and out in a couple hours, he booked a flight home for later that afternoon.

He got the shock of his life when the ag retailer took him out to the Young family farm for his first glimpse of no-tilled corn and double-cropped soybeans following wheat. The Young brothers (Harry and Larry) had hit on the idea of using paraquat as a burndown contact herbicide to control whatever vegetation was growing prior to no-tilling corn and soybeans.

The Chevron staff member canceled his afternoon flight. In fact, he was still in the area 2 days later, soaking up all the knowledge he could about no-till.

This spur-of-the-moment trip led to Chevron becoming a major player in the no-till market. They started working toward a paraquat herbicide label for no-tilled corn along full-season and double-cropped soybeans.

The rest is history, as Chevron and ICI soon became major players in the no-till movement. ✺

Welcome to a No-Till

MILESTONE

ENJOY THE JOURNEY

No-Till Gains Foothold

Year	U.S. No-Till Acres
1961	7/10 acre
1969	300,000

— *No-Till Farmer Tillage Practices Survey*

That field has been no-tilled every year since 1962.

With favorable results, Young quickly expanded the acreage of both no-till corn and soybeans that were double-cropped behind wheat and barley. By the early 1970s, the operation included the harvest of 2,150 acres of no-tilled corn, soybeans, wheat and barley from only 1,500 acres.

The rest is no-till history. His son, John, and grandson, Alex, carry on Young's longtime tradition of no-tilling 2,000 acres of corn, 2,000 acres of wheat and 2,000 acres of double-cropped soybeans ... all harvested from only 4,000 acres.

KEEN INTEREST IN NO-TILL. During the first 22 years of using no-tillage on the Harry Young farm, more than 11,000 visitors came to see how it was done.

November 1984

WHERE NO-TILL GOT ITS ON-FARM START

Sometimes photos can practically tell the whole story.

On a Sunday afternoon in the late summer of 1984, a historical roadside marker was dedicated at the Harry Young, Jr., farm near Herndon, Ky. Presented by Kentucky's Christian County Historical Society and, DuPont, the marker commemorates the beginning of no-tillage farming in the U.S.

It was appropriate that Shirley Phillips, the Cooperative Extension Service leader at the University of Kentucky, was on hand for the dedication. As leaders in the no-till movement, both men were pleased with the honor and the historical marker that still stands on the Young farm. ✽

John Harvey

WHERE NO-TILL GOT ITS START. In 1984, a historical roadside marker was dedicated at the spot where no-till was started on 7/10ths of an acre near Herndon, Ky. Harry Young (left) and University of Kentucky agronomist Shirley Phillips (right pioneered the commercialization of no-till in North America.

Industry Adoption

Based on the no-till successes on the Young farm, Allis-Chalmers, the full-line farm equipment manufacturer headquartered in West Allis, Wis., soon got in the game with a coulter-equipped no-till planter. Soon after, staffers at Chevron Chemical Co. saw the benefits of marketing paraquat as a burndown herbicide in no-till situations. While residual weed control in those days was pretty much limited to early applications of 2,4-D and atrazine, other chemical companies soon got into the no-till market. ❧

WHERE NO-TILL GOT STARTED

Here's what the no-till acreage looked like in 1969

State	No-Tilled Acres
Kentucky	150,000
North Carolina	30,000
Tennessee	35,000
Virginia	20,000
Ohio	8,000
Illinois	6,000
Arkansas	5,000
Iowa	3,000
Missouri	2,500
Indiana	2,000
Nebraska	900
Kansas	100

— 1969 No-Till Farmer Tillage Practices Survey

1960 DECADE
AMERICAN AG FACTS

- By 1960, 96% of the U.S. corn acreage is planted with hybrid seed.
- The new Gaines and Fortuna wheat varieties are introduced.
- Due to the severe financial conditions of the railroads in the 1960s, American agriculture sees an increase with cargo plane food shipments.
- The Trade Expansion Act leads to less than a 1% increase in agriculture's percentage of total exports.
- County Extension programs across the U.S. expand due to anti-poverty programs, such as the President's Committee on Rural Poverty and Child Nutrition Act.
- Agriculture sees an increase in unionization and fair labor standards, including the United Farm Workers Organizing Committee. This allowed the unionizing of California farm workers, while the Fair Labor Standards Act was extended to include agricultural labor.
- By 1968, 96% of cotton is harvested mechanically.
- Because of the on-going Vietnam War, the federal government uses food surpluses for the needy at home and aboard.
- Rachel Carson, a U.S. biologist, sparks an interest in the dangers of pesticide use in her book, *Silent Spring*.
- There are multiple movements throughout the 1960s to keep land in farming.
- The 1964 price of farmland in Iowa averages $291.77 per acre.
- The 1965 price of corn in Iowa averages $1.13 per bushel.
- The 1965 price of soybeans in Iowa averages $2.59 per bushel.
- The 1965 corn yield in Iowa averages 82 bushels per acre.
- The 1965 soybean yield in Iowa averages 26 bushels per acre.

"No-tillers never get plowed."
— Slogan on a No-Till Farmer hat

NO YIELD DIFFERENCES BETWEEN NO-TILL AND CONVENTIONAL TILLAGE SYSTEMS

One of the early scientific papers published on no-till was authored by Ohio State University agronomists Glover Triplett, and D.M Van Doren, Jr., along with ag engineer W.H. Johnson. With the title of "Non-Plowed, Strip-Tilled Corn Culture," it was printed in the 1964 proceedings of the American Society of Agricultural Engineers.

The paper summarized 3 years of research work from 1960 to 1962 in which corn was successfully grown without plowing or the

NO-TILL PIONEER. As an Ohio State University and Mississippi State University agronomist, Glover Triplett spent more than 55 years doing no-till research.

use of secondary tillage implements. The key was the control of non-crop vegetation with herbicides without injuring the no-tilled corn.

Summarizing 23 replicated experiments where corn was no-tilled into alfalfa sod, the researchers found no yield differences with corn between the no-till and conventional tillage systems on soils ranging from silt loam to clay.

A treatment of 2 pounds of 2,4-D that was applied either a month or a week before planting completely killed the alfalfa sod, but gave no control of grasses. Mowing the vegetation was completely unsatisfactory.

The paper pointed out that any successful no-till herbicide system must perform several key functions:

◆ Kill the vegetation initially present in the cropped area.

AMONG FIRST NO-TILL RESEARCH PAPERS. This research paper by Ohio State University scientists G.B. Triplett, Jr., D.M. Van Doren, Jr., and W.H. Johnson was printed in the 1964 proceedings of the American Society of Agricultural Engineers annual conference.

◆ Suppress growth of perennial and annual weeds from seed.
◆ Not injure the crop being planted.
◆ Not injure any following crops.
◆ Compete in cost with other crop cultural techniques.

A combination of dalapon and amitrole applied in April controlled both broadleaf and grassy vegetation so there was nothing growing on these plots when corn was no-tilled on May 15. ✳

No-Till Weed Control Treatments Compared
1961 corn yields on a megis silt loam soil in Ohio

Treatment	Corn No-Tilled in Grass Sod	Corn Grown After Corn
Plowed and disced	124 bu.	94 bu.
20 pounds sodium arsenate	86	79
8 pounds dalapon, 4 pounds amitrole	119	103
2 pounds atrazine plus 4 pounds amitrole, 2 pounds atrazine *	132	113
2 pounds 2,4-D, 2 pounds atrazine	127	113

* All herbicide treatments were made on April 27, except for one May 15 atrazine treatment.

— *Ohio State University*

No-Till Quickly Gained
Momentum in the '70s

"Tear out an old fence row and the soil tilth will be fabulous. But till this ground for 10 years and you'll never know where the fence stood..." — *Ed Hawley*

Back in the 1970s, I always tried to attend George McKibben's annual late-summer "No-Till Field Day" at the Dixon Springs Experiment Station in Simpson, Ill.

No other researcher in the country — either then or now — has ever put out over 420 unique yet practical herbicide comparisons for no-till corn, full-season no-till soybeans and no-till double-cropped soybeans as McKibben did in those days. I always came away from viewing his plots with fresh ideas on how to keep no-till weeds under control.

Where No-Till Got Started

In fact, it was McKibben's early-day plots in 1961 that encouraged no-till pioneer Harry Young to give no-till a try on the family's southwestern Kentucky farm in 1962.

Over the years, McKibben developed no-till systems that allowed growers to harvest high yields of corn and soybeans from land that's too steep for conventional tillage. Farmers, especially those living in the hilly areas of southern Illinois, eagerly adopted the practice.

He found no-till not only trimmed soil losses, but conserved soil moisture, reduced machinery, fuel and labor costs, permitted more timely planting and harvesting and saved time.

In fact, no-till's reduction in trips across the field has had a tremendous impact in reducing soil compaction caused by as many as a half dozen trips with intensive tillage equipment over problem soils.

No-Till Just Not Important

Sadly, most of these highly practical no-till plots were disbanded a few years after McKibben retired.

Years later, I was a member of an American Society of Agricultural Engineers conference panel with the dean of agriculture at the University of Illinois. I told him how the university had missed the boat by withdrawing funding from this highly practical look at no-till in favor of preferring to underwrite more basic ag research. As you'd guess, he didn't appreciate my comments.

In another academic incidence, I recall a University of Wisconsin extension agronomist who let growers know he had little use for no-till. After he became dean of agriculture at Wisconsin, he changed his tune and became a proponent of no-till when he saw the dollars available for no-till research and education efforts. ✴

May 1976

MAKING THE CASE FOR SOLID-DRILLED SOYBEANS

NARROWER SOYBEAN ROWS. Research trials at Purdue University in the mid 1970s showed a 12-bushel-per-acre benefit for solid-seeded soybeans compared to 32-inch rows.

Frank Lessiter

There's been lots of interest lately in drilling solid-stand soybeans. While most of the research work and farmer experience to date has been with conventional tillage, there's no reason this idea won't work with no-tillage.

Drills or row-crop planters with units mounted close together on a toolbar can be rigged for no-till to give a fairly solid stand of soybeans. The Bettinson drill, which is now available in the U.S. through the Melroe Division of Clark Equipment, could handle solid-planted soybeans.

No-tillers are already controlling weeds in soybeans with herbicides so this is not a major problem. However, weed control is a critical concern with solid-drilled soybeans.

It's possible to get a 10%-20% increase in bean yields by shifting from wider rows to rows spaced 6-7 inches apart, say USDA agronomists at Urbana, Ill. Besides higher yields, other advantages include less labor, lower equipment costs, reduced soil erosion and improved harvest efficiency.

In trials conducted by Purdue University agronomists, three varieties of solid-seeded soybeans averaged 82 bushels per acre under exceptional growing conditions at West Lafayette, Ind. Soybeans in 32-inch rows that relied to some extent on cultivation for weed control yielded 70 bushels per acre. ✴

TWO TRUE NO-TILL PIONEERS

Frank Lessiter

HARRY YOUNG. The first commercial application of no-till took place in 1962 on 7/10ths of an acre at the Young farm near Herndon, Ky. Note Harry's mailbox use for worn out no-till coulters that called attention to the farm's pioneering no-till efforts.

Frank Lessiter

GEORGE MCKIBBEN. The University of Illinois educator at the Dixon Springs Research Center, located in the far southern part of the state, typically put out over 400 herbicide test plots with no-tilled corn and full-season and double-cropped soybeans each year.

While Harry Young and George McKibben are deceased, both men made tremendous contributions to today's no-till success. Each received many awards and well-deserved distinctions, yet two unique chrome-plated plowshare awards from *No-Till Farmer* always had a special place in their hearts.

Young was named "No-Till Farmer of the Year" in 1973 and McKibben was named "No-Till Educator of the Year" in 1974.

Since the early 1960s, tremendous amounts of research work and further refinement of no-till has been done across the country by many educators and farmers. But both Young and McKibben played key roles in the advancement of no-tillage as we know it today.

From less than a single acre in 1961, no-till has certainly come a long way —to an estimated 100 million acres of no-till in the U.S. in 2017. ❋

January
1973

OVER 1,000 FARMERS AT EARLY-DAY WINTER NO-TILL MEETING

In December of 1972, four *No-Till Farmer* staffers attended one of the earliest big-time winter farmer meetings dealing with no-till. The Delmarva No-Tillage Conference attracted more than 1,000 farmers to the event held in Delmar, Md.

Sponsored by several agribusiness firms and the Cooperative Extension Services of Maryland, Delaware and Virginia, the event featured 19 researchers, extension staffers and farmers discussing many aspects of no-till. In addition, U.S. Secretary of Agriculture Earl Butz was on hand to cite the many erosion control benefits of no-till.

Ernest Fuchs told attendees the day might be coming when it will be mandatory to no-till. The Centreville, Md., grower got into no-till with 300 acres of corn and 200 acres of soybeans when bad weather didn't give him an opportunity to complete his conventional tillage tasks.

In the January 1973 issue of *No-Till Farmer,* we listed a dozen practical no-till tips from farmers, researchers and extension staffers who spoke at the conference.

1 If you have problems covering seed with a no-till planter, add disc coverers to throw soil over the seed. —*Tom Williams, University of Delaware.*

2 Use twice as much surfactant when applying paraquat with liquid fertilizer to enhance the killing action. — *Jim Parochetti, University of Maryland.*

3 The average response from using Furadan insecticide on no-till corn in 1972 was an additional 12 bushels per acre. — *Bill Mitchell, University of Delaware.*

4 There was no seed damage from knifing in anhydrous ammonia within 10 inches of no-tilled corn rows. — *Bill McClellan, Pennsylvania State University.*

5 With no-till, figure on doubling your labor efficiency. — *Bob Raver, ag agent in Dickerson, Md.*

6 Keep the soil covered since the more sun that hits the soil, the more fall panicum that germinates. — *Neal Hering, Westminster, Md.*

7 Cover crops will be essential with nutsedge problems. — *Ernest Fuchs, Centreville, Md.*

8 Anytime fall panicum is a concern, run soil tests as AAtrex and Princep aren't as effective once the soil pH goes down. — *Erskin Bedford, Bluemont, Va.*

9 With no-till, don't sample soils more than 4 inches deep. — *George Hawkins, Virginia Polytechnic University.*

10 Double-cropped no-till soybeans averaged 7 bushels more per acre in 20-inch rows compared to 30-inch rows at two locations in 1972. — *John Schillinger, University of Maryland.*

11 It costs $12.47 less per acre to plant corn with no-till. — *Richard Lester, St. Georges, Del.*

12 Lots of moisture made it easy to handle no-till last year, but remember you can still make mistakes. — *Joe Newcomer, University of Maryland.* ✾

HERE WE GO ... ONE MORE TIME

Daily newspaper headlines are once again proclaiming an energy shortage ... and the fact that we could have fuel rationing before long.

Up until yesterday, I figured it was just a lot of talk. But when I pulled into my favorite gas station last night, they were not using their self-serve pumps, they no longer pump gas 24 hours a day and I was limited to buying 10 gallons. All due to fuel allocation problems.

With fuel prices going up, experts say the problem is due to the Iranian situation, the severe winter and the diversion of fuels to more profitable markets, such as Europe.

So maybe it's once again a good time to review the fuel savings that less tillage can bring.

Save 3 1/2 Gallons per Acre

The newest fuel figures I've seen come from Michigan State University agronomists who compared nine tillage systems.

◆ They found conventional tillage — featuring the moldboard plow — required six trips across the field and 4.27 gallons of diesel fuel per acre.

◆ Minimum tillage required two, three or four trips across the field and 2.17 to 2.73 gallons per acre.

◆ No-tillage — with one trip for planting and another trip for spraying — required only 0.74 gallons per acre.

Using my calculator, a farmer with 600 acres of ground could save 2,118 gallons of diesel fuel by switching from moldboard plowing to no-tillage. At 60¢ per gallon, this would be a savings of $1,270 a year.

What About Herbicides?

While these fuel savings are substantial, one argument you always get is that the calculations didn't include the extra energy needed to manufacture herbicides.

Yet University of Delaware agronomists Tom Williams and Norman Collins have done so. They calculated that a no-till system that relies on paraquat and atrazine for knock-down and weed control requires the equivalent of only 3.06 horsepower hours per acre.

NO-TILL-AGE®

Why Bean Growers Haven't Moved to No-Till or Reduced Till *

Reasons for not switching	% Growers
Price of equipment	26%
Slow growth and lower yields	20%
Ground too wet in the spring	16%
Don't have proper equipment	15%
Doesn't fit rotation or cropping system	11%
Higher weed-control costs	9%
Ground too cold in the spring	9%
Poor soil-to-seed contact	6%
Concerns about soil compaction	4%
Roundup Ready seed not available	2%

* Reasons given by growers with more than 200 acres of soybeans who don't use no-till or reduced tillage to grow beans.
— 2001 American Soybean Association survey.

CHEMICAL WEED CONTROL SAVES ENERGY. Even factoring in the 3 horsepower equivalent for herbicide usage, no-till is a huge energy saver.

Adding this to the 13.61 horsepower hours needed for planting gives no-tillage a total of 16.67 horsepower hours per acre. By comparison, 86.28 horsepower hours are needed in a conventional tillage system with moldboard plowing.

So switching to less tillage can definitely help this country save energy — whether the current expected fuel crisis is real or not. ❀

Welcome to a No-Till
MILESTONE
ENJOY THE JOURNEY

No-Till Pushes Ahead

Year	U.S. No-Till Acres
2010	88,000,000
2019	106,700,000

— *No-Till Farmer Tillage Practices Survey*

1970 DECADE
AMERICAN AG FACTS

- Throughout the 1970s, inflation increases and economic growth slows down.
- Fertilizer usage increases from 32 million tons per year in the 1960s to 44 million tons per year in the 1970s, partially prompting the Clean Water Act in 1972.
- Paul Berg, a molecular biologist, begins transferring genes from one strand of DNA to another.
- The USDA and land-grant institutions establish the Council on International Science and Education.
- The Agriculture and Consumer Protection Act emphasizes increased production instead of government control of farming.
- In 1972, the grain market was rocked when the Soviet Union made $1.5 billion worth of deals with the five biggest U.S. grain companies, the largest ever grain deal between these two nations.
- Farm income skyrocketed from $2.3 billion in 1972 to $19.6 billion in 1973, as farm income pulled slightly ahead of average urban incomes for a short period of time.
- Consumer food prices rise, especially at the meat counter.
- Roundup, Monsanto's agricultural herbicide made with glyphosate, is released in 1974.
- The first twin-rotor system combine —created by Sperry-New Holland —is introduced.
- Norman Borlaug is awarded the Nobel Peace Prize in 1970 for his contributions to helping farmers increase the world's food supply.
- The 1974 price of farmland in Iowa averages $719 per acre.
- The price of corn in Iowa averages $2.66 per bushel in 1975.
- Soybean prices in Iowa averaged $5.23 per bushel in 1975.
- The 1975 corn yield in Iowa averages 90 bushels per acre.
- The 1975 soybean yield in Iowa averages 34 bushels per acre.

LETTER AVOIDS LEGAL SKIRMISH

With the title of "Energy Needed to Produce Corn Under Conventional, Chisel and No-Till Systems," the first "No-Till-Age" chart appeared in the March 1983 issue of *No-Till Farmer*. Appearing in every issue since that time, this chart has always shared a quick no-till fact or figure for subscribers.

But in the mid-1980s, there was a time when the streak appeared in jeopardy.

At that time, Shell Chemical Co. was a major player in the no-till market with Bladex herbicide. One day the first edition of a new Shell no-till grower newsletter showed up in my mailbox. I was shocked to see the name of their newsletter: "No-Till-Age."

Although we'd been using the No-Till-Age logo for a few years by this time, I figured we'd likely have to give up the catchy name. We were in no mood to finance a costly legal battle and I figured Shell had 30 attorneys on their staff just itching for a legal fight.

Our next step was to write a letter to Shell explaining how we had been using the No-Till-Age name since 1983 and felt they didn't have the right to use the name. I was shocked when their attorneys wrote back, said we were right and that future newsletters would carry another name.

After that happened, we quickly registered the No-Till-Age name with the federal government's copyright office to protect ourselves. ✺

DOUBLE-CROPPING CORN BEHIND HAY

*By Ed Hawley, as told to Darrell Smith, former No-Till Farmer Associate Editor and retired Farm Journal Agronomy Editor, in this article that appeared in the very first issue of **No-Till Farmer**.*

Erosion control, time and, as the ecologists remind us, animal life started me thinking about minimum tillage long before I finally tried it.

To begin with, it seemed to require too much work and expense to grow corn with conventional tillage. You can buy more machinery and more chemicals, but you can't put more days on the calendar.

I farm 355 acres at Stockton, Ill., and milk a herd of Holsteins with only my wife, two teenage sons and father-in-law for help. It keeps us moving, but you are beat these days if your income has to go for hired help.

Six years ago, I bought an Allis-Chalmers no-till planter and tried minimum tillage on 165 acres of corn where I just disced and then planted.

Two Crops in a Year

I soon discovered no-till would enable me to get an early spring cutting of alfalfa and then produce a corn crop on the same field in the same year. We usually chop that first cutting for silage, but last spring it was baled, since cool weather put us behind schedule.

Each year, we select 25 to 35 acres of alfalfa/brome that is due to be broken up. In mid-May we chop it for silage or bale it. Paraquat kills off the regrowth and we

"No-till lets us get a cutting of **alfalfa and grow corn on the same field in the same year...**"

no-till corn right away. AAtrex keeps the weeds down for the rest of the season.

Instead of our usual 110-day corn varieties, we no-till a 90-day variety into the alfalfa stubble. Corn yields

PLANTING GREEN IN 1972. Many of the early day reduced tillage pioneers started out by no-tilling corn into a growing grass crop. Northwestern Illinois veteran no-tiller Ed Hawley chopped silage or baled hay in mid-May from alfalfa/brome fields before immediately double-cropping no-till corn.

have always been comparable to the rest of our acreage, once getting up to 166 bushel per acre.

With no-tillage, the ground is much easier to drive on at silo-filling time. This makes the corn a dream to chop.

While the extra cutting of hay comes in handy, saving time is the biggest advantage. If I can plant corn a few days earlier — actually be planting instead of plowing — it pays off in extra yield and saves valuable time.

Timing is mighty important with our clay soils, as there are only a few days when that ground is fit to work. Plowing that soil when it's too wet will affect corn for the rest of the year. With no-tillage, you can stay out of the fields until the ground is fit to work and still get the crop planted on time.

NO-TILL-AGE®
Do You Apply Starter Fertilizer with No-Tilled Crops?

April 2002

Practice	Corn	Soybeans	Small Grains
Always	70%	11%	33%
Sometimes	14%	19%	15%
Never	16%	70%	52%

— Survey of 2002 National No-Tillage Conference attendees

"Too much tillage can **destroy** what nature put there for us..."

No-tillage allows us to get in the field a lot earlier, so there's not much conflict between chopping that first cutting of hay and no-tilling corn. We usually begin planting our full-season corn around the last week of April.

Last year, we planted corn in stalks without any tillage. We couldn't see any difference in yields between those 45 acres and where the stalks were disced before we planted.

More Below-Ground Critters

Even the soil microorganisms seem to benefit from reduced tillage. Last year, one of my fields was disced, treated with 5 pounds per acre of AAtrex and planted. Walking across it after a spring rain, my son and I no-

ticed literally thousands of earthworms; the holes were as thick as a pegboard.

But in a nearby conventionally-tilled field, which had suffered erosion from the storm, there wasn't a worm to be seen. That leads me to suspect a connection between the amount of tillage and earthworm numbers.

As for erosion, we've had rains hard enough to wash gullies in our graveled driveway without a bit of soil loss in contoured, minimum-tilled corn fields.

Building up organic matter is a long and slow process, but I believe cutting back on the number of tillage operations can help. For example, if you tear out an old fence row, the soil tilth there will be fabulous. But cultivate it for 10 years and you'll never know where the fence once stood. Too much tillage can destroy what nature put there for us.

Neighbors always look for trash in minimum or no-tillage fields. But by harvest time the residue is all back in the soil in the form of humus.

So far, we haven't encountered any disease or insect problems — we use a root worm insecticide, but so do area farmers who use conventional tillage. Minimum tillage is saving our soil, adding an extra hay crop and, perhaps more importantly, cutting down on man-hours per acre.

Avoiding hired labor is helping keep "Minimum Till Acres" a real family farm. ✿

No-Till Acres
Doubled in the '80s

"No bigger change has occurred during the past 15 years in agriculture than the ever-growing shift to less tillage..."

With innovative farmers across the country leading the way, no-till got its "roots" secured in American agriculture during the 1980s. During this 10-year span, the no-till acreage nearly doubled, expanding from 7.1 million acres in 1981 to 14.1 million acres.

After several decades of farmers cobbling together no-till equipment in farm shops, equipment manufacturers saw the light. This is when they finally got serious about the no-till market and came out with planters and drills that could make the seeding job under high residue conditions much easier.

While many of the original no-till planters couldn't guarantee a consistent and uniform stand, the newer no-till planters were very reliable — when properly adjusted and operated.

At the same time, farmers saw the weed control advantages offered by Roundup and other selective herbicides that led to improved weed control in no-till situations. Even so, the lack of consistent, complete weed control was still a major deterrent to more rapid acceptance of no-tillage.

GETTING STARTED WITH NO-TILL

October **1985**

Jim Kinsella has enjoyed 9 years of no-till experience on his own farm near Lexington, Ill. Besides the personal no-till experience he and his father have had, Jim also picked up many more no-till ideas from his travels as a district manager for BASF.

Based on all his experiences, he's outlined 17 ideas for anyone giving no-till a try:

NO-TILL SOIL IMPROVEMENTS. Illinois no-tiller and strip-tiller Jim Kinsella stressed the benefits of no-till in improving soil quality, organic matter and earthworm numbers.

Frank Lessiter

1 Correct soil compaction before you no-till. While no-till doesn't create compaction, it won't solve it. Used in the fall under dry conditions, Kinsella feels the Paraplow is an excellent tool (Page 74) that appears to be more effective after corn than soybeans.

2 Correct your soil pH or low fertility prior to no-tilling.

3 Start with a positive attitude. Make a personal commitment to leave your soils in good condition for future generations.

4 Start no-tilling on a limited basis, but don't give up if a few problems occur. Corn following untouched soybean stubble is the easiest place to start.

5 Don't be more critical of a no-till system than of your present tillage system. They both have problems.

6 Watch and learn from others. Adapt their ideas.

7 Don't be concerned if your fields don't look like your neighbor's. With current farm economics, you must farm for profit — not looks.

8 Work with landowners right from the start as you start to no-till. Work together and seek mutual goals. Be flexible, understanding and a little daring.

9 Choose the most cost-effective inputs — not necessarily the cheapest.

10 Leave untreated check strips and take yield checks. Don't expect miracles from no-till during the first year.

11 Become your own crop scout or hire a professional scouting service. With no-till, you will have more time for scouting. Buy an ATV if you don't like to walk.

12 Spraying weeds is the most critical operation with no-till. Purchase a good sprayer or find a reliable custom applicator with no-till experience.

13 Rotate crops if possible. A corn and soybean rotation is a natural for no-till.

14 Don't leave all of the needed nitrogen for corn on the surface since losses could be excessive.

15 Drain wet areas with tile. Since tile receives nearly the same tax incentives as equipment, consider trading some of your "excess equipment" for tile.

16 Question every pass across a field as to its benefits vs. its impact on the soil.

17 Reduce soybean row spacing to 8-10 inches. The quick canopy development with no-till solid seeding improves weed control, reduces soil erosion and increases yield potential. ✿

In addition, higher fuel prices and tough economic times for agriculture in the 1980s led many farmers to take a closer look at the advantages of no-till when it came to trimming production costs. This brings up the question as to what might have changed in regard to the popularity of no-till if the Iran oil crisis had never happened.

Farmers also saw new ideas that innovative no-tillers were adopting to not only trim costs, but also to protect the environment.

Others in the ag field were also starting to understand the benefits of no-till and started pointing out the flaws of conventional tillage. More government officials at all levels started to get onboard with the advantages of no-tilling on erodible ground.

No-Tilling a Growing Trend

Including no-till among the top 10 farming trends for 1983, *Successful Farming* editors and readers had some interesting things to say about the reduced tillage practice.

Stressing ways to increase profitability, they cited the example of farmers who use no-till to double-crop beans directly behind the combine at wheat harvest.

Another growing trend was the use of post-emergence herbicides, which is of extreme importance to no-tillers in tackling late-season weed pressures.

At the same time, no-tillage got some contradictory views in the magazine's survey. Some growers called no-till a major ag trend while others cited it as one of the biggest disappointments of the prior decade.

As the *Successful Farmer* editors pointed out, both groups may be right. But new planters and drills

1980 DECADE
AMERICAN AG FACTS

- ◆ The industry boom of the 1970s quickly turns into a bust in the 1980s because of grain surpluses, which led to farm protests.
- ◆ The average number of farmers shrink while the average farm acreage grows, and enrollment in collegiate agriculture programs drops during the farm crisis.
- ◆ Biotechnology starts to become a practical source for improving crop and livestock products.
- ◆ The U.S. and Canada trade accord starts free trade in all commodities
- ◆ In 1985, the Food Security Act lowers government farm supports and sets up the Conservation Reserve Program.
- ◆ Scientists at Monsanto create the first genetically modified plant cell, a tomato variety resistance to Roundup herbicide.
- ◆ In 1984, the Landsat 5 satellite launches into orbit and provides agricultural imagery for the first time.
- ◆ A severe drought in 1988 leads to poor crop yields, Congress approves $6 billion in aid for farmers and the corn supply shrinks by 31%.
- ◆ During the 1980s farm crisis, the U.S. farm debt doubles from its 1978 amount and grows to 15 times the amount of debt in 1950. The Federal Reserve estimates 19% of American farmers own 63% of the total farm debt.
- ◆ Net farm income plummets from $19 billion in 1950 to $5.4 billion in 1984.
- ◆ The 1987 price for farmland in Iowa averages $947 per acre.
- ◆ The 1985 for corn in Iowa averages $2.41 per bushel.
- ◆ The 1985 price for soybeans in Iowa averages $5.32 per bushel.
- ◆ The 1985 average corn yield in Iowa averages 126 bushels per acre.
- ◆ The 1985 average soybean yield in Iowa is 38 bushels per acre.
- ◆ Farm equipment sales increase at the end of the decade and more farmers begin using low-input sustainable agriculture (LISA) technology.

UNLIMITED CHOICES. With so many no-till planter and drill options, it may be difficult to decide which ones are best for your farming conditions.

CHOOSE FROM 864,000 AVAILABLE PLANTING COMBINATIONS

While there are many components to mull over in designing a no-till planter, the first step is correctly matching each soil-engaging part with the soil and crop residue on your own farm.

Actual performance of your no-till planter depends on the type of soil, moisture content, residue and how well these conditions interact with your machine, says John Morrison, USDA ag engineer with the Grassland, Soil and Water Research Laboratory in Temple, Texas.

Until now, matching a no-till planter to field conditions was largely a question of trial and error. But there's now more information available to help you decide what options you'll need and how they apply to the overall performance of your no-till planter.

Adapting the right components will improve planting efficiency and erosion control at the same time. Considering rainfall, water runoff, soil type, slope and steepness of grade is also essential.

Tracking Critical. How each component tracks or adapts to the surface of a field is another matter. Used on hillsides and curved rows, for example, a slot opener may not follow in the coulter slit or the press wheel may miss the seed slot. This can be caused by large fore and aft distances between no-till planter components. Closer-spaced components help overcome this problem.

Pivots placed between the coulter, furrow opener and press wheel will improve tracking with curved rows. Pull-type planters track better than mounted planters on curved rows, while mounted planters track better on hillsides.

Tackling Residue. At no-till planting time, surface residues usually cover much of the soil surface, says Ronald Allen, USDA ag engineer in Bushland, Texas. These residues can be coarse or thin, tall or short, chopped or long and attached or loose.

One problem is that residue can hairpin around soil-engaging components such as chisel shanks, supporting struts and frame members. Effectively cutting residue ahead of each soil-engaging tool will reduce hairpinning, says Allen.

Residue that catches on the no-till planter can be eliminated by staggering adjacent components. Using smooth-sided wheels, eliminating protrusions and bottlenecks between components is also helpful.

While rocks will slow down your no-till planting

speed, rolling coulters and disc openers will roll over obstructions with only a momentary loss of depth control. Protect rigid shank-type openers with trips or shear pins.

Active soil-engaging components perform different functions such as cutting residue, opening a seed furrow and pressing seed into contact with the soil.

Here are seven no-till planter components and how they function:

1   **Slicing Soil, Residue.** Rolling coulters cut through soil and residue. Coulters are sometimes not required on machines that have openers such as a staggered double-disc opener designed for cutting and opening seed slots.

9-ROW NO-TILL RIG. With a double toolbar frame, growers were able to no-till in narrower rows.

Smooth coulters generally cut through soil better and are easily sharpened. Rippled coulters tend to be self-sharpening and tolerate sticky soils. Narrow-fluted coulters and bubble coulters loosen soil in the immediate row area, but don't work well in sticky soil. While wide-fluted coulters accomplish strip tillage in friable soils, they throw too much soil out of the row when planting above 4 miles per hour.

Options for soil and residue cutting include: smooth coulter, notched coulter, coulter with depth bands, offset (bubble, ripped or fluted) coulter, straw straightener, powered blade or coulter, strip rotary tiller and dual secondary residue discs.

2   **Row Preparation.** These devices prepare the row area or deeply loosen soil ahead of the seeding units. Row clearing devices remove dry soil along with residues, which brings the no-till planter into contact with moist underlying soil.

Options include a sweep row cleaner, two-disc row cleaner, horizontal disc row cleaner, wide-fluted coulter, ripper chisel, subsoil ripper, packer roller, rolling basket, rotary cultivator, spring tines and S-tines.

3   **Accurate Depth Control.** This ensures even plant emergence. For no-till drills, rear press wheels are often used to provide depth control

because of space limitations. With this system, the opener and press wheel are mounted on a trailing arm arrangement or on the planter's parallel linkage.

4   **Soil Opening.** Components for depth control include rear press wheels, side gauge wheels, skid plate on each opener, front press wheels, frame lifting/gauge wheels, depth bands on the front leading coulter and disc openers.

Many no-till planters and drills use regular or staggered double-discs openers to open seed furrows. Components for opening the soil for seed placement include a double disc with or without a shoe, staggered double discs with or without a shoe, runner, stub runner, hoe, single discs, coulter, chisel, wide sweep, triple discs and powered blade or coulter.

5   **Seed Firming.** A seed-firming wheel is sometimes used to press seed into the bottom of the seed furrow. These devices are semi-pneumatic rubber wheels ranging from 1 by 6 inches to 1 by 10 inches, or solid-plate wheels as narrow as 0.25 inches.

6   **Seed Covering.** These covering devices must have loose moist soil available to place on top of the seed or must loosen the soil and move it over the seed. Components for seed covering include single covering discs, double covering discs, paddles, knives, loop or trailing drag chains and spring times.

7   **Seed Slot Closure.** Almost all seeders use press wheels to close and/or compact the seed slot. Seed slot closure devices include a wide semi-pneumatic or steel wheel, single rib wheel, double rib wheel, narrow semi-pneumatic or steel wheel (V-shaped or rounded), dual angled semi-pneumatic or steel wheels, split steel wheels and dual wide flat wheels.

No-till planter options such as chemical and fertilizer attachments may require additional weight for soil penetration and a stronger planting machine frame. Unfortunately, these options can reduce clearance between planting components and may reduce machine tolerance to heavy residues. ❁

FOUR REASONS TO NO-TILL

October 1983

When George Kapusta asked an audience, "why no-till?" he pointed to a list of four reasons they should consider when moving to this method of reduced tillage.

The agronomist at Southern Illinois University has been a long-time advocate of the many benefits of no-tillage and has always had interesting ideas to share on the subject. Here's what he told the farmers.

"We finally are beginning to realize we can no longer tolerate such losses and maintain high yield levels," adds Kapusta. "By no-tilling, farmers can reduce soil erosion almost to zero."

There they are...four key reasons as to why no-till will continue to grow in the years ahead. It's a pretty simple explanation of a question that's been asked many times over the years.

1 **Decrease Operating Costs.** No-till will reduce the amount of machinery needed to farm and let you use smaller horsepower tractors. This is because the no-till planter or drill is the biggest implement you have to pull through your fields. Since less trips are made over the field with no-till, labor costs are drastically reduced.

2 **More Timely Operations.** Since land preparation and herbicide incorporation are not necessary with no-till, planting can be more timely. This usually leads to higher yields with no-tillage.

3 **Moisture Conservation.** Retaining crop mulch on the soil surface and eliminating tillage can conserve soil moisture. This often results in higher no-till crop yields, which was proven a number of times this year in drought areas.

4 **Reduce Environmental Pollution.** Soil erosion due to wind and water has been a serious problem for many decades where the soil is plowed. In fact, 2 bushels of soil are lost due to erosion for every bushel of corn produced using conventional tillage. ✻

SAVE ON HORSEPOWER NEEDS. With no-till, considerably less tractor power is required since the biggest implement to pull is the planter or drill.

along with better chemicals have made no-till much easier than it was 10 years ago.

More Full-Season No-Till Needed

Holding back the growth in no-till acres in the early 1980s was the lack of acceptance for no-tilling more full-season crops. As an example, a 1985 survey in Illinois showed only 6% of the corn and 1% of the full-season soybeans were no-tilled.

But by the end of the 1980s, that had changed. Data from the *No-Till Farmer* Tillage Practices survey showed the no-till corn acreage had increased from 2.9 million acres in 1980 to 5.3 million acres by1989.

During that same time period, the no-till soybean acreage increased from 2.6 million acres to 5.1 million acres. And by the end of the decade, 48% of the no-till soybean acres were full-season beans. ❈

FARM THREE TIMES MORE LAND WITH NO-TILL

March 1984

Without no-till, Vernon and Rick Foster figure they'd only be able to plant one-third as much corn on their Baltimore County, Md., dairy farm. A big time-saver, the father and son team also credits no-tillage with reducing erosion by 50%-80% over conventional tillage.

Since 1973, they've used no-till with the same contour strip-cropping system laid out nearly 4 decades ago. Along with no-till, they alternate grass and nitrogen fixing legumes with row crops to rebuild soil tilth and fertility. ❈

October 1983

Frank Lessiter

NO-TILL SOLVES EROSION CONCERNS. With no-tilled wheat, barley and lentils, 71% of Idaho farmers experienced less erosion compared to using more extensive tillage.

WHY GROWERS ARE NO-TILLING

With no-tillage turning in a sizeable increase in Idaho, officials of the state's Department of Water Resources recently surveyed farmers regarding their attitudes about this reduced tillage practice.

Back in 1980, only 16,000 acres were no-tilled in Idaho. This increased to an anticipated 140,000 acres for 1983 in our *No-Till Farmer* Tillage Practices Survey.

No-Tillage Works. Some 58% of the surveyed farmers no-tilled, with wheat, barley and lentils being the primary crops. Some 21% of the farmers reported higher yields with no-till while 33% saw no change.

This left 46% of the farmers who saw a drop in yields, with 55% of these farmers seeing a yield decrease of 20% or more. This is at least partially due to remaining plant residues, which keep the soil cool, slows germination and also increases weed and insect problems.

Some 85% of these no-tillers reported a reduction in fuel consumption — with the majority having a reduction of up to 39%.

Soil Is Saved. An overwhelming 71% of the farmers experienced much less erosion with no-till.

Before 1981, 29% of the farmers had already tried no-tillage, 29% had tried minimum tillage and 18% had used both reduced tillage practices.

Idaho farmers seemed pleased with the results of trying less tillage, although they are definitely concerned about rising production costs and changing yields — just like farmers everywhere. ❈

NO-TILL OUTSHINES THE PARAPLOW

BRITISH WORK. The Paraplow was developed by ag engineers in Great Britain.

Howard Rotavator

We've been hearing a great deal about the benefits of using the Paraplow from Great Britain, but recent Ohio research shows no-tillage is still a better bet. In fact, tests with the Paraplow over 3 years in Ohio studies indicate no benefit when compared to no-tilling.

Studies conducted by Ohio Agricultural Research and Development Center agronomist Dave Van Doren, Jr., and ag engineer James Henry showed an average yield of 132 bushels of corn with no-till and 127 bushels for the Paraplow treatment. These figures are based on results from a half dozen experiments representing 11 cropping years.

In one comparison, use of the "bare" no-till technique, where as much crop residue as possible was removed from the ground prior to planting, averaged 127 bushels per acre. Moldboard plowing gave a yield of 99 bushels while combining moldboard plowing with a Paraplow treatment yielded 105 bushels per acre.

"For the soil, climatic and timing of operations pertaining to these experiments, use of the Paraplow did not produce sufficiently improved soil conditions to cause greater corn yield than was produced from a stable, no-tillage situation," says Van Doren. "One would have difficulty recommending the extra expenditure of energy and time to Paraplow."

The researchers say the Paraplow might show more promise if used when the soil was drier — such as in the fall. Another good use might be where excessive compaction occurs. ❋

DOES PARATILLING REALLY PAY OFF?

(4-year averages)

Tillage	Corn Yield
Annual paratill, no-till	150 bu.
Biannual paratill, no-till	146
No-till	135
Annual paratill, moldboard plow	156
Biannual paratill, moldboard plow	155
Moldboard plow	154

— *Purdue University*

MORE COST, MORE TIME WITH PARAPLOWING. While there was keen interest among a number of U.S. growers, especially with serious compaction conditions, tests by university scientists in a number of states found there was little benefit to combining paraplowing with no-till.

USDA

NO-TILL WORK. Smaller firms, like Jon Kinzenbaw's startup company, led the way in selling the many benefits of no-till.

NO-TILL MAKES ROUGH GROUND PAY

March 1984

When a rough 40-acre field came on the market 9 years ago, Charles, Ed and Matthew Clavin weren't overly enthused about taking it on.

"But the price was such that we couldn't refuse it," says the Rosamond, Ill., farmer. "But what to do with it was something of a problem.

"The field was so rough that some spots weren't farmed at all. Other places were such that you couldn't pull a silage wagon for fear of turning it over."

After the Clavins bought the 40 acres in 1975, they decided the only thing they could do would be to try a rotation of continuous wheat and no-tilled double-crop soybeans.

Wheat averages 50 bushel an acre and double-crop soybeans make 30 bushels. Best of all, this program lets them conserve and improve their valuable soil. ✿

January 1987

WE'RE 15 YEARS OLD!

Looking back over the first 15 years of this publication, many of our pioneering articles have revealed, and in many cases, created new trends and new ways of farming with less tillage. They've also pinpointed new directions and possibilities for future growth in what was the still relatively young no-till area of crop production.

Over the past 15 years, the *No-Till Farmer* editors have risen to the challenge by reacting clearly, objectively and forcefully in providing needed no-till information.

From today's advantage point, no-till has dramatically matured with new innovations in equipment, pesticides and crop production practices. No bigger change has occurred during the past 15 years in agriculture than the ever-growing shift to less tillage.

With the tremendous savings no-till has to offer in reduced machinery, labor and other costs, it's definitely a practice whose time has come. ✿

6 ROWS NO-TILL CORN, 6 ROWS NO-TILL BEANS

The cropping program Dan Stadtmueller follows usually draws a crowd of curious people. It's because the Monticello, Iowa farmer alternates no-till corn and beans every half-dozen rows on about one-half of his 1,100-acre farm.

Corn is planted, then 6 rows of beans are planted between the corn strips 2 or 3 weeks later. These soybean strips give the corn more sunlight while the corn helps shade the beans. Stadtmueller says the yield increase with corn more than offsets any yield loss in the shaded beans.

Since planting is done on last year's ridges, there is no problem in finding the alternating rows. By banding herbicides and fertilizer, there's no chemical interference.

Corn is planted with a John Deere Max-Emerge and beans are planted with a Hiniker planter. Corn is cultivated once and beans twice.

"We think this system has promise as an erosion control system because you never have the whole field covered with just soybean residue," says Stadtmueller. "Soybean residues often don't provide adequate protection for the soil during heavy rains. With these alternating strips, half the field is always covered with corn stalks, which gives excellent erosion protection."

Dry fertilizer application is completely mechanized since Stadtmueller pulls a 5-ton cart behind his planter. This set-up lets him fill fertilizer boxes on the go.

While most no-tillers apply herbicides either before or after planting, Stadtmueller is convinced there are advantages to applying herbicides with the planter.

"I think it's crucial to apply herbicides as you are planting," he says.

"This way, the herbicide goes right to work even without rain. Saddle tanks mounted on the tractor carry herbicides that are sprayed in a band directly over the row area — cutting chemical costs right in half.

"The cultivator is the pressure point," he adds. "So I may broadcast more herbicides in the future.

"I'm a believer in sidedressing ammonia. I've done part of it as a separate trip and also done part as I cultivate. I've done that on every other row for 6 years."

Speaking about fall tillage, Stadtmueller recommends only one practice — tilling fields that need improved tillage to make no-tillage work. ✺

ALTERNATING CROPS, ROWS. Dan Stadtmueller found that alternating no-till corn and soybeans resulted in higher yields, more sunlight for the corn and increased shading for the soybean crop.

SLICED $150,000 IN EQUIPMENT COSTS

April 1985

Terry Schneider says equipment savings are probably the most important economic aspect of switching to no-till.

"We've eliminated $150,000 of equipment since starting to no-till," adds the veteran no-tiller from Shirley, Ill. "When you take that times 14% interest divided by 1,020 acres, you see a savings of more than $20 per acre."

Schneider credits no-till with maintaining yields while sharply reducing costs. He's found no-till has also freed up more time to spend making management decisions like marketing or walking the fields rather than riding through them on a tractor. ✸

NO-TILL-AGE®

February 1986

Growers See No-Till Disadvantages

No-Tillage Disadvantages	Continued No-Till Users	Former No-Till Users
Increased chemical costs	58%	65%
Weed control concerns	40%	59%
No-till equipment costs	27%	43%
More difficult to manage	25%	35%
More precise planting needed	17%	28%
Need to keep conventional planter	13%	28%
Spraying residue	8%	24%
Yield variability	7%	22%

— *University of Tennessee survey of 509 west Tennessee farmers*

LESS WATER NEEDED

November 1984

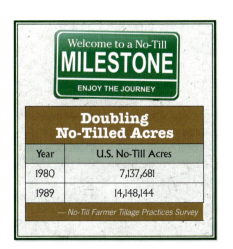

Welcome to a No-Till
MILESTONE
ENJOY THE JOURNEY

Doubling No-Tilled Acres

Year	U.S. No-Till Acres
1980	7,137,681
1989	14,148,144

— *No-Till Farmer Tillage Practices Survey*

OCEANS OF WATER. A change in his herbicide tankmix program allowed Mac Baggett to save $100 per day in fuel and labor costs needed to haul water for weed control in double-cropped soybeans. ✸

KNOW SOIL CONDITIONS. The amount of no-till residue and soil moisture conditions can have a major impact on how planter attachments work.

Frank Lessiter

IMPROVED SOILS. No-till will improve soil quality, which can reduce erosion, increase organic matter and provide more effective planting.

CHECK HOW SOIL CONDITIONS IMPACT NO-TILL

April 1988

When selecting no-till planter attachments, keep these performance conditions in mind:

1 Heavy, wet, poorly drained soils tend to be adhesive and leave seed furrows that are difficult to close over the seed.

2 Heavy, dry soils tend to be difficult to penetrate with planter openers, produce clods if disturbed by tillage tools and are difficult when it comes to closing seed furrows.

3 Crusting soils are susceptible to excessive amounts of compaction over the seed row, which may reduce plant emergence.

4 Friable, medium-textured, well-drained soils may be planted over wide ranges of moisture content with satisfactory results.

5 Naturally consolidating soils are difficult to penetrate at low moisture contents. They are often susceptible to excessive compaction that is caused by the planter's gauge wheels and press wheels when soils are wet.

6 Soils with consolidated subsoil layers often can only be planted when both the topsoil and subsoil properties are amendable to planter attachment disturbance. ✼

November 1972

NO-TILL LOOSE ENDS...

NO-TILL GOES TO THE DOGS

Quite often, we hear of new crops being grown under no-tillage conditions. But one favorable comment I've heard that was really different dealt with the promotion of no-tillage corn for beagle clubs.

The author of this article in the *National Beagling News* was suggesting beagle club managers promote no-tillage corn because it can improve dog training conditions. The dead vegetation from planting corn in sod would remain on the soil surface and make running easier for the beagles. (And probably also for the rabbits.)

The author also told beagle owners that no-tilling corn and soybeans will reduce both crop costs and erosion. Could this be the end of dogs falling into 3-foot deep gullies in conventionally-tilled corn fields? ✼

Great Plains Mfg.

No-Till Growth Goes
'Hog Wild' During the '90s

"Farmers are going to shift faster than ever before to no-till in the next 20 years to save soil, boost profits, cut costs and do a better job of protecting the environment..."

In the years since *No-Till Farmer* got its start in 1972, we've seen many progressive things happen as our editors visited with no-till farmers around the country.

No longer was "no-till" a dirty word and no longer were farmers using this reduced tillage practice only in "back 40" fields away from the highway so nobody could harass

them about all the surface trash. No-till had definitely come of age.

After seeing the many benefits of no-tilling, farmers in the 1990s were no longer waiting for better crop prices or lower machinery prices before moving to less tillage.

Instead, the switch to no-till was stronger than ever before. The same was also true among farmers

moving away from moldboard plowing to various forms of minimum tillage.

No-Till Had Arrived

The 1990s was the era when no-till came into its own. During this 10-year span, there was an astounding 296% growth in U.S. no-tilled acres. By 1999, farmers were no-tilling

NO-TILL TECHNOLOGY HAS PLACE, BUT FIRST FOCUS ON MOTHER NATURE

February 1999

Despite all the talk about the latest technologies that claim to boost yields for no-tillers, Ray Rawson says the key to healthier crop life and higher yield is often overlooked by over-zealous farmers.

First Things First

"The first step you should take is to evaluate your immediate environment," explains the developer of the zone-till system and a large-acreage grower at Farwell, Mich. "Each of us needs to know what we have to work with and where we're at."

Evaluating the environment, according to Rawson, means assessing three key elements.

"All living things must have oxygen, water and nutrition," he says. "If we're going to do well, we have to have a good environment to live in."

Also important is the mindset of the farmer.

"Everything we do out there is going to be directly related to our income," explains Rawson. "We must have an excellent stand and improvement of soil tilth. The question is: How do we get there?"

Turning your attention toward soil quality is a step in the right direction, says Rawson.

> "Use a no-till crop rotation system to **replenish the soil...**"

"Labs tell us that air and soil water are critical soil concerns. Most list it as the number one and two areas to focus on," he says.

One way of setting your sights on soil improvement is to concentrate on carbon dioxide.

"They tell us there is not enough carbon dioxide to produce quality crops," explains Rawson. "We can solve this problem by allowing decay to occur in the soil."

This is good news for the no-tiller, but perhaps not such good news to the conventional tilling farmer. Their soils lose 80-90 times more carbon dioxide than is the case with no-till.

Increasing the amount of carbon dioxide in the soil has a cycling effect, according to Rawson. Since plants depend on carbon dioxide for growth, they will have heavier root mass with the increase of carbon dioxide in the soil. The heavier the root mass, the more carbon dioxide will be released. The plants will only grow healthier and stronger, as more water and more holding capacity will make them less likely to be damaged from disease.

Keep It in Balance

It's a delicate cycle, but as Rawson says, one that can be understood.

NO-TILL NUTRIENT NEEDS. With improved soil quality occurring with no-till, farmers find that their fertilizer needs often change dramatically.

"Don't just focus on one crop," he says. "Use a no-till crop rotation system to replenish the soil."

Finding the right system is imperative. The one that works best for Rawson includes aerating the soil with deep slots, utilizing the available crop culture system and focusing on harvestability and water management.

"This is not recreational tillage," he says. "We're all business people, with big dollars involved. Anytime we go out there with something that isn't necessary, it's a loss. Everything we do, every tillage pass needs to be profitable."

One way to check your soil type is by reverting back to the old-fashioned way of simply digging a hole.

NO-TILL-AGE®

March 1992

Should You Rotate No-Till Soybeans?
(1975-1991 average yields with 30-inch rows)

Tillage System	Soybean Yield Per Acre
No–till soybeans after corn...	
Fall plow	53 bu.
Fall chisel	51
No-till	49
No–till soybeans after soybeans...	
Fall plow	48 bu.
Fall chisel	47
No-till	46

— *Purdue University*

NO-TILL BOOSTS SOIL QUALITY. Ray Rawson maintains conventionally-tilled soils lose 80-90 times more carbon dioxide than is the case with soils that are no-tilled.

Frank Lessiter

"The first thing I do on a new farm is dig holes," explains Rawson. "I dig in the highest elevation in the field and then in the lowest. This way, I get a good idea of soil makeup and what needs to be done for the healthiest plant life."

Understanding the soil is extremely important to no-tillers like Rawson. But he also takes into account location, weather and climate.

"It's about going back to the basics — water, oxygen, carbon dioxide and soil warmth," he says. "Everything we do relates to timing, relieving, lag time and accelerating plant growth. We must take into consideration the difference in location. We plant differently in the north than in the south."

But no-tillers should not play down the value of sunlight.

"Some 95% of everything we get is from sunlight," says Rawson. "We need to have the capacity to utilize that. We can do it with massive plant leaves if we keep them green. But we also need to have a massive root-structure underneath that plant to store it.

"It's the things we get for free that make the difference. It's your job to always understand what matters." ✳

"As growers gained experience, they moved toward 100% no-till..."

Frank Lessiter

NO-TILL FINANCIAL BENEFITS. Most no-tillers quickly learned they could profit from a sharp reduction in cropping, equipment, fuel and labor costs. Others found no-till offered a more effective way to utilize already available manpower by expanding their operations to even more acres.

nearly 50 million acres, up from 16.9 million acres in 1990.

So what led this dramatic expansion? A number of things, including the realization that no-till could be used with any crops. There was no longer any reason to no-till one crop but then use minimum tillage on other crops in the rotation.

As growers gained experience and success with one or two no-tilled crops, they developed the confidence to move toward 100% no-till. But they quickly recognized a change in cropping strategies and management abilities were needed if they were to make no-till successful year after year.

Refinements and innovations (particularly among shortline farm equipment manufacturers) with planters, drills and air seeders made life much easier for no-tillers. With these commercial innovations, growers were no longer spending long hours over the winter months in their farm shops modifying equipment.

New herbicides and insecticides offered no-tillers many fresh opportunities for combating weed and insect concerns. Thanks to new chemistries, weed control was no longer the excuse holding back further no-till growth.

Cold Soil Concerns

During the 1990s, the strip-till acreage also started to take off. Our staff recognized strip-till's promise in some no-till operations and starting publishing strip-till articles in the late 1980s.

In fact, two speakers at the very first National No-Tillage Conference in 1993 zeroed in on the pros and cons of strip-tilling. One was Cliff Roberts of Kentland, Ind., who was among the first growers to strip-till by leaving a small ridge in the fall where corn would be no-tilled the following spring. Like many growers, Roberts was looking for a new way to overcome early-spring concerns with no-tilling corn into cold soils.

While "pure no-tillers" turned up their noses —and continue to do so today — at the value of strip-tilling, it became another option for some no-tillers.

Another major reason leading to the huge increase in no-tilled acres were the financial benefits. No-tillers recognized they could profit from a sharp reduction in crop costs and labor needs. Others saw no-till as a way to utilize available labor to take on even more acres.

By 1994, no-till was used on 14% of the nation's cropland. Yet conventional tillage was still the rage, being used on 62% of the country's crop acreage.

Adoption of no-till is among the best ideas we've seen in years and years of change in farming. It will continue to grow as more farmers see it as a way to save soil, boost profits, cut investment costs and do an even better job of protecting the environment. ❋

FIRST INVITED, THEN UNINVITED

This story indicates what can happen when herbicides become a critical issue in promoting no-tillage.

It all started in March when the Hiniker Co. of Mankato, Minn., a major ridge till and no-till equipment manufacturer, was invited to participate in a no-till seeding demonstration at the North Iowa Area Community College (NIACC) in Mason City, Iowa.

By mid-April, they had been "uninvited."

Who Uninvited Them?

In a letter sent to the *Mason City Globe-Gazette,* Hiniker officials claimed a chemical company insisted that the firm's equipment not be included in the seeding demonstration because it reduces the amount of herbicide needed for weed control.

However, Dana Dinnes, director of the Agronomic Demonstration center at NIACC maintains additional weed control with Hiniker equipment comes from cultivation. That is inconsistent with the strictly herbicide weed control feature of their no-till system.

"The demonstration of equipment is not meant to be an endorsement, but just a chance for the kids to see different systems in use," Dinnes told Kevin Baskins, a reporter for the *Mason City Globe-Gazette* newspaper. The no-till seeding demonstration project is a partnership with NIACC, Iowa State University and BASF.

The letter to the newspaper editor stated Hini-

NO THANKS. Because cultivation was used to help control weeds, the ridge till system wasn't welcome at a northeast Iowa no-till demonstration.

ker equipment leaves residue from the previous year's crop on the soil surface and tills the soil under the residue, which helps removes weeds.

Not A Pure System

Dinnes says it's precisely this soil disturbance via cultivation, which prompted the decision to "uninvite" Hiniker from participating in the true no-till system demonstration.

"There's no hidden agenda," Dinnes says. "We just wanted to stay with a no-till system in as pure as possible a form. The Hiniker system has more soil disturbance than what we are comfortable with for our demonstration project."

While relying on herbicides is definitely the way most no-tillers control weeds, you'd be surprised at the USDA data that shows a sizable number of no-tillers also rely on high residue cultivators to take out escaped weeds.

Mike Kelly, assistant manager for the BASF Agronomic Development Center in Lexington, Ill., says the burndown herbicides needed prior to crop emergence as part of the NIACC project aren't even manufactured by BASF.

"With either system — no-till or Hiniker — BASF does not really have anything to gain in regard to the specific herbicides used for weed control prior to planting," he says.

Regarding the reasons behind withdrawing the invitation to participate in the no-till demonstration, you can draw your own conclusions. ❋

ADDRESSING THE COLD SOIL CURSE: NO-TILLING SOYBEANS IN NOVEMBER?

June
1999

"No one's ever accused me of being completely normal," is a claim often made by no-tiller Dave Savage. And there's good reason.

This crop consultant from Farley, Iowa, has tried almost every possible scenario when it comes to early planting dates, earning him the "distinguished" honor of being the butt of many no-till jokes.

Believe it or not, there's a method to his madness — and the metamorphism of snickers and sneers to asking questions and learning has begun. All of this is thanks to his persistence in pushing the no-till envelope just a little further.

Confused? Let Savage Explain

"The philosophy I take with early planting dates with no-till is we have this cold soil curse that seems to dog us with corn yields," he says.

"I came to the conclusion that if I can back my corn planting date off until it warms up a little bit more and

> ## "November planted soybeans stayed viable until the middle of April..."

get into that first week of May, we can get picture-perfect, picket-fence stands of corn. So I thought I'd rather sacrifice some stand to get the planting date benefit.

"We have the unique opportunity with no-till to take advantage of cold soils."

However, Savage was more interested in working with the planting dates of no-tilled soybeans. He says the premise for early planting is based on the idea that if soybeans could be no-tilled into cold ground, it would act like a cold storage unit that would preserve the seed until the soil warmed up enough for germination and growth. Theoretically.

You Want to Plant When?

In 1996, Savage was surprised to see soybeans planted on April 18 out-yielded soybeans planted on April 27, May 22 and June 16. It surprised him because the weather was exceptionally disagreeable.

"I thought I would be early with mid-April," he says. "But I put them in and got scared to death because it

MANAGEMENT CHALLENGES WITH EARLY NO-TILLED SOYBEANS

David Savage offers this advice for no-tillers thinking of aggressively seeding soybeans earlier.

♦ **Be Ready to Go.** You may only get a 2-day window of opportunity in which to no-till your soybeans.

♦ **Keep The Acreage Small.** Don't go overboard the first time you try this idea.

♦ **Soil Conditions.** You don't always get a chance to find the right planting conditions.

♦ **Planting Depth/Cutting Trash.** If you're no-tilling soybeans following Bt corn and you've got a lot of foliage, it's tough trying to cut that residue.

♦ **Harvest Loss.** There's some loss due to short plants and low bean pods.

♦ **Seed Treatment vs. Non-Treatment.** Definitely use a seed treatment.

♦ **Selection of Soybean Varieties.** There isn't

any big difference in varieties, so use one that's adapted to your area.

♦ **Managing the Burndown.** Put some thought into what you're going to do with a herbicide program. Roundup Ready is going to make it awfully easy, but there are other options. Avoid marestail or fields where you have used 2, 4-D.

♦ **Risk Assessment.** Review your soybean seed replant policy, because you will likely have to replant at some point. Can you afford to cover the same acres twice?

"In a 4-year period, even if one of the years is a complete disaster, you're still going to make an additional $10.88 per acre per year," says Savage. "And that's with 10% overseeding and paying more for additional seed." ❋

"If you want to go earlier than that, consider yourself a researcher and failure at that point **is a good learning experience...**"

not only rained but it also snowed. I thought those beans were dead. They came out with 54 bushels an acre. This has even been replicated."

The second year, Savage pushed the planting date ahead to April 3. Again, Savage was nervous.

"The ground froze pretty good after that date," he says. "People were spreading manure on top of the soil to make it warm. I kept digging up seeds and putting them in the kitchen window, and sure enough, they were germinating."

Since the seeds had swollen, he feared they might not germinate. But he reports an average yield of 66 bushels per acre, even after the swollen seeds sat in frozen soil for 7 days.

Thanksgiving Beans?

The third year was the one that earned Savage his "nutty" reputation. It was the year he planted soybeans on November 22. Savage also planted soybeans one week before Christmas as well. Other planting dates included March 7 and May 14.

"The November beans stayed viable until the middle of April," he says, "but the earthworms did the harvesting for us.

"The Christmas beans got a hard freeze right after planting, so they basically sat in cold storage until spring. They were our only factory-planted treated seed. We got 66.7 bushels per acre."

But the yield difference between the March planting and the May planting really made the neighbors talk.

"We didn't plant any special bean variety or use any different seed coating with the March date and we got a 3-bushel advantage over the May planting with 65.8 bushels an acre," says Savage. "It was hard to believe, but it was there."

So perhaps Savage isn't so crazy. He's made valiant efforts for no-tillers everywhere to take advantage of "the cold soil curse." But he does so with a word of caution.

"I picked the first part of April because that's when I thought we would get maximum results with the least risk," he says. "If you want to go earlier than that, consider yourself a researcher and failure at that point is a good learning experience." ❁

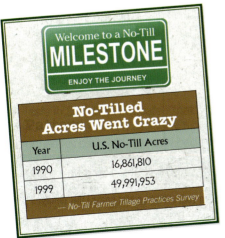

No-Tilled Acres Went Crazy

Year	U.S. No-Till Acres
1990	16,861,810
1999	49,991,953

— No-Till Farmer Tillage Practices Survey

20 YEARS OF U.S. NO-TILL PROGRESS

(1972 to 1992 acreage changes)

Crop	Change
Total no-till acres	674%
No-till corn	399%
No-till soybeans (all)	1,004%
No-till soybeans (full-season)	1,951%
No-till soybeans (double-crop)	579%
No-till small grains	1,390%
No-till grain sorghum	519%
No-till forages	240%
No-till cotton	4,984%
Total minimum-till acres	236%

— 1972 and 1992 No-Till Farmer data

1990 DECADE
AMERICAN AG FACTS

- The information age and Green Revolution lead to drastic technological improvements in American agriculture and the expansion of precision farming techniques using satellite technology.
- In response to rapidly growing technology, the Food and Drug Administration grants approval of the first biotechnology produced food, the Flavrsavr tomato, declaring it safe for human consumption.
- With a rise in the use of no-till and conservation tillage, methods to counter erosion leads to a 40% drop in cropland erosion between 1982 and 1997.
- The 1990s mark the availability of the first weed and insect resistant biotech soybeans, leading to the fastest adoption ever of any new ag technology. It went from 4.3 million acres of biotech crops in 1996 to 98.6 million acres in 1999.
- Technology advancements bring a substantial decrease in labor needs for farmers. The farm population drops from 6 million (3.4% of the labor force) in the 1980s to 2.9 million (2.6% of the labor force) by the end of the 1990s.
- The new North American Free Trade Agreement (NAFTA) in 1993 increases trading and leads to an agricultural export record of $60.4 billion in 1996.
- In 1996, Monsanto and Bayer release the first herbicide resistant corn varieties genetically altered to express proteins from the bacterium *Bacillus Thuringiensis (Bt)*.
- At the end of the 1990s the USDA releases the first official organic standards and seal of approval.
- The U.S. Environmental Protection Agency sets a precedent by registering Monsanto's new beetle-killing *BT* gene potato, New Leaf Superior, as an insecticide.
- Increased productivity leads to crop surpluses and a price slump in the late 1990s.
- The 1997 price of farmland in Iowa reaches $1,697 per acre.
- Diesel costs in 1995 average $1.10 per gallon.
- The price of corn in Iowa averages $2.47 per bushel in 1995.
- Soybean prices in Iowa averaged $5.74 per bushel in 1995.
- The 1995 corn yield in Iowa averages 123 bushels per acre.
- The 1995 soybean yield in Iowa averages 44 bushels per acre.

NO-TILL LOOSE ENDS...

SOUTH AMERICANS RUSHED TO NO-TILLING

January 2006

Although no-tilling has experienced steady U.S. growth, the adoption lags far behind South America. Since 1987, no-tilling has exploded by 47% in South America, compared to only 6% in the States.

There are a handful of reasons for no-till's quick adoption in South America, according to Rolf Derpsch, a noted no-till advocate from Paraguay.

- No-till provides efficient and economically viable erosion control in countries where climatic conditions favor high erosion and soil degradation.
- Widespread use of cover crops aided weed suppression, organic matter buildup and biological pest control.
- No-till was the only conservation tillage method recommended to South American growers by all parties.
- Public and private sectors voiced the same consistent message about the benefits of no-till without any disagreements about the reduced tillage practice.
- The necessary knowledge for no-till adoption was widely available and shared between researchers and farmers.
- Farmers aggressively spread the positive word about no-till among themselves.
- Publications with adequate, practical information were made available to farmers and extension agents. ✿

New Day for No–Till as the World
Moves Into the 21st Century

> "A number of ag editors in the early days of the new century indicated no-till was the 'hot topic' for reducing costs, **expanding farm income and finding ways to earn Farm Bill incentives**..."

As American agriculture moved into a new century, no-till started to come into its own. When *No-Till Farmer* got its start in 1972, no-till was used on less than 2% of the U.S. land.

By 2000, the no-till acres had jumped to an astounding 52.1 million acres, the total conservation tillage acreage was well over 100 million acres and no-till made up 17.5% of all cropped land.

By the end of the new decade, the no-till acreage had reached 86.5 million acres.

Up to $50 More Per Acre

A survey conducted of 2001 National No-Tillage Conference attendees indicated 55% credited no-tillage with adding an extra $20-$29 per acre profit. Another 19% of attendees indicated that no-till had boosted their

"Extensive tillage on CRP ground destroyed the soil quality benefits..."

Allen Berry

CHANGING NO-TILL CONDITIONS, CHANGING NEEDS. Illinois grower Allen Berry recognizes different no-till crops and conditions require many refinements and adjustments to make no-till work.

returns by over $50 per acre.

While no-till will never be the favorite tillage system for all producers, a number of farm publication editors in the early days of the new century cited no-till as the "hot topic" among farmers for reducing costs, expanding income and finding ways to earn Farm Bill incentives.

At the start of the new century, no-till was playing a key role in the advancement of American agriculture. With 40 million acres enrolled in the federal government's Conservation Reserve Program (CRP) that was designed to reduce grain surpluses and trim the erosion that was returning to crop production, no-till was the key to keeping future soil losses under control. Extensive tillage, unfortunately, destroyed much of the long-term build-up of soil quality benefits on much of this CRP land.

Many other new developments came of age during the new decade. These included the development of polymer seed coatings that led to earlier and more profitable no-tilling, refinements in no-till seeding equipment, the rapid acceptance of GPS and increased use of cover crops. ✳

NO-TILL CORN OR SOYBEANS FIRST?

June 2005

At least in the Corn Belt, tradition says you'll probably no-till corn before no-tilling soybeans. Yet more growers today are questioning this conventional wisdom.

Steve Prochaska says there are numerous benefits to no-tilling either crop early into dry soils. Even so, the extension advisor in Crawford County, Ohio, says you need to recognize how the risks of early-season no-tilling vary from crop to crop.

Timeliness is Critical

Prochaska says both corn and soybeans are being no-tilled earlier than ever before. And he's found numerous reasons to consider no-tilling soybeans before corn.

"With farm size increasing and many farmers working off the farm, it is imperative to effectively utilize all available planting days," he says. "One of the impediments to large or small-farm operation success is the lack of planting time on dry soils."

Prochaska says seeding corn and soybeans into wet soils is more of a concern with extensive tillage. "Planting in wet soils is a disaster that not only robs crop yields, but forces additional tillage to rectify the damage," he says. "And more tillage creates the potential for greater soil erosion and reduced farm profit.

"While there are additional windows of opportunity to replant soybeans or repair poor stands, you realistically only have one chance to get a good stand with corn. As a result, soybeans might be a better option for very early planting and may outweigh the risk of possible later cold weather and/or wet soils suitable to compaction." ✳

NO-TILL-AGE®

How Soil Organic Matter Changes After 20 Years

(Tons per acre measured at a 0- to 12-inch depth)

Nitrogen Rate Per Acre	Conventional Tillage	No-Tillage
0 lbs.	37.6 tons	42.6 tons
75	43.3	44.9
150	43.4	45.2
300	47.2	51.1

— University of Kentucky

OUCH, A $60,000 NO-TILL SPEEDING TICKET

"If I can convince you to slow down your no-till planting speed, I can make you some easy money," maintains Gregg Sauder of Precision Planting in Tremont, Ill., and a veteran no-tiller who over the years has pioneered numerous precision ag developments.

A strong believer in testing planter meters and reducing planter speed, Sauder worked extensively with one farmer who liked to "race" through the planting season.

"When yields were checked, reducing planter speed from 7 down to 4 1/2 mph boosted corn yields by 24 bushels per acre," says Sauder. "With 1,500 acres of corn, that's an extra $60,000 bonus for slowing down."

In another test, Sauder checked seeding results in 89-foot long strips. At 5 mph, there were eight skips. When the planter speed jumped to 7 mph, there were 18 skips.

Speed also has an impact on depth control since bouncing over

Frank Lessiter

HIGH PLANTING SPEEDS ARE DANGEROUS. Don't calibrate your no-till planter for 4 1/2 mph and end up seeding no-till corn fields at 7 mph.

no-till residue makes it difficult to place seeds at the same depth.

Why Speed Kills

"No-tillers usually start out planting at the right speed," he says. "With a 30,000 corn population in 30-inch rows at 4 1/2 mph, your planter is designed to do a good job.

"But with a 6 1/2 mph planting speed, your neighbors are dropping 22 seeds a second. If you want seeding perfection, you're going to miss the mark at this speed.

"Plant at 5 mph or less as that's what most planters were designed to run at. It's too easy to overdrive your planter meter capacities."

Even with 2,000 acres of corn and soybeans to no-till in a short period of time, Scott Dollen has found slowing down pays. The Persia, Iowa, finds skips and doubles are much more evident at faster planting speeds.

Purdue University agronomist Bob Nielsen found significant yield differences between 4-5 mph and 6-7 mph planting in 23% of surveyed corn fields. Losses ranged from 1.9-4.7 bushels per acre for every 1 mph increase over 4 mph.

"The effect of excessive planting speed is more likely seen with older, poorly maintained or misadjusted no-till planters," he says. "If a farmer questions the capability of the planter to accommodate faster speeds, check the planter's response."

Steve Butzen, agronomy information specialist at Pioneer Hi-Bred International, says seedbed condition, weather stress, insect concerns and disease problems impact yields. Yet he says no-tillers have more control over planter-induced variables such as planting speed.

Doubles, Triples, Skips

Illinois agronomist Ken Ferrie says finding the right planting speed is critical when no-tilling. If you plant too fast, doubles, triples

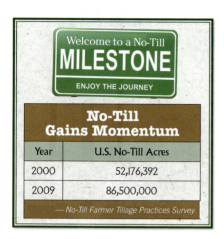

Welcome to a No-Till
MILESTONE
ENJOY THE JOURNEY

No-Till Gains Momentum	
Year	U.S. No-Till Acres
2000	52,176,392
2009	86,500,000

— *No-Till Farmer Tillage Practices Survey*

and skips are more likely since seed will ricochet down the seed tubes. A no-till corn planter that's bouncing up and down can lead to uneven planting depth.

The Heyworth, Ill., crop consultant found a 6 bushel-per-acre yield boost from reducing planter speed from 6 down to 4 mph. He says you'll run into serious no-tilling problems at over 5 mph.

In one test, Ferrie found medium rounds planted at 4 1/2 mph resulted in a 100% seed drop. When planter speed increased to 6 1/2 mph, there was a 3.9 error rate due to doubles and skips. With large flats, there were 4.7 errors at 4 1/2 mph.

When calibrating a no-till planter, Ferrie recommends including whatever seed treatments, graphite or other material you'll be adding to the seedboxes. He suggests calibrating the planter for your most commonly used seed size. With other seed sizes, experiment to see how speeding up or slowing down changes plant uniformity.

Finally, Ferrie encourages farmers to be honest about the speed at which they typically plant. "Don't calibrate your no-till planter for 4 1/2 mph and drive 7 mph," he says. ❧

2000 DECADE
AMERICAN AG FACTS

- Golden rice is engineered to alleviate vitamin A deficiency in 2000.
- In 2001, John Deere equips new tractors and mobile machines with GPS capabilities.
- Advancements in software and mobile devices offer farmers greater productivity and immediate access to data.
- The sale of organic products grows 20% since 1990 and continues to climb throughout the decade.
- In 2002, 27% of farm operators are women and 11% are the principal farm operators.
- U.S. production of tobacco drops by 60% between 1997 and 2003.
- The federal government's first electronic data reporting system begins in 2001 with the Cotton Ginning report.
- The National Agricultural Statistics Services publishes statistics on organic agriculture for the first time.
- The 2002 Farm Security and Rural Investment Act increases projected subsidy payments by 74% over 10 years.
- U.S. grain prices plummet after export handling capabilities fell 63% due to damage from Hurricane Katrina.
- Greenpeace launches an attack in 2007 on genetically modified corn produced by Monsanto, claiming rats developed liver and kidney problems from eating the corn.
- In 2009, the pharmaceutical product ATryn is the first biological produced by a genetically engineered animal approved by the Food and Drug Administration.
- The 2005 price of farmland in Iowa averages $2,760 per acre.
- Diesel costs in 2005 average $2.36 per gallon.
- The price of corn in Iowa averages $1.90 per bushel in 2005.
- Soybean prices in Iowa average $5.88 per bushel in 2005.
- The 2005 corn yield in Iowa averages 173 bushels per acre.
- The 2005 soybean yield in Iowa averages 52.5 bushels per acre.

WAS THE FEDERAL GOVERNMENT WRONG? No-tilling Conservation Reserve Program (CRP) ground may be essential for feeding the world.

WHEN THE CONSERVATION RESERVE PROGRAM MISSED THE BOAT

After Frank Martin talked about cover crops at the 2004 National No-Tillage Conference, we discussed the idea of utilizing no-till and cover crops on Conservation Reserve Program (CRP) land to pump needed dollars back into rural America.

The veteran no-tiller from Hallsville, Mo., maintains the CRP project has had a negative effect on rural communities. Instead of sitting idle, much of these 40 million CRP acres could be producing corn, soybeans, wheat and other crops to feed America's people and livestock, to export grains and to produce ethanol and biodiesel fuels.

"As a result of the present program, rural communities have lost hundreds of dollars per acre of economic activity," says Martin. "Counties with a substantial percentage of CRP acres have seen businesses close and their infrastructure suffer."

Martin is convinced that no-tilling CRP ground would sharply reduce erosion losses. Among his ideas for getting CRP ground back into production would be waiting until after no-tilling corn or soybeans before killing a wheat or rye cover crop that's grown up to 3 feet tall. This would

> *"The **economics** of using no-till to bring CRP acres **back into production are staggering...**"*

keep needed residue on the soil surface to soak up excessive moisture and provide needed soil protection.

Two Primary Reasons

The federal government's CRP program was established in the 1985 Farm Bill to ease the overproduction of U.S. farm commodities and to reduce erosion. While CRP has definitely reduced erosion, Martin says the government's policy gave South American farmers the green light to clear millions of previously uncultivated acres.

How to Get the Job Done

Martin says landowners with CRP acres could be given the option of receiving a reduced annual government payment in exchange for the right to produce a crop with no-till and cover crops.

"The economics of doing this are staggering," he says. "If 30 million CRP acres were put back into production and produced $200-$400 per acre in income, that would be $6-$12 billion of new economic activity in America's rural communities. It would also reduce CRP government payments by $1 billion per year for 30 million acres."

COATED SEED HOLDS PROMISE FOR EARLIER NO-TILLING

While many seed companies have long maintained almost all of their hybrids are ideal for no-till conditions, it may not be long before you'll be able to get more specific answers to this question. That's because Landec Ag is expanding its temperature-sensitive seed polymer research efforts to determine which hybrids will give no-tillers the best yields when planted 2-3 weeks earlier than normal.

The early plant system being developed by the Monticello, Ind., firm relies on its patented Intellicoat polymer technology to keep water out of the seed until conditions are ideal for germination. Some 160 Midwestern corn growers tested the early plant system on 3,000 acres in 2001.

The temperature-activated polymer should provide a wider planting window that will lead to more efficient use of labor and machinery, which should result in higher corn yields, says company president Tom Crowley.

No-Till Much Earlier

As soil reaches a preset temperature (55 degrees, for example), the polymer allows the seed to take on water and begin germination.

He cites an Iowa State University study that indicates the optimum corn planting time in Iowa is April 20-May 5. If you no-till before this time, you may lose yield due to cold temperatures. After May 10, yields start dropping because you're not getting enough growing degree days.

Finding No-Till Traits

Landec Ag is determining which corn hybrids offer the greatest chance for success regardless of when, where or how you no-till.

"We're choosing hybrids that have cold tolerance, early season vigor, strong emergence, strong root systems and disease resistance. You must have the right hybrids," says Crowley.

"We're studying each hybrid's water uptake and how these seeds respond under extreme cold weather conditions. We're looking carefully at this area because no-till soils are colder and wetter than normal."

To determine how corn responds to extreme no-tilling conditions, researchers subjected seed to cooler temperatures for several weeks before conducting germination tests. Crowley says seed corn must be able to withstand such conditions for the polymer system to be commercialized.

In these extreme condition tests, the polymer-coated seed had germination percentages of 85%-95% compared to 40%-70% for uncoated seed. Crowley believes researchers will eventfully raise the germination rate above 95% percent for coated seed.

Editor's Note: *In recent years, the seed coating technology seems to have disappeared from the programs of seed suppliers.* ❋

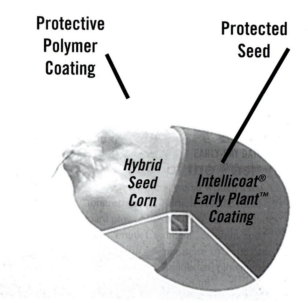

Protective Polymer Coating

Protected Seed

Hybrid Seed Corn

Intellicoat® Early Plant™ Coating

HOW THE COATING WORKS. The Intellicoat polymer coating regulates seed corn water uptake, which delays germination. This offers a predictable germination delay over a range of spring weather conditions.

TACKLING SERIOUS RESIDUE CONCERNS. Kentucky no-tiller Jerry Peery changed the knives and choppers on his combine in order to distribute thick small grain straw much more evenly. This has allowed his no-till planter to more effectively slice through the straw with the help of row cleaners.

NO-TILL USED TODAY TO CROP CIVIL WAR BATTLEFIELDS

May 2001

When thousands of troops walked around the huge rocks and out of the fields on the morning of Sept. 17, 1862, in the Civil War's bloody Battle of Antietam, little did they realize 19,510 soldiers would be wounded and 3,600 killed on this gruesome day.

Battlefield Crops

Many of the battlegrounds at the Antietam National Battlefield near Sharpsburg, Md., are no-tilled by Dale and Terry Price. They grow 1,400 acres of corn and soybeans and also custom plant 1,500-2,000 acres.

"We're probably 80% no-till, mainly because of the rocks in the fields we farm," says Dale. "All of our corn is no-till, 70% of the soybeans are no-tilled and 70% of the wheat is no-tilled."

The Prices no-till corn in 30-inch rows with a 6-row John Deere 7200 planter. Soybeans and wheat are no-tilled with a 24-row John Deere 1560 drill.

"We no-till over the battlefield rocks very slowly," says Price. "Our normal planting speed is in the neighborhood of 4 1/2-5 mph, but it drops as low as 2 mph when we're no-tilling over the rocks."

Because of the rocky conditions, the Prices attach no-till coulters to the planter frame, well out in front of the row units.

"When coulters are mounted directly to the row unit, the force of the rocks will destroy the coulters," he says. "The coulters can't handle the excessive force from the rocks when mounted close to the row units." ✺

LATE CORN PLANTING A MAJOR CONCERN

January 2006

According to a survey of 13,250 corn growers in 20 states conducted by Landec Ag, 32% of corn plantings over the past 5 years were not wrapped up by the grower's ideal date, regardless of the tillage system being used.

Wet soil conditions were the primary cause. For instance in 2005, timely planting was a particular challenge in Wisconsin, Ohio, Michigan and Pennsylvania where 30%-36% of farmers missed the ideal planting windows.

Some 58% of farmers counted on finishing planting corn in 3 weeks or less, while 16% hoped to finish in 10 days or less. ✺

"No-tilling into sod **is easy with the right equipment and management..."**

CASH IN ON THE BIOLOGICAL BENEFITS OF NO-TILLING CRP GROUND

March 2008

The first impulse of many growers may be to plow under Conservation Reserve Program (CRP) ground when converting it back into cropland. While there is a dense network of fibrous material beneath the soil surface, what really gets the attention of producers is the "low-flying jungle" that is in plain view.

Yet no-tilling into sod can be easy, says John Baker, a no-till researcher from New Zealand. In working with the Cross Slot system, he says the original machine's design intent was to find a way to successfully seed into sod.

"Sadly, the problem of physically handling the biomass often compromises the issue of making the best use of the soil's biological benefits," Baker told attendees at the 2008 National No-Tillage Conference in Cincinnati, Ohio. He says there are two main considerations when no-tilling into CRP ground.

The first is coping with the jungle of grasses, weeds, legumes and pests without plugging the seeders or destroying all the good biological things that have happened with the CRP program.

The second is creating a biological environment that favors and protects the newly sown crop and nurtures the biological assets that developed during 10 years of no production.

"The worst possible thing that can be done to CRP is to till it," he says.

Here are Baker's 11 suggestions for effectively no-tilling CRP ground.

1 **Kill It.** When sprayed with an effective herbicide, the CRP ground should behave and be treated as a heavy crop residue. With untouched CRP land, handling residue should be easier than dealing with sliced and pushed-down cereal grain residues.

2 **Harvest Helps.** Where cutting grass silage is allowable before CRP land is cropped again, residue handling will be greatly simplified, especially for non-disc-type openers. You may be able to spray

NO-TILLING CRP GROUND. Bringing CRP ground back into production with no-till offers the chance to create a new biological environment.

the grasses with glyphosate and harvest 3-4 days later before no-tilling a cash grain crop. However, this can test your confidence due to glyphosate translocation to the grass roots, since grass removal will disguise any die-back for a period of time.

3 **Control Weeds.** This involves identifying weeds and possibly tankmixing two or more herbicides for effective weed control. For that reason, knowledge of herbicide compatibility is important.

4 **Avoid Cereal Grains.** Sowing cereals as the first crop after CRP should be avoided. Many pests will have flourished for 10-plus years under CRP, so be sure to identify these pests before no-tilling.

5 **Watch for Pests.** Slugs and rodents that thrive in dense sod can become an issue when seeding into CRP residue.

NO-TILL-AGE®

How Corn Yields Have Increased Since 1975

Tillage System	% Yield Increase
Moldboard plow	4%
Chisel plow	8%
No-till	18%

— *Tony Vyn, Purdue University agronomist*

> "The worst possible thing with CRP ground is **to till it...**"

6 **Add Nitrogen.** Decomposing the CRP biomass will likely lock up large quantities of soil nitrogen after spraying herbicides, which can lead to poor initial new growth and yellowing of plants. Add nitrogen only if it can be safely banded separately from the seed and avoid broadcasting fertilizer since results are usually disappointing.

7 **Select Openers Carefully.** Most disc-type openers will hairpin heavy residue. No-till opener choices are limited, which is why many people reluctantly go back to extensive tillage with CRP ground.

8 **Horizontal Slots Best.** Baker says it's best to avoid vertical slots when seeding into sod. Horizontal slots allow double-shooting fertilizer and handling residue without compromising either function.

9 **Wait to Drill.** Openers perform better if you in-crease the interval between spraying and seeding since waiting makes the soil more crumbly.

10 **Don't Wait Too Long.** The downside of waiting too long to seed after spraying is that weed regeneration can occur. Remember that the weed seed pool found within CRP soils is vast.

11 **Plant Later.** When seeding CRP ground in the spring, excessive soil water will be removed quickly by the growing sod. If dead or dying sod is preventing the soil from drying out or warming up, Baker favors a shorter interval between spraying and seeding, no-tilling later or planting later-maturing hybrids or varieties.

"It seems like everyone is astonished when you tell them no-tilling into sod is easy, but it can be with the right equipment and management," Baker says. ❁

CHECKING RESULTS. Dan Towery (in hat) measures the water infiltration level with corn no-tilled into a burned down stand of annual ryegrass as Mike, Jeff and Nick Starkey look on at the family's Brownsburg, Ind., farm.

REMEMBERING NO-TILL WRITER, EARLY ADVOCATE GAYLIN MORGAN

Back in 1962, Harry Young Jr., began experimenting with no-till on 7/10ths of an acre on his Herndon, Ky., farm. It led the family to no-till 1,800 acres of crops each year from only 1,200 acres thanks to double-cropping.

Ag Communicator Pioneer

By the mid-1960s, farm writer Gaylin Morgan was spending lots of time on the Young farm reporting on this new-fangled no-till movement. Morgan also did considerable writing and public relations work to further the acceptance of no-till for pioneering companies such as Chevron Chemical, Allis-Chalmers, DuPont and Velsicol Chemical.

Passing away in 2005 after a 4-year bout with cancer, Morgan (everyone called him by his last name) dreamed up the publication concept that became *No-Till Farmer*.

When I came to Milwaukee in mid-1972 to launch this magazine at Reiman Publications, Morgan became my good friend and mentor when it came to learning about no-tillage. And he encouraged my wife and me to buy this publication in 1981.

He was an important part of the team that put together the first farmer-friendly text written on no-tillage. Co-authored in 1973 by Young and University of Kentucky agronomist Shirley Phillips, the 224-page *No-Tillage Farming* book, edited by Morgan, remains one of the most impressive and practical texts ever written on the topic. The book was later translated into Portuguese for South American growers.

An Iowa State University ag journalism graduate, Morgan was known for his gentle nature, creative ideas and wry wit. Survived by his wife, Carol, a son,

NO-TILL BELIEVER. Gaylin Morgan (center) was among the first farm editors to report on the no-till movement that got underway in Kentucky in the 1960s. He played a key role in the early-day acceptance of no-till among U.S. farmers.

> ## "Gaylin Morgan became my good friend and no-till mentor."

Charlie, two daughters,Kim and Gaylee, and grandchildren, he formed his own public relations firm in 1973 and co-founded another, Morgan & Myers, in 1976. Heading the firm that was named to Inc. magazine's 500 list of fastest-growing companies in 1987, he retired 10 years later.

A gentle, ethical man who created an employee-friendly culture, co-workers credited him with coming up with numerous marketing innovations that contributed to the success of many programs and projects. Receiving many awards, he was named Ag Communicator of the Year by the National Agri-Marketing Association in 1998.

Morgan played an early-day key role in the acceptance of no-till across the U.S. He would have enjoyed a sense of accomplishment in seeing no-till expand from 3.3 million acres in 1972 to an astounding 62.4 million acres by 2004.

The Teacher

While I've learned a great deal about no-till over nearly 5 decades, much of my early education came from the many words of wisdom received from Morgan. While it's sad to see the passing of an editor who played such a critical role in communicating the no-till practice to farmers, Morgan should be proud of his role in helping develop one of today's most widely accepted cropping practices — a critical farming development that's caught on faster than hybrid corn ever did.

For years, I've continued to have a number of old-time no-tillers ask about Morgan's friendly smile and what he was doing. Knowing Morgan, he's got the heavens abuzz with no-till farming and many other thoughts on how to improve global agriculture. ❊

Trimming Costs, Trying New Ideas Help
No-Tillers Grow in the 2010 Era

"No-tilling and seeding cover crops **is saving $80 to $100 per acre on fertilizer**..." — *Dan Towery*

With more growers enjoying the benefits of leveraging cover crops with no-till, the no-tilled acres will grow by an anticipated 21% during the 2010 decade.

Thanks to the development of new high-tech ideas, more emphasis on soil quality and the ups and downs of grain, milk and livestock markets, no-tillers are more profit-conscious than ever before.

During this decade, many new cropping ideas and refinements of existing practices were being adopted by innovative no-tillers. This has been the case even as economic conditions made it tougher for growers to take risks with new practices.

Several examples that we'll describe later in this chapter include no-tilling corn, soybeans or other crops into living cover crops and the impact green bridge (See page 100)

may be having on no-till yields, such as with continuous corn.

No Guarantees

Dan Towery says farmers still using extensive tillage often expect a first-year guaranteed return on investment when switching to no-till. However, weather, mismanagement and other factors usually don't lead to any yield drop during when transitioning to no-till, says Towery, the

"They leave out how important **preserving soil, or reducing soil erosion, is to soil health...**"

HIT WEEDS RIGHT AWAY. A lack of early-season weed competition is among the biggest contributors to unseen yield losses in herbicide-tolerant corn. In most instances, no-tillers can't afford to wait to spray until the weeds come up, as a lack of control will lead to late-season weed concerns.

head of Ag Conservation Solutions in Lafayette, Ind. Despite the lack of immediate returns, there's plenty of incentive to encourage more growers to switch to no-till.

"Growers who are no-tilling and seeding cover crops are saving $80-$100 per acre per year on fertilizer," he says. "If you factor in the increased yield no-till gives you in a dry summer, the profits can add up to another $100-$200 per acre."

Yet it often takes 4-5 years of no-till and improved management to fully enjoy the economic benefits. Plus, there's still no good soil test to verify how much no-till can reduce fertilizer rates.

"Common nitrogen, phosphorus and potassium recommendations today don't account for the high biological soil activity with no-till," says Towery. "Over the years, these recommendations have been accepted as accurate, which they are not.

"If the impact of biological activity on these nutrients could be accepted, no-till could be a real game-changer."

New Buzz Words?

While there's increasing emphasis on "sustainable agriculture" and "soil health," these four buzzwords can ruffle the feathers of veteran no-tillers. It's because no-tillers recognize that they were the original true innovators behind these "not-so-new" concepts as far back as when no-till arrived on the scene in the 1960s.

The terminology for some old ideas appears to have suffered from overuse. It seems we've reached the point where sound, important ag principles must be described with new names in an effort to stir up excitement and bring about change.

One example is the steady decline in soil conservation interest. Today, the

term seems to have fallen out of favor among many folks in agriculture.

David Lobb, a soil scientist at the University of Manitoba, senses that ag folks need to move on and come up with a new look and different terminology for some of the older practices.

Tom Bauman shares a similar story. The head of Agren, an ag consulting firm in Carroll, Iowa, recalls a time in the 1980s when a decision was made among conservation agencies to change the popular soil erosion terminology and instead emphasize what they felt was the more exciting water quality term. Working for NRCS in Iowa at the time, Bauman recalls voicing his frustrations since the U.S. was not even close to solving its soil erosion concerns.

"A very wise conservationist, a mentor for many Iowa conservationists, reassured me," says Bauman.

"Lyle Asell always had a unique perspective and this time was no different.

Many programs tend to reach the end of their lives, Lyle told me very convincingly. At some point in time he said it's helpful to repackage the old problem, develop new policies and get new funding to continue to tackle the issue at hand.

Bauman says we still have the same soil erosion and water quality concerns today. Yet once again in the past few years we've pivoted the terminology one more time, turning "soil health" into the latest buzzword.

He understands that conservation agency folks are tired of talking about soil erosion and understands why they dreamed up "soil health" as a replacement. But Bauman finds it disturbing that today's so-called "soil health" experts are only telling half the story.

"They leave out the part about how important preserving soil, or reducing soil erosion, is to soil health," he says. "It's like they think we have solved the soil erosion problem and we are moving right into building soil health. It can't be done, as it would be like trying to stabilize climate change without addressing carbon emissions."

He's adamant that soil erosion can't be left out of the equation for improving soil health, water quality, carbon sequestration, national and global food production or long term-profitability.

"If you are weary of talking about soil erosion, get over it," says Bauman. "Bundle the problem of soil erosion into a pretty package like soil health if you must. But don't forget that the foundation of soil health is based on the condition that soil exists in place. No amount of repackaging will change that." ✳

THE 2010 DECADE
AMERICAN AG FACTS

- Crops like corn increase drastically from the 256 million tons produced in 2003 to 354 million tons in 2013. During the same years, milk production increases by 18%.
- Wheat production drops between 2003 and 2013 from 64 million tons to 58 million tons, while cattle numbers drop from 12 million to 11.7 million head.
- Data continues to revolutionize farming, as growers begin using data to better understand what's going on in their fields.
- The number of agricultural scientists grows 9% between 2002 and 2012.
- In 2010, cold weather in Florida destroys 70% of the state's tomato crop, causing a severe national shortage.
- Agricultural sensors contribute to farm automation and become financially viable in 2015.
- Livestock biometrics, new GPS developments and RFIC collars become scientifically viable in 2017, allowing farmers to track vital information remotely.
- In 2014, the Agricultural Act ends the direct payment program, the countercyclical program and the Average Crop Revenue Election program. It also expands crop insurance and adds Agricultural Risk Coverage and Price Loss Coverage programs for expanding subsidies.
- The Supplemental Nutrition Assistance program quadruples in size from $18 billion in 2000 to $74 billion by 2016.
- Okanagan Specialty Fruits releases the Arctic Apple, the first U.S. genetically engineered apple that doesn't turn brown from oxidation.
- The 2015 price of farmland in Iowa averages $8,200 per acre.
- Diesel costs in 2015 average $2.64 per gallon.
- The price of corn in Iowa averages $3.67 per bushel in 2015 after reaching its high point several years earlier.
- Soybean prices in Iowa average $9.37 per bushel in 2015, well below the higher prices farmers enjoyed a few years earlier.
- The 2015 corn yield in Iowa averages 192 bushels per acre.
- The 2015 soybean yield in Iowa averages 56.5 bushels per acre.

September 2006

Jim Cook

GREEN BRIDGE ROOT DISEASE. Here's a severe case of barley no-tilled into barley residue.

NO-TILL YIELD ROBBERS MAY TRAVEL ALONG 'GREEN BRIDGES'

Plant diseases crossing the so-called "green bridge" might be stealing yields from no-tillers who don't even know it.

"If the whole field is infected, you might not know there was a problem, or you might think you need a little more phosphorus or nitrogen," warns Jim Cook. The retired Washington State University plant pathologist deserves much of the credit for closing the green bridge gap in the Pacific Northwest, the only region in the U.S. where growers are widely aware of the costly problem.

Cook was hired by the USDA's Agricultural Research Service in 1965 to study plant diseases. He retired in late 2005, having devoted his career to understanding and controlling soil-borne pathogens and root diseases with wheat.

Defining the Problem

Cook defines green bridge as the weeds, volunteer crops or cover crops that grow in fields between harvesting one crop and no-tilling the next. He says disease pathogens feed on the roots of volunteer plants, surviving long enough to infect the roots of the next season's crop. Wind-spread pathogens and insect pests are also a green bridge source.

"The pathogens in the soil don't care if you planted a crop or if it's a volunteer crop or weeds," he says. "Roots are roots, and they can be the source of infestation for the next crop. In the Pacific Northwest, we count on soil biological activity to decompose the food that the pathogen gets from the roots. That biological activity is a key to sanitizing the soil.

"But when it's warm enough for soil biological activity, it's usually too dry. And when it's wet enough for biological activity, it's usually too cold."

Cook says no-till fields are particularly vulnerable because plowing kills the pathogens, which can't stand to be stirred. For years, some no-tillers solved the problem by burning the stubble, a practice that's out of favor today.

Monoculture systems, such as wheat on wheat or corn on corn, face the greatest green bridge risk. However, some pathogens can't be controlled through crop rotations and must be eliminated with a plant-free period that deprives them of a life-sustaining host.

Among the most common and difficult to control soil pathogens is *pythium*, which is likely present in every cup of agricultural soil in the U.S. This root rot strips off root hairs, robbing the root's absorption capacity and hindering its efficiency in obtaining nutrients.

Rhizactonia is another common pathogen that pinches off the roots, often leading to a plant growing with only half its root system.

"With a disappointing crop, green bridge may be the cause..."

There's been little research, but Cook says it would be interesting to know to what extent it's a problem in corn country, especially with continuous corn.

Mistaken Identity

Cook says the green bridge problem went unrecognized or misdiagnosed for many years. "At one time it was called Roundup injury because it happened where weeds or volunteer crops had been sprayed with that herbicide," he says. "In reality, when the infected weeds or volunteer crops were sprayed, the plant's defense system collapsed, and the pathogen thrived on the dying plant's roots, then jumped to the newly planted crops."

In the case of Roundup, green bridge was able to shut down the shikimic acid pathway that plants use as a defense against most pathogens. As an example, the *rhizoctonia solani* that is already found in plants as a pathogen can take complete possession of the roots. This can occur in time to attack the roots of a newly seeded crop.

If cereal grains are planted too soon after an initial burndown with Roundup, root diseases can be transferred from the dying roots of volunteer plants or grassy weeds to the young roots of the following crop.

Noting the increasing popularity of cover crops among no-tillers, Cook says the green bridge threat with cover crops needs much more study in different environments.

"Whatever it is that you are doing with cover crops, you're being successful," he says. "But if you do have a wreck or a disappointing crop, green bridge could be the cause." ❋

January 2013

MANAGING THOSE 'UNDERGROUND' CRITTERS

For the nearly 5 decades I've been tracking no-till, soil biology has never received as much attention as today. As no-tillers refine their systems, more are recognizing the importance of doing a better job of managing the millions of critters living under the soil surface.

During a panel discussion at the 30th anniversary celebration of the Conservation Technology Information Center, the increased value of soil biology was mentioned numerous times.

Start With the Soil

As no-tillers become more efficient, Jerry Hatfield says taking a closer look at soil biology is critical to boosting efficiency. The director of the USDA National Laboratory for Agriculture and the Environment in Ames, Iowa, says farmers must consider how increasing climate variability will affect their soil management decisions.

Hatfield says the key is fitting the pieces of the no-till puzzle together. "There are a lot of different pieces in today's technology, but we haven't yet figured out how to make it all work together effectively," he says. "Yet it's not so much new technology that we still need, but rather better ways of putting it all together."

Underground Livestock

Early in his farming career, Dan DeSutter became interested in managing the critters living below the ground. As a result, the Attica, Ind., no-tiller wants a crop or cover crop growing throughout the year, so the millions of microbes found in the soil are never starving.

He maintains every acre of healthy agricultural soil contains tons of fungi, bacteria and organic matter that lead to higher crop production. "We already have livestock on the farm," he noted. "It's just that they're located underground."

DeSutter has found a 2% increase in organic matter is worth $40-$50 per acre in reduced input costs. "When we focus on organic matter, then everything else comes together," he says. "This includes providing drainage so the microbes can breathe and do their job."

DeSutter has managed his underground livestock so effectively that manure has become the primary source of fertility in his no-till operation. ❋

10 YEARS OF NO-TILL CHANGES

Here's a look at changes that have occurred over the past 10 years among *No-Till Farmer* readers. This average data comes from a comparison of the No-Till Operational Benchmark study conducted each year by the *No-Till Farmer* editors. (Figures used in the 2008-2011 column represent the data from the first year that the data was collected during this 4-year period.)

Practice	2008-2011	2017
GENERAL...		
Cropped acres	1,191	1,163
No-tilled acres	917	922
Operating expenses per acre	$336.81	$344.73
EQUIPMENT		
Planter options...		
6-row	34%	23%
16-row	18%	26%
24-row	7%	13%
Planter attachments...		
Coulters	53%	48%
Closing wheels	78%	87%
Pop-up applicators	35%	44%
Down pressure	36%	42%
Sprayer ownership...		
Pull-type sprayer	43%	33%
Self-propelled sprayer	37%	44%
CORN		
No-tilled acres	61%	67%
Per acre population	30,535	31,794
Sidedressing nitrogen	58%	73%
Corn yields...		
No-tillage	156 bu.	181 bu.
Strip-tillage	166 bu.	203 bu.
Minimum tillage	163 bu.	187 bu.
Vertical tillage	—	186 bu.
SOYBEANS		
Acres no-tilled	80%	80%
Seeding options...		
Planter	64%	70%
Drill	52%	38%
Air seeder	—	10%

Practice	2008-2011	2017
Bean fertilization...		
Applying N	30%	23%
Applying P	72%	57%
Applying K	80%	70%
Applying micronutrients	50%	45%
Varieties...		
Roundup Ready	95%	68%
LibertyLink	8%	25%
Dicamba-tolerant	—	36%
Soybean Yields...		
No-tillage	45 bu.	55 bu.
Strip-tillage	—	58 bu.
Minimum tillage	46 bu.	54 bu.
Vertical tillage	—	58 bu.
WHEAT		
No-tilled acres	83%	87%
COVER CROPS...		
Growers seeding covers	49%	83%
Species...		
Cereal rye	44%	74%
Annual ryegrass	26%	26%
Radishes	35%	50%
Oats, barley, wheat	34%	51%
Clovers	15%	37%
Peas	12%	21%
Hairy vetch	6%	14%
Seeding options...		
Drills	—	54%
Rotary spreaders	—	27%
Air seeders	—	16%
Aerial seeding	—	12%
High boys	—	7%
Interseeding devices	—	4%

— 2008-2018 No-Till Operational Benchmark Surveys

NO-TILLING INTO LIVING COVERS LEADS TO PICKET-FENCE STANDS

Frank Martin quit raising hogs in 2001 and turned his focus to doing a better job of raising crops — with an emphasis on reducing soil loss. For the Halls-ville, Mo., farmer, this meant shifting to no-till and integrating cover crops. Initially, Martin killed his cover crops before planting — until a sprayer skip pointed him in a different direction.

"The first year I tried cover crops, I had a near-perfect stand of corn on one side of the field I missed with the sprayer," he recalls. "Right next to it where I had sprayed the cover crop, the stand was pretty bad. That got me thinking about waiting until the crop was up and growing to kill the cover crop."

66% Rye, 33% Wheat

Martin no-tills 850 acres of corn and soybeans into a cover crop mix consisting of two-thirds rye and one-third wheat.

He knifes anhydrous into the cover crop immediately before planting and uses RTK guidance to no-till corn in rows between the anhydrous bands. When he accidentally planted on top of a few anhydrous bands he saw excellent crop response.

The wheels and knives on the anhydrous application equipment lay down some of the growing cover crop. Then, aggressive Yetter Shark Tooth floating row cleaners do a good job of creating a nice planting path without plugging, he says.

Manages Moisture

Martin has never replanted when no-tilling corn or soybeans into a living cover crop and normally gets a nearly perfect, picket-fence stand.

He credits this success to the fact that a living cover crop helps manage soil moisture, an issue where 8 inches of topsoil covers a clay layer — a bad situation for water infiltration.

"When you plant into a living cover crop, you're letting it manage moisture levels until the crop gets up a few inches and can manage the moisture on its own," Martin says.

He finds cover crops suck up extra moisture in wet years while residue holds moisture around the seed longer in dry years. He kills off the cover crop sooner in a dry year, so it doesn't compete with the growing crop.

"We never have a crusting problem and there's always enough moisture to get the seed germinated," he says. "One thing I love about this system is that I know I'll have a good stand."

Though he seeds both wheat and cereal rye, Martin prefers cereal rye. "It can grow 3-6 feet tall and when you lay it over, you get a lot of material on the ground for better soil protection and residue levels," he says. "Plus, cereal rye is easy to kill."

Down the road, Martin expects to add cover crops that can also fix nitrogen. He's even mulled over how cover crops may allow farmers to add live-stock to their no-till systems.

"I've made some mistakes in the past, but it's important to experiment and keep trying things to see what works best," Martin says. ✺

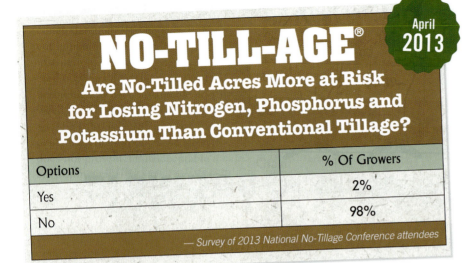

NO-TILL-AGE®

April 2013

Are No-Tilled Acres More at Risk for Losing Nitrogen, Phosphorus and Potassium Than Conventional Tillage?

Options	% Of Growers
Yes	2%
No	98%

— Survey of 2013 National No-Tillage Conference attendees

Welcome to a No-Till
MILESTONE
ENJOY THE JOURNEY

No-Till Pushes Ahead

Year	U.S. No-Till Acres
2010	88,000,000
2019	106,700,000

— No-Till Farmer Tillage Practices Survey

AFTER SELLING $180,000 OF EQUIPMENT, NO-TILL LET HIM STILL KEEP FARMING

March 2001

Roy Pontzius owes a great deal of gratitude to what no-till has done for his farming operation. Without it, the Columbia City, Ind., farmer says he would be out of farming.

Nearly 2 years ago, Pontzius watched an auctioneer sell off farm equipment he had paid more than $180,000 for. After the crowd of farmers had gone home, all Pontzius had left was a sprayer and an old tractor that had sat unused in a shed for 3 years.

Less than 3 weeks after the machinery sale, Pontzius leased a no-till planter, borrowed his brother-in-law's tractor and started over with most of his previous acreage — this time as a no-tiller.

Cash Flow Better with No-Till

Today, he owns practically no machinery. But thanks to no-till, he's made his farm cash flow.

Like many farmers, Pontzius had continued to buy more land, more machinery and figured he couldn't do anything wrong. With a "Cadillac-style" equipment system, he came up short when interest rates soared and land prices dropped in the mid '80s.

Lenders Got Tough

When he couldn't make the equipment payments, his lenders got tough. The next step was the auction.

Even though he was selling his equipment, Pontzius convinced some landlords to keep renting him land. And his idea of moving to no-till has paid off big.

In fact, he turned out the best corn yields he's ever harvested the first year he tried no-till. Yields averaged 110 bushels per acre, up from the farm's typical 90-bushel average. Last year, he rented a no-till drill and no-tilled beans into standing corn stubble that averaged 40 bushels per acre.

While machinery costs were cut sharply, you might expect chemical costs to soar with no-till. But that hasn't been the case for Pontzius. As a result, he is totally sold on no-tilling and says he'll never go back to conventional tillage.

CHECK OUT ADS FROM EARLY DAYS OF NO-TILL FARMER...

Companies running these advertisements in our early '70s issues were among suppliers who saw the value in being part of the no-till movement.

Nobody knows the ins ad outs of NO·TIL farming better than ALLIS·CHAMERS…we pioneered it

Back in 1966 we introduced the Allis-Chalmers *No-Til* Planting System. Today, more acres of no-tillage crops are put in with our planter than with all others. That says a lot about our planter…farmers know it's still the most versatile no-tillage system on the market.

And, after the crop is in, you can look to Allis-Chalmers experience to help you get it out…*right* and on time. This year, *Gleaner* combines will harvest over 45,000,000 acres of wheat, corn, soybeans and other crops. More custom operators are using *Gleaners* than any other combine. That says a lot, too.

So, when you turn to no-tillage, turn to the company that pioneered the practice. See your Allis-Chalmers dealer.

We can help you grow.

ALLIS-CHALMERS AGRICULTURAL EQUIPMENT

NO-TIL is an Allis-Chalmers trademark.

Who'd plant rye in a stand of corn?

Ans: The man who's looking ahead to reducing operating costs next spring.

With the big switch to reduced tillage practices, corn growers are finding that a fall cover crop pays off in more ways than one. A rye cover is most easily established but either small grains are good, too. You can plant rye right after harvest or seed by air directly into the unharvested standing corn. It takes fast, winters well and is easily killed with Paraquat prior to planting.

Consider the advantages to your land. The rye sod helps prevent soil erosion and conserves considerably more soil moisture than would otherwise be available. This is true in spite of moisture uptake by the rye. Soil compaction is reduced. It lets you get into the fields earlier even in a wet spring.

But the biggest benefits are to you personally. You can plant directly into the Paraquat-killed rye by opening a seedbed with a notched or fluted coulter. It's essentially a once-over operation. The savings to you in time, labor and equipment usage can be tremendous. And at least fuel is used as well. Tests prove that you get yields at least equal to those achieved by conventional planting methods.

Reducing tillage in corn is a well established practice that works. If you're thinking of trying out reduced tillage next year, get that cover crop in as soon as possible, so you will have a mulch to plant into come spring. By the way, the Paraquat you'll need next spring is not only an effective herbicide, but an environmentally sound product. It's inactivated on contact with the soil so there's little danger of stream or lake pollution. For best results, apply Paraquat using Ortho X-77 Spreader. How about talking the whole matter over with your Ortho Dealer or your seed dealer?

ORTHO Chevron Chemical Company
Paraquat

THIS (ORTHO, CHEVRON AND DESIGN) — REG. U.S. PAT. OFF.   AVOID ACCIDENTS: READ THE LABEL AND USE ONLY AS DIRECTED.

QUESTIONS AND ANSWERS ABOUT TRI-LEVEL-BED FARMING.

Q: WHAT IS TRI-LEVEL-BED FARMING?
A: Originated by a farmer in Nebraska and introduced by Orthman Mfg., Inc., several years ago, TRI-LEVEL-BED farming is the only "no-tillage" system of row crop farming under furrow irrigation.

Q: WHAT DOES "TRI-LEVEL-BEDS" MEAN?
A: Just what the name implies…the bed is formed in 3 distinct levels. Each level has an important function as shown on diagram.

The ridge (top level) increases the water carrying capacity of the furrow and helps protect young seedling from wind and frost damage. During growing season it stores moisture. At bed splitting time it provides the soil to cover old crop residue.

The ledge (middle level), where soil is not disturbed, provides a firm seed bed so necessary for fast germination and emergence. Because it's a drained area, seedling damage, due to crusting or excessive rainfall, is practically eliminated.

At planting time, the furrow (bottom level) provides drainage in event of excessive rainfall. Because it is formed at the same time crop is planted, you can start irrigating immediately. Between crop cycles it provides the cavity for covering old crop residue.

Q: HOW DOES IT WORK?
A: Here again we ask that you refer to the diagram.

1. During Growing Season:
This illustration shows crop growing on TRI-LEVEL BED.

2. After Harvest:
This illustration shows how one pass with the TRI-LEVEL-BED splitter partially buries old crop residue and leaves a trash free area that's been worked up black for early spring warm-up. Field is corrugated for winter irrigation.

3. At Planting Time:
This illustration shows how beds are shaped, furrows are formed and seed is planted—all in the same pass. Note how crop residue is completely buried in ridge where it can be left to deteriorate for a full crop cycle.

Q: WHAT ADVANTAGES DOES TRI-LEVEL-BED FARMING HAVE?
A: Because of limited space, we will mention just a couple. With TRI-LEVEL-BED farming you'll make far fewer trips. In the spring the Orthman TRI-LEVEL-BED Shaper (Photo A) lets you shape the bed, form the furrow, plant the seed and apply herbicide—all in one pass. (this lets you start irrigating immediately.)

After harvest, one pass with the Orthman TRI-LEVEL-BED Splitter (Photo B) eliminates all soil fitting operations. Your field is left corrugated with a trash free furrow for winter irrigation.

Both the Orthman Shaper and Splitter are available for 20" rows on 40" beds, 30" rows on 60" beds, 36" rows on 72" beds.

It's impossible, of course, to cover the full story of Orthman TRI-LEVEL-BED Farming in this ad. If you're interested in learning more, we invite you to call us or mail the coupon below. We'll send you detailed literature together with the names of farmers in your area who are now farming this revolutionary new way.

ORTHMAN MANUFACTURING INC.
P. O. Box 8, Lexington, Nebraska 68850
Phone: 308-324-4654

Name _____
Address _____
City _____
State _____ Zip _____
Phone _____

PARAQUAT— THE HERBICIDE THAT MAKES NO-TILL FARMING POSSIBLE.

Simple forms of reduced tillage farming have been practiced over the years to control erosion, conserve soil moisture and reduce soil compaction. But the ultimate, No-Tillage or Zero-Tillage, wasn't achieved until five years ago with the introduction of a chemical called Paraquat.

Paraquat is a highly efficient, non-selective contact herbicide that provides exceptionally quick kill and fast knock-down of weeds and grasses. On corn for example, mixed with a residual herbicide like atrazine, it gives season-long control of all vegetation competing with your crop. In effect, *Paraquat substitutes for the plow and disc while the residual chemical performs the function of the cultivator.*

Naturally, a herbicide as effective as Paraquat has a multiplicity of uses. These are known as ORTHO-TIL Farming Systems using Paraquat. These cultural practices include many different chemical weed control techniques and are particularly pertinent to such crops as corn, soybeans, sorghum and cotton. Field tests prove that the various techniques hold working-of-the-soil to a minimum while delivering better results than the various mechanical weed control methods. Beyond that, ORTHO-TIL Farming Systems conserve soil moisture, eliminate much unwanted soil compaction, improve soil structure and often make heretofore unfarmable land farmable.

To acquaint you with the many different uses of Paraquat, ORTHO has prepared a wealth of informative literature on each of the various systems. Any part or all of it is yours for the asking. Just see the attached coupon.

ORTHO
Chevron Chemical Company

I am interested in ORTHO-TIL Farming Systems using Paraquat and would appreciate receiving the following free information:

- [ ] New "How To" Book (Describes the many ORTHO-TIL Farming Systems using Paraquat currently in use and those being researched.)
- [ ] For Corn (In all No-Till Situations.)
- [ ] For Cover Crops
- [ ] Fall Panicum Control
- [ ] Paraquat & Liquid Nitrogen Soybeans
- [ ] Double Cropping, Preplant, Stale Seedbed
- [ ] Stubble Planting —Cocklebur Control
- [ ] Direct Spray Program

Name _____

Address _____

City _____ State _____ Zip _____

Please mail this coupon to: Dept. OT, ORTHO DIVISION, CHEVRON CHEMICAL COMPANY, 200 Bush Street, San Francisco, Calif. 94120

From the DEKALB Stress Breeding Program come hybrids bred to yield big under most stresses.

High yielding hybrids must be tough. That's why DEKALB research men put source populations, inbreds and hybrid crosses through various forms of torture. High plant populations. Diseases. Insects. Just about every stress known to corn. The weak are eliminated. Superior inbreds are developed that help us make new DEKALB hybrids with the potential for high productivity under all kinds of conditions.

The screening continues all through the breeding process. Hybrids are tested under varying cultural and climatic conditions at many widely spread locations. Performance is checked at planting rates ranging from 14,000 to 30,000 plants per acre. Only after a hybrid can pass our tests can it wear our name.

DEPEND ON DEKALB

Injecting plants with stalk rot is one tool DEKALB uses otherwise in our Stress Breeding Program. The result is stronger, more productive hybrids for your farm.

BANVEL® herbicide gets tough broadleaf weeds <u>after</u> your corn is up

In spite of pre-plant weed knock-down, it is now well-known that certain tough, resistant, late germinating broadleaf weeds are a serious problem in many no-till corn fields.

Banvel, a translocating herbicide, is absorbed through leaves, stems and roots, moves with plant fluids from leaf tip to root tip, destroying as it goes. Banvel is being used increasingly to kill such broadleafs as: smartweed, Canada thistle, cocklebur, pigweed, lambsquarter, many others. In tank mix combination with 2,4-D it even takes care of the supertough velvetleaf.

Banvel is not a contact, quick knockdown herbicide. It may require several days to bring tough weeds down. But, because it kills from within, it does its job thoroughly. Fortunately, Banvel does not cause brittleness in corn.

Late germinating weeds need not lower your corn yield. Use low cost per acre Banvel to eliminate them. At your Ag Chem Dealer.

BANVEL from VELSICOL

Velsicol Chemical Corporation. 341 East Ohio St., Chicago, Illinois 60611 A Subsidiary of Northwest Industries, Inc.

Let Paraquat be your plow.

So you've decided to give reduced tillage farming with Paraquat a try. Then there's no longer any need to plow under unwanted weeds and grasses. Paraquat does that job. It knocks down competing vegetation fast, in a way that no residual herbicide can.

ORTHO Paraquat CL is a broad spectrum, fast-acting herbicide that burns back unwanted weeds and grasses in corn and soybeans. Mixed with a residual herbicide, it provides extended weed control without endangering your crop. Paraquat has been called the key to reduced tillage farming and reduced tillage means significant savings in time and labor. And, because reduced tillage farming with Paraquat controls erosion, you can often farm hilly land which you couldn't farm before.

Of great importance, too, ORTHO Paraquat CL is an environmentally sound product. It's inactivated on contact with the soil so there's little danger of lake or stream pollution from runoff.

If your program of weed control has been largely mechanical tilling, the use of Paraquat can save you a whole lot of labor, time and equipment usage. It can free you for countless other duties. Why not talk to your ORTHO Dealer about it? For best results, use Paraquat with ORTHO X-77® Spreader.

ORTHO PARAQUAT
Chevron Chemical Company

TM'S: ORTHO, CHEVRON AND DESIGN – REG. U.S. PAT. OFF. AVOID ACCIDENTS: READ THE LABEL AND USE ONLY AS DIRECTED.

Circle 72 on Reader Service Card

Frank Lessiter

New Coulter Designs
Got No-Till Started

"You want a coulter that will slice through residue **without pushing it down**..."

Without the development of the no-till coulter by Allis-Chalmers engineers, this conservation tillage system would never have caught on as fast as it has. While only half of no-tillers use coulters on their planters today, outside the box thinking in regard to coulter design led to commercialism of no-till in the late 1960s.

Prior to that time, wheel-track planting had confirmed that only a small area of fine, firmed soil was needed for seed germination. As a result, planter engineers were soon asking the next logical question:

"Why do we need to turn over 500 to as much as 1,000 tons of soil per acre with a moldboard plow when we only need 8% of the ground for a seedbed?"

That question led to the 1960s development of the A-C No-Til coulter by Maynard Walberg, an implement engineer at Allis-Chalmers, a full-line farm machinery manufacturer that later became part of AGCO.

As explained in *Allis-Chalmers Farm Equipment 1914-1985*, a book that was authored by long-time company employee Norm Swinford, the initial product was a deep-fluted

No-Till Coulter Usage

Type of Coulter	% No-Tillers Using Coulters on No-Till Planters	% No-Tillers Using Coulters on No-Till Drills
Narrow fluted	36%	29%
Wide fluted	15%	10%
Bubble	20%	3%
Ripple	15%	13%
Straight	8%	21%
None	6%	24%

— 1991 No-Till Farmer subscriber survey

17-inch diameter coulter that was mounted on the A-C 500 series toolbar during the 1960s and 1970s.

It was designed to slice through crop residue or sod while preparing a 2 1/2-inch wide seedbed ahead of the row unit for planting. A heavy rubber torsion-spring penetrated tough soils and protected the coulter when going over rocks.

To accommodate the coulters, Allis-Chalmers engineers designed no-till planter frames with three toolbars to handle 20- to 40-inch rows. The first toolbar held the coulters, the second toolbar carried the fertilizer boxes or tanks plus openers and the

Four Styles of Rolling Coulters

	Plain Edge	Ripple Edge	Fluted Edge	Notched Edge
Side				
Edge				
Pattern				

DISTINCT COULTER DIFFERENCES. Four coulter styles demonstrate how soil is moved when forming the furrow. In 1986, Montana State University ag engineer Bill Larson told no-tillers they may need a rolling coulter ahead of the row openers on their drills, which adds to the cost and weight.

— Mid-May 1986, No-Till Farmer

planter units were attached to the third toolbar.

By the early 1970s, the company offered three coulter blades for varying soils and field conditions. Its full 2 1/2-inch fluted coulter proved to be the most popular model for most no-tillers.

Over the years, new coulter styles have come along to work with single, double and triple openers. The key with no-till is to have a coulter that slices through the residue rather than simply pushing it down or rolling over residue left in the seed trench. Failure to do so aggravates the residue problem by tucking residue (called hairpinning) into the furrow along with the seed.

NO-TILL-AGE®

April 1986

How Coulters Influence Surface Residue

Coulter Type	Residue Left After Corn	Residue Left After Soybeans
No coulter or smooth coulter	90%	95%
Narrow ripple coulter	85%	90%
Wide fluted coulter	80%	85%
Sweeps or double disc furrowers (till-plant)	60%	80%

— Univ. of Nebraska

Some growers felt bubble coulters were ideal for no-till. Others argued that the development of the bubble

"It is important to match the **width of the coulter** with the **row opener width...**"

Pros, Cons of Coulter Styles

In late 1991 and early 1992, *No-Till Farmer* published a thought-provoking six-part series on the pros and cons of using coulters with no-till planters and drills. Based on survey data, it was evident that no-tillers, educators and ag supply firm representatives had sharply different viewpoints on the use of coulters under no-till conditions. In the early 1990s, 96% of growers were equipping no-till planters with coulters.

Yet the big complaint in those days was that many coulters didn't effectively slice through heavy residue left in the row area, didn't properly form the seed trench, tossed too much dirt to the side of the seed trench and led to excessive soil compaction.

coulter (shown at right) was the worst thing that had ever happened because of their impact on soil compaction and cutting depth.

Among no-tillers using fluted coulters in the 1991 survey, two-thirds preferred the narrower models to the 2-inch wide models.

The biggest objection to straight coulters was that they didn't do a good job of cutting residue. Others preferred this style of coulter, as it tended to leave the soil more intact, which leads to fewer weed concerns.

Tests conducted at Iowa State University in the mid 1980s showed few differences between four styles of coulters. The ag engineers found no significant impact on the amount of draft force or the percentage of corn stalks sheared by smooth, fluted,

rippled or notched coulters. They determined dry corn stalks are easier to cut than wet stalks.

In fact, these researchers determined coulter sharpness was more critical than coulter style. They also found that having enough weight to get adequate coulter penetration is essential, which was confirmed decades later with the addition of down-pressure units on many no-till planters.

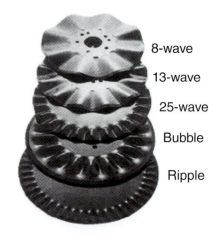

8-wave

13-wave

25-wave

Bubble

Ripple

TAKE YOUR PICK. These were the five most commonly used coulters that were found on a number of no-till planters in the early days.

Match Coulter Depth to Moisture Conditions

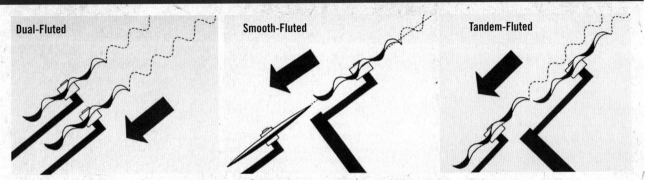

Dual-Fluted

Smooth-Fluted

Tandem-Fluted

MATCH COULTERS TO SOIL RESIDUE. When selecting coulters for use with no-till planters, you'll want to match them to soil conditions and seedbed residue. Dual fluted coulters (left) work a wider seedbed. Placing a smooth coulter ahead of a fluted coulter (center) aids in cutting through residue. Tandem-fluted coulters (right) produce more loose soil, which leads to better seed coverage.

— *No-Tillage Farming, Minimum Tillage Farming*

Tracing a Half-Dozen Coulter Tracks

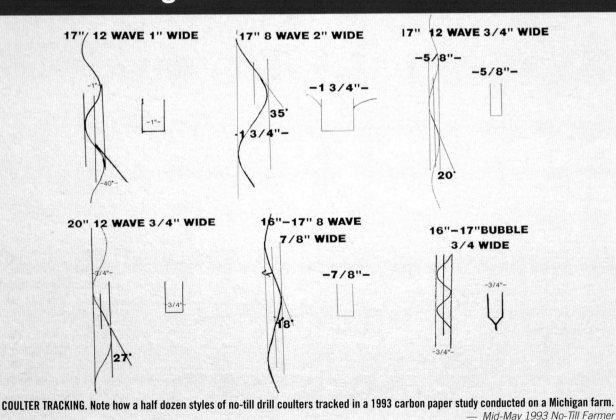

COULTER TRACKING. Note how a half dozen styles of no-till drill coulters tracked in a 1993 carbon paper study conducted on a Michigan farm.

— *Mid-May 1993 No-Till Farmer*

Tracking Coulter Tracks

In the early 1990s, Joe Malburg evaluated the action of several types of no-till coulters on a Tye Series V drill. The Ray, Mich., grower had been no-tilling for 20 years, but still wasn't sure how different style coulters moved through the soil.

To find an answer, he rolled different style coulters across pieces of carbon paper to determine the actual width and angle of its path. After establishing a center line, he measured the angle from this line to the outside of the waves with a protractor.

He reasoned that the greater the angle, the more aggressive a coulter will be and the more dirt it will lift. (See chart on page 111.)

When Malburg ran his no-till drill at 6 mph, the 12-wave coulters lifted more dirt than the 8-wave coulters. Even at slower speeds, the 12-wave coulters worked the soil in the seed zone slightly more than did the 8-wave coulters.

At higher speeds, the 8-wave coulters loosened just as much

NO-TILL-AGE®

March 1995

Planter Attachment Impact on 1994 Corn Yields

Attachment	Yield
No attachment	180 bu.
Row cleaner (1)	176
Combo at seed depth (2)	181
Combo below seed depth (3)	178
Combo plus 2 ripple coulters (4)	185
Combo plus 2 wavy coulters (5)	181
Soil finisher (6)	185

(1) Residue manager only.

(2) Residue manager plus a 13-wave coulter running at seed depth.

(3) Residue manager plus a 13-wave coulter running 1-inch below seed depth.

(4) Residue manager plus a 13-wave coulter running at seed depth and two ripple coulters running hub deep.

(5) Residue manager with a 12-wave coulter running 1-inch below seed depth and two wavy coulters at hub depth.

(6) For comparison, two passes were made with a field cultivator equipped with discs and cultivator shanks and corn was seeded with no attachments on the planter.

— *Calmer's Agronomic Research Center*

NEW NO-TILLERS OFTEN PICK
WRONG COULTER

March 1993

Ed Meek says growers often select the wrong coulter when making the switch to no-till.

The former University of Missouri area agronomist at Memphis, Mo., says it's natural for a first-time no-tiller to want to till a narrow area where seed will be placed. As a result, first-time no-tillers often choose a 3-inch or 2 1/2-inch wide wavy coulter, which tosses moist soil out of the seed zone, leaving a shallow open trench. ✿

dirt, but the 12-wave coulters didn't throw as much soil since they took smaller bites of dirt.

In general, he found smaller and sharper coulter blades were a positive development. With 24-inch diameter blades, there was less pushing on the soil and more hairpinning of residue in the seed trench. There were less trash clearance concerns with the 17- or 18-inch diameter blades.

In his 500-acre no-till operation, Malburg preferred the wavy blades. He found bubble coulters pushed the soil down and to the side, which led to difficulty in closing the V-trench. He determined that it was also important to match the width of the coulter blade to the row opener width in order to place the seed deep enough in the trench while avoiding sidewall compaction and excessive pressure on the back of the opener.

Malburg's conclusion was that the selection of coulters depends on whether you want to cut a path through residue, remove residue, create a proper seedbed or use the coulter to "knife in" fertilizer.

Soil conditions must also be considered when choosing the right coulter. This means looking at soil type, the amount and kind of residue, crop root structure, size of the field, tractor horsepower and the particular planting or drilling speed.

One thing Malburg noticed from the carbon paper tracing evaluation was a discrepancy between actual and advertised coulter widths. In some cases, 2-inch wide coulters were only 1 3/4 inches wide, and a number of 3/4-inch coulters were found to be only 5/8-inches wide.

In other work with coulters, Paul Schaffert found the double-disc openers and bearings on no-till

drills should last twice as long when a coulter is run ahead of the discs. A veteran no-tiller and the owner of Schaffert Manufacturing at Indianola, Neb., he says a coulter provides a softer seedbed for no-tilling.

Not All Coulter Ideas Worked Out

In the mid 1970s, University of Tennessee ag engineers experimented with a vibrating coulter on a no-till planter. While the vibrating no-till coulter improved plant stands by 5% over fluted coulters when no-tilling corn into sod, it was a different story with a bare dry silt loam soil. Under these conditions, the vibrating blades tossed up more soil and the resulting clods led to a lack of uniform seed placement and depth.

However, the vibrating coulters never caught on among manufacturers. ❃

NO-TILL'S BIGGEST HEADACHES

Every spring, Rick Tallieu says two problems frequently spell the difference between no-till success and failure.

The reduced tillage agronomist with Alberta Reduced Tillage Linkage in Olds, Alberta, says farmers generally start out seeding at the right speed.

"Unfortunately, they increase their tractor speed toward the end of the day or when rain is expected. Pretty soon they're no-tilling too fast for conditions and the crop suffers," he says.

Some growers spread straw over the full combine width, yet don't have a chaff spreader, says Tallieu. He maintains dealing with chaff is critical.

"If you don't spread the chaff, it creates a seeding nightmare," he says. "It's even worse when the next crop is canola, as that crop won't grow in a chaff row."

Another option is collecting chaff in a wagon pulled behind the combine. Still another option is to blow chaff over the top of windrowed straw for baling. ❃

PARKING LOT SCENES. A few growers have invested in personalized license plates to promote their use of no-till, as seen on vehicles at mid-summer no-till field days.

Frank Lessiter

Frank Lessiter

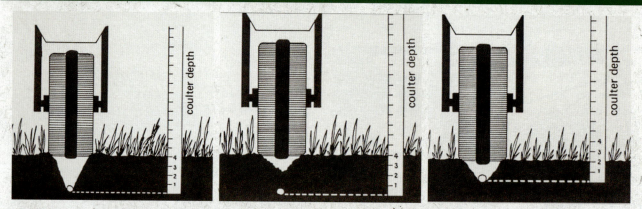

Run Shallow, Run Deep

ADJUST NO-TILL COULTER DEPTH. Running coulters too deep in wet soil (left) can result in seed placement being too deep. Running the coulter at the same depth in dry soil (center) results in proper seed placement. Proper coulter depth adjustment for wet conditions is shown at right.

— Mid-May 1993 No-Till Farmer

NO-TILL LOOSE ENDS...

BIG HORSEPOWER VS. NO-TILL

January 1977

You'd need a 260-row no-till planter to fully utilize this 650-horsepower, two-engine articulated Steiger tractor that was displayed at the 1976 Farm Progress Show. With 30-inch rows, the no-till planter would stretch out 650-feet wide behind this monster tractor that had a price tag of around $100,000 in 1977 ($423,581 in 2018 value).

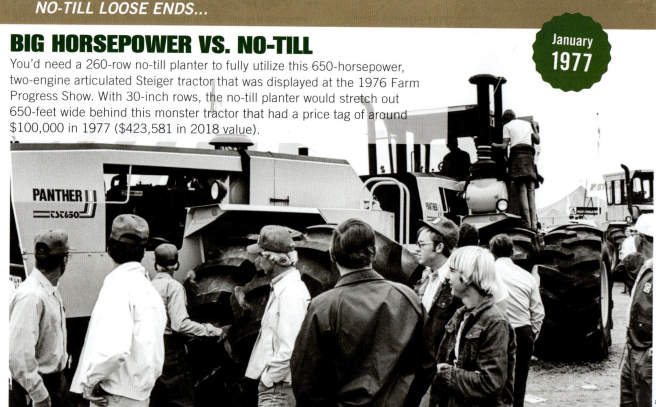

Frank Lessiter

Plowing is a Practice
from the Past

"Nobody has ever advanced a
scientific reason for plowing…" — *E.H. Faulkner*

John Deere was a typical central Illinois blacksmith in the mid-1800s turning out hayforks, horseshoes and other essentials for farming on the prairies. Then a broken steel blade led him to fashion a highly-polished moldboard from which sticky prairie soils would slide off easily while turning under the sod.

The innovative moldboard plow created in 1837 by the blacksmith in Grand Detour, Ill., revolutionized farming to the point where huge plows and high horsepower tractors are now used to turn under the soil. In fact, it got to the point where multi-bottom plows are capable of turning up thousands of tons of dirt in a single pass.

But with the increased use of no-till and minimum tillage, fewer acres are plowed each year. In 1972, 85% of the farmed ground in the U.S. was conventionally tilled, with almost all moldboard plowed. By 2012, the use of conventional tillage had dropped to 38%.

In 1973, no-till was used on only 1.6 % of the U.S. cropland. This increased dramatically to 34.6% of the ground by 2012.

MOLDBOARD PLOW.
The original moldboard plow that made it possible to tame the prairies and started the tillage revolution in 1837 was created by central Illinois blacksmith John Deere.

Deere & Co.

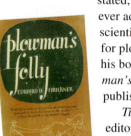
IS THE PLOW BEING BURIED?

Edward Faulkner Calls Out Plowing

On July 5, 1943, Edward Faulkner stated, "no one has ever advanced a scientific reason for plowing" when his book, *Plowman's Folly*," was published.

Time magazine editors said the book was "one of the most revolutionary in agricultural history, and the *Associated Press* quipped, "His crops are the wonder of his neighbors and the despair of the conventional farmer."

Reader's Digest said, "Probably no book on an agricultural subject has ever prompted so much discussion in this country."

Faulkner was honored in 2011 by *No-Till Farmer* as being one of "40 Legends of the Past 40 Years" for no-till practices, noting the book was, "a revelation on understanding and appreciating the value of reducing soil disturbance."

A Tool of the Past

Over the years, we've written a few "tongue in cheek" articles on the moldboard plow. Much of this chapter includes a few "Frank Comment" columns and other *No-Till Farmer* items over the years that poked fun at moldboard plowing and its lack of a role with no-till. 🌾

There's no doubt about it…and there certainly hasn't been any doubt for a long time. The trend is definitely toward reduced tillage as our exclusive No-Till Farmer "Tillage Practices Survey" has shown since we started this publication way back in 1972.

Farmers are changing tillage practices to hold costs down, to keep soils in their fields and to secure government price supports, says Michael Hirschi, a University of Illinois ag engineer.

Results of a survey conducted among Illinois farmers in 1989 indicate only 22% of the state's farmers used the moldboard plow last year. In a similar survey 7 years earlier, 59% were moldboard plowing.

"Growers are recognizing they don't have to plow every year, but it's still a new way for many and they're cautious," Hirschi says. "They know extensive tillage is a consistent performer and there

MOLDBOARD PLOWING TRADITION. Generations of farmers maintained that decent-yielding crops couldn't be grown unless the sod or remaining crop residue was turned over in the fall or spring with a moldboard plow. Not turning over the soil each year is something they didn't believe could work.

Minneapolis-Moline

seems to be a perception of risk with other tillage methods."

Hirschi says this perception may be unfounded since researchers compared yields with a half dozen no-till, reduced tillage and intensive tillage systems and found only a slight yield advantage for plowing.

"Whatever reduction in yields we've seen from reduced tillage and no-till was insignificant most years and was offset by less soil erosion and lower costs compared to the plowed plots," he adds. "That's not a bad trade off."

Hirschi says plowing is expensive in terms of power. A moldboard plow system requires four or more trips across a field. This cost adds up so that by the end of the year per acre costs among different tillage systems are equal despite difference in yields.

New Equipment Needed?

Hirschi says one common misconception among growers thinking about changing tillage practices is that new equipment will be needed. As an example, a move to a rotation tillage strategy such as discing after corn and no-tilling after beans should not require new equipment on most farms.

"The moldboard plow is a farming tool that has a place beside many other farming tools," concludes Hirschi. "But growers should be aware that there are many tillage options in addition to plowing." ✺

THINGS THAT WILL SOON DISAPPEAR FOREVER

At the end of 2016, the *Kiplinger Today* newsletter out of Washington D.C., listed 10 items Americans will be saying goodbye to over the next few years. Among your car's clutch pedal, dial-up Internet, the incandescent light bulb and keys, #7 on the list was the predicted demise of the moldboard plow.

Here's what they had to say about the coming demise of the plow.

"Few things are as symbolic of farming as the moldboard plow, but the truth is, the practice of 'turning the soil' is dying off.

"Modern farmers have little use for it. It provides deep tillage that turns up too much soil, encouraging erosion because the plow leaves no plant material on the surface to stop wind and rainwater from carrying the soil away.

"It also requires a huge amount of diesel fuel to plow, compared with other tillage methods, cutting into farmers' profits.

The final straw: It releases more carbon dioxide into the air than other tillage methods.

"Deep plowing is winding down its days on small, poor farms that can't afford new machinery. Most U.S. cropland is now managed as no-till or minimum-till, relying on herbicides and implements such as drills that work the ground with very little disturbance. Even organic farmers have found ways to minimize tillage, using cover crops rather than herbicides to cut down on weeds." ✺

Case IH

FUEL BURNERS. Plowing just 8 inches deep means a farmer is turning over 1,000 tons of soil per acre. By switching from a moldboard plow system to no-till, a grower can trim his yearly diesel fuel bill by 1,684 gallons on 400 acres. No-till also offers sizable labor and machinery savings.

WHY FARMERS WON'T CHANGE

Plowing and smoking cigarettes may have more in common than you think. And the excuses for not giving up either sound pretty similar.

January 1991

Wes Robbins of the Soil Conservation Service in Burlington, Colo., offers a dozen excuses that deal with not giving up smoking and plowing.

1. Some farmers still don't consider plowing to be a dirty habit. Instead, conventional tillage is designed to keep farmers busy, be respected in their community and make them happy.
2. Even though it isn't "cool" to plow, many farmers still do it.
3. Despite being bad for your farm's economic and productive health, some people still plow. The farmer who worked hard, plowed, had bad luck and went broke is still treated OK in most communities. But if a farmer went broke no-tilling, he couldn't even go into town.

   Ask a conventional till farmer how much profit he earned last year and he may tell you to get lost. Yield — not profit — is normally what counts to the man who plows.
4. Like a chemical deficiency, an addiction to tillage is hard to break. What to do with their time when not plowing is a problem for some farmers.
5. Many suppliers maintain plowing is not a filthy habit. Too many farmers listen to people who say you must till the soil to make it produce.
6. Tradition still counts. Plowing must be the thing to do since it's been done by generations of farmers.
7. Peer pressure is still there. At the local coffee shop, no-tillers may sit at a small table in the back with other no-tillers instead of sitting up front talking with big horsepower farmers who think no-tillers are a bunch of trashy, lazy growers.
8. Lender acceptance can be a concern. Some lenders tell farmers they can't no-till if they want to borrow money.
9. Too many landowners still believe you've got to work the soil. If you won't plow their soils, they may rent to someone who will.
10. Free time is a concern. What are you going to do with your free time if you quit plowing? Get an off the farm job? Or drive the neighbor's tractors?
11. A farmer may watch soap operas for 2 days, then get so bored that he will go out and plow.
12. If the corn yield goal in your area is 180 bushel per acre, you're considered a bad farmer if you fall below it and a liar if above it. This is why no-tillers don't talk about yields.

How many conventional tillage farmers in your area are guilty of these dozen "bad habits?" Maybe it's time for you to help by forming a "Plowers Anonymous" group!

'JUST BURY THE RESIDUE'

In the early days of no-till, plant pathologists were certainly not fans of the new practice. In the 1960s and 1970s, their answer to plant disease concerns with no-tilled crops was to plow under the residue. By doing so, they maintained most plant disease concerns would go away.

Even today with numerous fungicides on the market, there are still plant pathologists who believe you can plow away many plant disease worries.

To many of them, "no-till" is still a dirty word.

SO LONG EARTHWORM POACHERS. With the soil remaining covered with residue in a no-till system, glimpses such as this photo of seagulls and other birds finding earthworms on top of the soil in freshly-plowed fields are a sign of the past. Higher earthworm numbers are a big benefit of no-till.

'JUST CALL ME IRRESPONSIBLE'

February 2000

That's a typical comment we've heard numerous times in recent weeks from *No-Till Farmer* readers regarding the latest John Deere advertising campaign that emphasizes the need for intensive tillage.

In these ads, John Deere states:

◆ "Tillage is the responsible way to manage residue."
◆ "Tillage creates the most favorable seedbed."
◆ "Tillage is an environmentally sound farming practice."
◆ "Tillage helps reduce plant and insect problems."
◆ "Tillage boosts reliability in weed management."
◆ "Tillage is a reliable means of incorporating herbicides."
◆ "Tillage effectively breaks up soil compaction."

The Deere ads go so far as to suggest that buying a moldboard plow is a low-cost alternative for controlling pests. These ads suggest field cultivators, discs, mulch finishers and other tillage tools offer profitable solutions to reducing compaction, weed, residue and pest concerns.

Not once do they mention no-till in these ads.

Promoting soil and chemical incorporation along with ways to handle high residue and poorly drained fields with field cultivators, discs and moldboard plows doesn't make sense to no-tillers. They wondered whether John Deere officials are serious about promoting no-till while protecting the environment.

Maybe they're just interested in selling big horsepower tractors and more iron. ✳

HOW TO SPOT 'HABITUAL TILLERS'

March 1984

In talking with farmers, you can spot the real "died in the wool" conventional tillers by listening for one or more of these 10 telltale phrases:

1 "I would love to have a big tractor like…"
2 "That black field sure looks pretty…"
3 "All the neighbors are out…"
4 "I would rather be on the tractor than doing this…"
5 "It's sure a nice day – I think I'll plow a few rounds…"
6 "The neighbors will think I'm lazy if I don't plow…"
7 "Just one more discing to smooth up the…"
8 "That should bury those little…"
9 "If I don't bury the trash, it will plug…"
10 "Fall fertilizer always has to be incorporated…"

— Jones County, Iowa, Soil Conservation Service

$1.5 MILLION FINE FOR PLOWING?

June 2000

Results of a recent California judicial ruling indicate plowing and deep ripping wetlands may be against the law. As no-tillers have maintained for years, the cropping and environmental perils of plowing may finally be recognized when it comes to controlling erosion and protecting our land.

The Clean Water Case

A federal judge has fined Galt, Calif., farmer Angelo Tsakopoulos $1.5 million for "deep plowing." The judge ruled deep plowing violated sections of the federal Clean Water Act.

Even so, many ag people are concerned this ruling may set a dangerous precedent since one of farming's basic tillage tools could be designated as an instrument of water pollution.

U.S. District Judge Garlan Burrell, Jr., ruled Tsakopoulos committed 358 violations of federal environmental law. Burrell gave him the option of paying a $1.5 million fine or spending $500,000 to finance a 4-acre environmental restoration project on his ranch. The judge reasoned that Taskopoulos must pay a stiff civil penalty or make appropriate restoration for "depriving the nation of wetlands that support wildlife and endangered species."

The judge rejected the farmer's challenge of the U.S. Army Corps of Engineers and the Environmental Protection Agency (EPA) power to regulate deep ripping on the 8,348-acre ranch. Farming interests are closely watching the EPA's counterclaim for water law violations, the first action of its kind in the nation.

Costly Violation

Government agencies argue the deep ripping destroyed sensitive wetlands and violated federal law. "The amount of the penalty is sufficient to take the profit out of the violations," says Edmund Brennan, an assistant U.S. attorney. "'People should take from this result the lesson that it costs more to violate the Clean Water Act than to comply with it."

On the other hand, Taskopoulos contends regulators have no right to tell farmers how and what ground to plow. He maintains the idea that a plow represents a point source of pollutants is totally wrong. When the case first went to court 3 years ago, he vowed to make this a national test case and take it all the way to the U.S. Supreme Court if necessary. He hasn't changed his mind.

"Two farmers were hit with fines over $4 million for plowing so-called wetlands..."

Pasture Conversions

Deep ripping was used to deal with a hard, relatively shallow layer of soil before converting pasture land to vineyards and orchards. By using deep rippers without a permit, the judge ruled Tsakopoulos violated the Clean Water Act by causing soil to be deposited into environmentally sensitive wetlands, swales and intermittent drainage from 1995 to 1997.

The judge maintains the rancher violated the federal act by discing a vernal pool that was likely home for fairy shrimp, a threatened species under the Endangered Species Act.

Editor's Note: *The U.S. Supreme Court in 2002 upheld the lower court's verdict against Tsakopoulos who had accused the government of going overboard to protect wetlands. He had earlier been ordered to pay $500,000 in fines and restore 4 acres of wetlands.*

Another Similar Case

Farther north in California, there's another related case. In 2012, John Duarte, who owns Duarte Nursery in Modesto, Calif., bought 450 acres near Red Bluff in northern California. His intention was to grow wheat.

Because the property had some wetlands, Duarte hired a consulting firm to map out areas that were not to be plowed since they were part of the drainage system for nearby creeks. Even though a small amount of the wetlands were accidentally worked to a depth of 4 to 7 inches, there was no significant damage.

Duarte was shocked when hit with a $2.8 million fine for plowing before seeding wheat. The Army Corps of Engineers and the California Central Valley Regional Water Quality Control board claimed Duarte had violated the Clean Water Act by failing to obtain a permit to discharge dredged or fill material into seasonal wetlands, which are considered waters of the U.S.

In addition to the enormous fine, Duarte may have to repair the damage to the wetlands, including leveling the soil and replanting native plants. He may also be required to purchase other wetlands to compensate for the alleged property damage.

Need a Permit for Plowing?

Duarte sued the two agencies, maintaining they violated his constitutional right of due process by issuing cease and desist orders without a hearing that kept him from harvesting the wheat. The U.S. Attorney's Office counter-sued to enforce the Clean Water Act violation, despite the fact that farmers plowing their fields are specifically exempt from the rules.

"The case is the first time we're aware of that says you need to get a U.S. Army Corps of Engineers permit to plow to grow crops," says Anthony Francois, a lawyer for the Pacific Legal Foundation that fights for private property rights and limited government. Some environmental groups felt the Army Corps lawyers should also have claimed that Duarte had violated the Endangered Species Act by destroying fairy shrimp or their grassland pool habitat.

It may be an important ruling as the case could set a precedent for requiring farmers to obtain costly and time-consuming permits to work their land.

Editor's Note: Faced with immense legal and business risks and a government threat of up to $45 million in penalties, Duarte reached a settlement with the justice department in August of 2017.

Under the agreement, Duarte admitted no liability, agreed to pay a $330,000 civil penalty, buy $700,000 of vernal pool mitigation credits and perform additional work on the site where plowing took place. ❀

SHOULD YOU NO-TILL OR PLOW?

August 1993

A couple of issues back, we looked at 10 scientific reasons for plowing instead of no-tilling. Put together by no-tiller Piers Woodriff of Somerset, Va., the list has brought many letters from readers.

"I find all 10 of his reasons to plow to be disadvantages for using the plow," wrote Gene Vincent, Soil Conservation Service conservationist at Austin, Minn. "I hope he made his list with tongue-in-cheek."

Vincent offers another viewpoint to each of the 10 previously listed reasons for moldboard plowing.

1 *Facilitates root growth by loosening the soil.* Plowing loosens 6-10 inches of soil, but corn roots grow at least 5 feet deep. By planting time, the soil has recompacted.

2 *Increases water intake by loosening compacted soil.* The reasons for not plowing are the same as for the previous item.

3 *Lets more oxygen into the soil for plant roots to use.* Spring plowing and discing, plus tractor traffic, recompacts the soil.

4 *Aids earthworms.* In recently dug ridge-tilled fields, there were earthworms all over the place. There were few earthworms in a nearby fall plowed field.

5 *Covers the trash.* Plowed corn stalks rot slower when buried 6 inches under the soil.

6 *Gets rid of weed seeds by burying them.* Buried weed seeds return to the surface to germinate in plowed fields.

7 *Covers and kills unwanted vegetation and cover crops.* By covering the vegetation, you lose erosion protection and reduce the speed of residue decomposition.

8 *Provides loose soil to later cover weed seedlings.* Not true since most conventional tillage farmers no longer cultivate.

9 *Kills harmful over-wintering insects.* Farmers have been plowing for more than 100 years and still have plenty of insects.

10 *Aerates the soil, providing nitrogen release from accumulated crop residues.* It's doubtful that plowing increases decomposition or increases nitrogen release. ❀

DOING WHATEVER IT TAKES

Back in the late 1970s, I was visiting with an Illinois farmer who was on his second or third year of giving no-till a try. Having both grain bins and flat grain storage, he told me where he stored his first newly-harvested corn.

"The first three loads of corn are dumped in the southwest corner of my large machinery storage building," he says. "That's where I keep my six-bottom moldboard plow.

"I dump those first few loads of corn over the top of that plow, so I'm not tempted to get the plow out when all of my neighbors are fall plowing."

THE NEGATIVES OF MOLDBOARD PLOWING

Many folks talk about the negatives of no-tillage. But we'd bet you can't recall anyone talking about the negatives of moldboard plowing, as it's seldom done.

February 1980

Here are 10 negatives to plowing as worked out by veteran ridge tiller and former Soil Conservation Service area conservationist Ernie Behn of Boone, Iowa.

1 It takes too much fuel.
2 It takes too much equipment.
3 It causes water erosion.
4 It causes wind erosion.
5 It takes too much time, plus the time needed for conventional tillage comes at a critical time in the spring.
6 Soil compaction is high.
7 It does a lousy job of conserving soil moisture.
8 It forces you to pull volunteer corn plants.
9 It makes weed control more difficult. Despite two discings, two rotary hoeings and two treatments of herbicides, weeds keep coming back.
10 Insect problems can be great.

Now that we stop to look at them, Behn's list of moldboard plowing negatives makes a lot of sense.

But there's still one thing that bothers me about all of these plowing negatives. With all of these disadvantages, why haven't we been able to talk more people out of moldboard plowing and conventional tillage?

'MAKE THAT A DEAD FURROW, PLEASE'

During the 1972 World Plowing Contest held in Mankato, Minn., bartenders at the Holiday Inn came up with a special drink that they called "the dead furrow."

Here's how the drink with this plow-related name was made:

- 2 ounces of 151 proof rum.
- 1 dash of Triple Sec.
- 1 tablespoon of sugar.
- Juice of one half lemon.
- Splash of sparkling water.
- Serve over crushed ice.

January 1973

Farmers who are still moldboard plowing may need a strong drink like this one when they think of how many dollars they've wasted by not no-tilling.

If you're not familiar with the "dead furrow" term, that's great. It means you've likely never sat on a tractor pulling a moldboard plow.

A "dead furrow" is a double-wide furrow that is 8-10 inches deeper than the surface as soil has been tossed in opposite directions as plowing is wrapped up in an area of a field.

On the other hand a "back furrow" is when dirt from the first two back and forth moldboard plowing trips in the opposite direction are tossed together in an area of a field.

SO LONG TO THE PLOW

By John N. Landers

Throw your plough through the window
And revel in the crash
Of broken glass and paradigms
And non-organic trash

History had many reasons,
Learned from going wrong
But when there is a breakthrough
Don't listen to the throng

Inventions break the ancient mould
And open up the doors
To spread a new technology
That sets a better course

Man spent a lot of energy
Breaking down the soil
But ploughs and heavy tractors
Give less and less for toil

Horse and ox did not compact
The way a steam plough could
And later, heavy tractors
Turned our soils to pudd.

We've burnt the saintly particles
That held our crumbs as bread
And lost organic matter
So our soils are nought but dead

The devil takes the hindmost
And those who don't evolve
Will pay the price of blindness
And their problems never solve

Conservation is for soils, sir
Not for clinging on
To now outdated practice
In an everlasting con

So throw your plough through the window
And revel in the crash
Of broken glass and paradigms
And non-organic trash

Brazilian no-till enthusiast John N. Landers wrote this poem for the 2016 World Congress on Conservation Agriculture in Rosario, Argentina. Living in Lago Sul, Brasilia, Landers serves as an international conservation tillage consultant.

IF HE'D JUST NO-TILLED...

There are folks around Matteson, Ill., who wish Ray Dettmering would give no-tillage a try. They aren't happy with his plowing tactics.

In fact, all Leonard Reh wants is a good night's sleep. But what he is getting instead is the noisy drone of tractors as they plow through the night behind his suburban home.

Reh and other neighbors in their Matteson subdivision have to turn their lawnmowers off at 7 p.m. because of village ordinances against excessive noise. And they wonder why a farm tractor should be any different.

On the other hand, Dettmering, who was arrested for plowing his fields too late at night last spring, sees it differently.

"The way I feel, he moved next to the field — the field didn't move next to him," says Dettmering, who resides in Peotone, Ill. "The only question I have is when I go to harvest this fall, what happens then?"

Reh, a postal employee whose property abuts the land Dettmering rents, says the late-night forays must stop. The once-rural village (30 miles southeast of Chicago) has seen a significant influx of middle class commuters and an upswing in luxury housing in recent years.

All Day, All Night

"I don't want to stop him from making his living," says Reh. "But farming at midnight is just a little bit outrageous."

Around 11 p.m. on April 26, Matteson police, responding to Reh's complaint, tromped through muddy fields to arrest Dettmering as he, his wife, Kathy, and a hired hand finished plowing the land he leases. He planned to plead guilty to two counts of disturbing the peace – each carrying a maximum $500 fine and 6 months of jail time.

Dettmering has a day job and farms rented land on nights and weekends. He'd like to be a full-time farmer, but can't afford it, depending on his second job as a farm equipment dealership mechanic to make ends meet.

A wet spring put him behind in preparing corn ground. The days when it wasn't raining meant even longer days — and nights — in the field.

Peace and Quiet

"I have to get up at 4 a.m. to go to work," Reh says. "Whoever owned this land before used to work the land during the day rather than at night and never bothered anybody."

So, what's this got to do with no-tilling? Let's dream a little.

If Dettmering were to no-till, he wouldn't have to farm at night. He could trim his machinery investment enough to farm more land...become a full-time farmer...and so forth.

Then Reh would be able to sleep better. ✺

TOO MANY FARMERS STILL LOVE TO PLOW

March 1979

Is moldboard plowing a recreational activity? Just asking such a question shakes the very foundation of much of agriculture.

The Enjoyment of Plowing

Iowa State University ag engineer Wes Buchele says many farmers simply enjoy plowing and do it because plowing:

1 Buries crop residue.
2 Buries fertilizer.
3 Buries insects and their eggs.
4 Buries weed seeds.
5 Eliminates last year's mistakes that led to poor stands, weeds and erosion problems.
6 Gets the tillage job done in the fall when there is plenty of good weather.
7 Minimizes the threat of delayed spring field work.
8 Maximizes yields.
9 Allows the planter and cultivator to work best in clean land.
10 Is fun.

Plowing's Social Advantages

Buchele also maintains plowing offers certain "tongue-in-cheek" social advantages, such as:

- Plowing improves the stature of the farmer in his neighborhood.
- The cleaner the soil surface, the higher the status of the farmer.
- Farmers who plow are more religious.
- Banks loan money to farmers who plow.
- Farmers who plow are able to pay back their loans.
- Farmers who plow are more industrious.
- Farmers who plow are pillars of strength in their local community.

Buchele believes there's something wrong with what has been said in this "somewhat humorous" listing of why farmers like to plow.

We agree. Many farmers who plow still have nagging fears about their operation and how the soil erodes during rainstorms and blows away during windstorms.

These farmers have heard of minimum tillage, no-tillage and conservation tillage. Yet too many seem afraid to give reduced tillage a try. ❀

WHY NO-TILL WON'T WORK

April 1979

Last month, we reported a number of "tongue-in-cheek" ideas on why many farmers still plow.

Going further, Wes Buchele has created a list of what farmers who plow for fun and higher social standing in the community think of no-tillage, minimum tillage and conservation tillage.

These farmers think people get in trouble with reduced tillage because:

1 They have poor germination and stand problems with less tillage.
2 They have poor crop growth.
3 They have insect problems due to cutworms that reduce stands and corn borers that produce high first brood populations.
4 They have weed and grass problems.
5 Their fields look ragged throughout the winter and spring.
6 They always seem to be behind their neighbors with spring field work.
7 They must rely on chemical weed control rather than cultivation, and sometimes that doesn't work.
8 They appear to be lazy.
9 They have cold, wet soils in the spring.
10 Crop residue interferes with operation and efficiency of the planter.
11 Crop residue interferes with their cultivator.
12 Down deep, most reduced tillage farmers still would like to plow.

That seems to be the way "clean-tillage" farmers look at the reduced tillage picture. And you probably know neighboring farmers who would "worship" this list of reasons on why not to do less tillage.

Hopefully, these farmers will change. And as Buchele says, "The best thing they could do is to trade in their four-wheel drive, 150 horsepower tractor on a 150-horsepower motorboat.

By switching to less tillage, they would have the time in the fall to plow the lakes and rivers of this nation when the fishing is good."

Yes, plowing "water is definitely a recreational activity!" ❀

NO-TILL-AGE®

Two Reasons to 'Park Your Plow'

October 1983

Tillage System	Nutrient Loss Per Acre	Soil Loss Per Acre
Fall plow	186 lbs.	23 tons
Fall chisel	44	4.4
No-till	21.6	0.5

— Iowa State University

WHY PLOWING IS BETTER

May 1994

Unfortunately, a small scale, simple living publication called *Plain Magazine* recently devoted a special issue to agriculture. I say unfortunate because several pages were devoted to a story, which carried the above title.

The article by former *Farm Journal* writer Gene Logsdon directly attacked no-till, something he's done in several of his books that have promoted the anti-chemical market. After berating no-till for "repeated and heavy application of herbicides," Logsdon called for a return to plowing.

Fat Chance!

"I'm beginning to believe the attempt to abolish the plow is just another way the new oligarchy hopes to push small farmers out of business," he wrote. "The most irritating aspect of toxic, so-called no-till farming to me is its attempt to pose as the savior of farm soils."

Everyone Feels His Wrath

Logsdon blasts the chemical industry for promoting no-till, manufacturers for selling no-till planters, no-till drills and sprayers and takes after the government for going along with this increased expense.

While the author admits the stupid use of the plow can certainly lead to farming and erosion problems, he maintains "intelligent plowing" of suitable land can result in less erosion than no-till along with less energy and expense.

Logsdon goes so far as arguing that plowing can be done with small horsepower tractors or even horses. He believes Amish farmers definitely see the ways of the plow.

Yet it's hard for me to see bigger labor savings available anywhere else than by watching progressive Amish farmers making only one trip across the field with a no-till planter or drill pulled by a team of Belgians instead of walking five, six or more times behind a big team of horses with a plow and several tillage tools to get a crop in the ground.

Few Facts, More Baloney

Logsdon calls on his own small farm experience in preaching the virtues of the plow. He maintains plowed fields are completely protected from erosion all year long and are areas where abundant wildlife thrives.

Yet numerous studies show wildlife numbers increase and erosion is practically halted with no-till. But he doesn't let the facts confuse him as he insists no-till is bad, is toxic to your health and that the plow will be the savior of American agriculture.

Unfortunately, too many city people reading that small scale, simple living publication will get the wrong viewpoint. ✺

SHOULD NO-TILLERS EVER PLOW?

June 1980

For 8 years, we've been encouraging farmers to give up the moldboard plow while moving to less tillage. We've told readers how this idea can save them time, fuel, money, soil and much more.

You Must Decide

We've talked with no-tillers over the years that figured they would have to plow every third or fourth year to help control weeds — using no-till during the other years. And we've always told them to go ahead.

Yet most of these no-tillers later found moldboard plowing wasn't necessary. And a few no-tillers haven't plowed in 10-15 years.

We've always said each farmer has to work out the best tillage system for his farm and his or her own management ability. If that means plowing every fourth or fifth year, then they're still ahead during the other years.

Does Plowing Have Value?

That's why a recent Stauffer Chemical Co. News release suggesting that no-tillers should plow every 3-5 years doesn't bother us in the least. In fact, the idea has plenty of merit.

This news release quoted Jim Parochetti, a USDA pesticide specialist, as saying occasional plowing of no-till or reduced tillage fields can keep perennial weeds in check, reduce weed seeds on the surface and improve herbicide performance.

He goes on to say fall panicum, yellow nutsedge and triazine-resistant redroot pigweed are becoming increasingly difficult to control in no-till fields. By plowing every 3-5 years, Parochetti believes farmers can bury these weed seeds below their germinating zone.

Fall plowing can also make perennial weeds — like johnsongrass — more vulnerable to winter kill.

Parochetti also sees other benefits for plowing no-till fields, like mixing the soil. Since organic matter content increases and soil surface pH decreases in no-till fields, mixing of the soil can improve herbicide performance.

In conclusion, the moldboard plow might still have an occasional place in a no-tiller's production program. But it's a decision each no-tilling farmer has to make for himself. ✳

NO-TILL LOOSE ENDS...

SKIP ROW SOYBEANS FOR PARKING?

April 1992

After harvesting small grains in June, the staff at the University of Tennessee Milan Experiment Station no-tills double-crop soybeans in this interesting pattern.

During the station's late-July No-Till Field Day, this single-row, skip-row pattern of double-cropped no-tilled soybeans marks parking places for visitors. The idea not only helps promote no-till, but serves a valuable purpose while marking parking stalls in the recently-harvested small grain fields. ✳

Frank Lessiter

PLOWBOY PETE VS. NO-TILL NED

Known by how they tilled their farms, these
*two characters were part of the early issues of **No-Till Farmer.***

In this "tongue in cheek" look at tillage practices, it seemed like Plowboy Pete was always in trouble while No-Till Ned never seemed to do any wrong.

In 1972, 1973 and 1974 when we produced a magazine (before switching to a newsletter format due to limited no-till advertising support), these two characters appeared in most issues.

Readers who wrote the winning captions were treated to a free dinner for four at the restaurant of their choice with *No-Till Farmer* picking up the tab.

Our cost for most of the free dinners in the early '70s ranged from $28-$60. However, we always felt one winner took advantage of the free dining experience when his bill topped the $325 mark. ❋

"PETE, I certainly hope your soil that's washing over into my no-till field has a lot of fertilizer in it."

"PETE, With all the weeds growing in your fence row, it looks like you've created your own version of the Berlin Wall."

"**PETE,** you were just voted the 'man of the year' by the local farm equipment dealers."

"**PETE,** even your corn is rooting for no-till."

"**PETE,** we could sure use a nice rain shower, right?"

PETE, you always seem to plow the year around, making dead furrows wherever you go."

"PETE, you'd better head to the no-till meeting in Hawaii, so both you and your corn fields can wear a grass skirt next year."

"PETE, you need a 150 horsepower tractor to pull both your 8-bottom plow and combine."

"PETE, don't sweat the fuel shortage with all your farm machinery as you can switch to steam and use all those weeds for fuel."

Nothing's More Critical Than
Managing No-Till Residue

"I wouldn't sell residue from my no-tilled fields **even if the price was $100 per ton**…"

— *Ray McCormick*

No-till veterans recognize success starts with how residue is handled at harvest. Proper distribution of residue across the full width of the combine is critical to managing concerns with weeds, moisture, sunlight, disease, insects, planting and even next year's harvest.

With more acres of no-till continuous corn being grown, excessive residue is becoming a worry to some growers. While some growers may bale or graze a portion of the residue from their fields, other no-tillers insist on leaving all the residue in the field as part of their overall crop management package. Still others in the past burned off residue to insure better no-till seedbeds.

Ray McCormick is among many die-hard no-tillers who place a high value on the nutrient and other benefits that residue contributes to his success. In fact, the Vincennes, Ind., grower and soil conservation leader maintains he wouldn't sell residue from no-tilled fields even if the price climbed as high as $100 per ton. ✾

SPREAD STRAW ... AND CHAFF

If you expect to be successful with no-till, make sure you properly distribute the straw and chaff. In fact, it's just as important to effectively spread corn stalks when no-tilling.

John Smith, a University of Nebraska ag engineer, says 24- to 30-foot combine grain platforms and 8- to 12-row corn headers complicate residue distribution. With crop residue gathered over a 30-foot width and distributed only in a 6-foot band or windrow, you can expect residue problems when no-tilling, he says.

WHAT NOT TO DO. This combine is not spreading straw and chaff over the full width of the combine. Bunched-up residue can lead to serious seeding concerns when no-tilling.

As an example, take a look at a wheat crop that produces 100 pounds of above-ground residue for each bushel of grain. Thus a 40-bushel-per acre wheat yield would produce 4,000 pounds of straw and chaff.

Depending on the height and cut, about 50% of this residue will pass through the combine. Some 40-70% of the total residue is chaff, which normally does not reach the straw spreader or chopper — instead dropping from the combine chaffer and sieve section.

Proper Spreading Critical

Here's how different spreading devices measure up in terms of distributing crop residue:

◆ If the combine has no straw or chaff spreader and if the residue drops in a 6-foot windrow behind the unit, you will have the equivalent of a concentration of 12,000 pounds per acre in that area.

"This situation is nearly impossible to seed into without removing the windrow or using some form of tillage," explains Smith.

◆ Combines with a 30-foot grain header equipped with straw spreading devices are capable of spreading straw over an 18-foot width, yet often do not come with chaff spreaders. With this setup, the 6-foot wide strip where chaff falls will have a total residue accumulation equal to nearly 9,000 pounds per acre.

"This amount of residue is still very difficult to no-till through and demonstrates that the distribution of both straw and chaff is necessary for acceptable results," says Smith.

◆ What happens when you have a combine equipped with both straw and chaff spreaders capable of spreading 24-30 feel wide? The maximum amount of residue is usually less than 5,000 pounds per acre and is acceptable for success with most no-till planters or no-till drills.

Can Create Problems

Smith says a number of problems can result from inadequate residue distribution. These include: weed seed concentration, interception of herbicides, pest infestations, insulating the soil from the sun, poor tillage performance, a need for more tillage, increased demands on planting equipment, poor seed-to-soil contact and excess residue directly over the seed furrow.

"If the planter or drill can't cut through or separate heavy accumulations of residue, it may punch the residue into the soil with the seed," says Smith. "Known as 'hairpinning,' this can cause poor seed-to-soil contact and poor germination. Planting equipment with discs or coulters is most vulnerable to this problem."

If there's too much crop residue directly above an emerging plant, various degrees of plant injury may result. While this phenomenon, often termed phytotoxicity or allelopathy, is not completely understood, poor stands and poor yields often result.

Smith says a general rule is to avoid leaving the equivalent of more than 5,000 pounds of residue per acre directly over the emerging no-till plants. ✸

FOR TOP NO-TILL YIELDS, DON'T HARVEST STRAW, CHAFF OR STALKS

With current economic pressures, some no-tillers are wondering whether they should sell straw rather than leave it in double-cropped soybean fields.

While selling straw can be an attractive source of income, it's a valuable source of organic matter plus minerals and helps in growing no-tilled soybeans.

University of Delaware agronomist Bill Mitchell says farmers need to consider the long-term benefits from cutting wheat or barley at a height of 10-12 inches and leaving the straw in the field.

Worth $20 per Ton

He says the nutrients in 3,000 pounds of wheat straw have a fertilizer replacement value of about $20. The organic matter contained in the same amount of straw may be worth even more.

"This alone," says Mitchell, "is enough to question the wisdom of selling straw — especially from farms where the soil is sandy and the humus level is low."

With no-tilled soybeans, stubble and straw residue left on the ground shades the soil and holds back weeds. It can also conserve soil moisture.

Mitchell adds that herbicides such as paraquat and Roundup perform much better when small grain straw stubble is left tall.

Tests in Delaware indicated that improved weed control due to leaving a relatively high barley stubble (10-15 inches) boosted double-cropped soybean yields by 5 bushel per acre. In some cases, a drop in bean yields due to weed competition caused by low stubble has been more than 5 bushels.

Mitchell urges farmers to add up all the costs for removing grain straw from double-cropped fields. Make

sure these costs are more than covered in any price you get for straw at harvest time.

Besides leaving straw in double-cropped bean fields, it's also important to evenly spread the straw. This usually means a straw chopper is needed on the combine.

Spread Straw Evenly

The failure to spread straw evenly can lead to difficulty with no-till planter and drill openers penetrating the heavy residue. As a result, seed will not be properly placed.

Heavy layers of residue can also cause changes in fertility and moisture, which impact yield and crop maturity.

Present-day straw choppers are able to spread straw over a 13-20 foot width. However, ag engineers in western Canada doubled the spreading width of straw choppers by equipping them with relatively inexpensive deflector vanes. They also found new double rotor combine straw choppers can increase the straw spreading width to 60 feet.

Corn Head fitted with Calmer BT Choppers

Corn Head fitted with Rotary Blades

The BT Chopper simultaneously explodes stalks and exposes pith for maximum decomposition

Rotary blades leave stalks in tact, resulting in slower decomposition

STALKS EXPOSED. A 2014 central Illinois side-by-side comparison demonstrates how the Calmer BT Chopper explodes the top of the corn stub. This exposes the stalk's pith for much faster stalk decomposition. It also results in less tire wear compared to corn heads with rotary blades that leave the corn stalks intact.

Bank More Dollars

Paul Unger, a USDA soil scientist at Bushland, Texas, says leaving straw on the soil surface is money in the bank.

By leaving straw on the soil surface during a 10-month fallow season, Unger increased grain sorghum yields by up to 1,960 pounds per acre.

Without any mulch on a bare field, grain sorghum yielded 1,600 pounds per acre over a 3-year period. Yields increased to 2,100 pounds with a half ton of mulch, went to 2,000 pounds with 2 tons of straw mulch, to 3,200 pounds with 4 tons of mulch and to 3,560 pounds with 6 tons of mulch.

Besides getting extra yields from leaving straw on the soil surface, weed control was also cheaper. ✿

THREE 'WILD AND CRAZY' IDEAS FOR TACKLING NO-TILL RESIDUE

Effective residue management is among the cornerstones of any successful no-till program. And back in the '70s and '80, no-tillers were often forced to head to their farm shops to come up with innovative ways to deal with residue concerns.

'Up and Over' Mulching

Not satisfied with commercially-available no-till equipment, Charles Monty combined a flair chopper and no-till planter for tackling residue concerns in soybeans and cotton.

The Clarksdale, Miss., grower's one-trip rig chops existing stubble or vegetation, no-tills the crop, applies herbicide, lays the chopped residue down as a protective mulch and controls mulch placement behind the planter even in high winds.

MODIFIED FORAGE HARVESTER. This rig blows residue into raised slots cut by a chisel in order to improve no-till water-holding capacities.

The flair chopper has an adjustable deflector hood, and 48 cutter blades that are each 6 1/2 inches wide. Running at twice the normal speed, they clip the vegetation at ground level and chop the residue into a consistent sized mulch.

A coulter runs in front of each planter unit and the chopped stubble or vegetation is tossed back over the no-tilled seed to serve as a valuable mulch. Flood nozzles directly behind each planting unit offer complete herbicide coverage.

Slot-Mulch Residue Machine

The development at Washington State University of a slot-mulch residue system that could be easily combined with no-till helped increase water usage and reduced soil runoff and erosion concerns.

Developed by USDA ag engineer Keith Saxton, the concept may help eliminate compaction, crusting and sealing problems where tillage is not used.

The rig stuffs a continuous slot with straw that runs below the usual depth of soil freezing. The removed soil is placed in a narrow ridge along the downside of the slot and compacted straw and chaff is placed in the slot. The 2-4 inch wide slots that are 8-10 inches deep are repeated every 12-24 feet while going up steep slopes in the Palouse area of eastern Washington.

When combined with no-till, Saxton says the system offers a low energy form of tillage that allows for the removal of crop residue, which often leads to drilling and plant growth concerns. It also reduces runoff, increases

CHOP, PLANT IN ONE TRIP. By combining a flail chopper with a coulter-equipped planter, Charles Monty protects the soil while no-tilling soybeans and cotton.

SLOT-MULCHING RIG. Straw or crop residue is forced into 8-10 inch deep subsoiled slots to improve water usage and soil runoff.

available water for plant growth and the standing stubble offers a more uniform snow catch.

Saxton believes the system has great potential for no-till double-cropping in drier areas.

'Stuffing' the Residue

Saxton also modified a forage harvester to blow crop residue into a slot formed by a chisel. Spaced 20 feet apart, these 9-inch deep V-shaped slots are 3 inches wide at the base and 5 inches wide across the top.

"Water can infiltrate the soil even when the ground is frozen," he says. The residue insulates the slot so the bottom portion remains unfrozen."

NO-TILL-AGE®

November 1986

Wheat Straw Impact on Soybean Yields

Treatment	Yield Per Acre
Burn straw, no-till	49 bu.
Burn straw, conventionally-till, plant	50
No-till in high stubble	51
No-till in low stubble	52
Till, plant after wheat	50
Kill growing wheat, no-till	49
Disc, fallow, plant	48

— Milan Experimental Station, University of Tennessee

July 2007

CHECK NO-TILL HARVEST-TIME REMINDERS

If you don't want a mess next spring in your no-tilled fields, Gordon Hultgreen suggests paying close attention to harvesting this year's no-tilled crops.

The researcher at the Prairie Agricultural Machinery Institute, in Humboldt, Saskatchewan, suggests following a half dozen rules when it comes to managing no-till residue:

1. **Check stubble height.** For no-till seeders with hoe or knife openers, stubble height should not exceed the row width. For a seeder with 8-inch row spacing, this means the stubble height should be less than 8 inches. When damp in the spring, taller stubble often causes bunching and plugging problems with some no-till drills.
2. **Spread straw, chaff evenly.** Chaff that is concentrated in rows may tie up nitrogen and have a toxic impact on the next no-tilled crop. Spread chaff to the full width of the combine.
3. **Dense straw is bad news.** Crop residue can hairpin under disc drills and lead to poor or late germination when straw is pushed down into the planting V instead of being properly sliced.
4. **Spread the chaff.** Post-harvest harrowing will spread some straw, but the chaff normally remains untouched. Chaff spreading is best done with a combine-mounted unit. Buy this combine attachment before spending dollars on a no-till drill or air seeder.
5. **Deep banding fertilizer.** Do this as a separate operation in the fall. When done in the spring, deep banding disturbs the soil and leads to costly moisture losses.
6. **Get a jump on weeds.** Consider fall spraying of winter annuals and perennials if they are still actively growing after harvest. Mark major annual weed infestation areas on a field map to aid in planning future weed-control programs.

NO-TILL-AGE®

July 2004

When No-Till Residue Turns to Ashes
(Amount Of Nitrogen Lost To Burned Corn Stalks)

Corn Yield Per Acre	Nitrogen Lost Per Acre
120 bu.	40 pounds
130	44
140	47
150	50
160	54
170	57
180	60
190	64
200	67

This data is based on the previous year's 15.5% moisture corn yield. The nitrogen contained within the corn residue after harvest would not yet have been released and made available for next year's crop.

— Ohio State University

WHAT'S THE VALUE OF YOUR NO-TILL CORN RESIDUE?

Nutrient	Pounds per ton	Per ton value of nutrients
Nitrogen	17 lbs.	$6.80
Phosphorus	4	$1.80
Potassium	34	$8.50
Sulfur	3	$0.75
Total value per ton		**$17.85**

With a 200-bushel corn crop, 5 tons of corn residue per acre is produced. Considering nutrient removal, lime value, yield loss, soil loss, increased irrigation needs, raking, baling and transporting round bales, total costs could be as high as $75.60 per ton to harvest the residue. ✳

— *University of Nebraska*

NO-TILL LOOSE ENDS...

CONTROLLING JOHNSONGRASS NEVER EASY

March 1979

A Corn Belt no-tiller asked for advice on the best way to control Johnsongrass in no-tilled corn and soybeans.

In tests at the Dixon Springs Agricultural Center near Simpson, Ill., agronomist George McKibben found a combination of 7 pounds of Dowpon plus surfactant applied on June 10 and 5 pounds of Princep applied on June 22 gave the best yield with corn that was no-tilled on June 22.

"Johnsongrass is still a problem, in areas where plants have produced seed in previous years, and will likely continue to be a problem for several years," says one of the early-day pioneers in no-till research.

With soybeans double-cropped on June 22 into seriously Johnsongrass-infested ground, yields averaged 17 bushels per acre. To get that yield, McKibben applied 2 quarts of Roundup on June 10 along with an application of 2 pounds of Surflan and 1.5 pounds of Sencor when no-tilling the beans on June 22. ✳

No–Till Acres Jumped Dramatically
Once Roundup Arrived

"I'm looking forward to Roundup approval since **I'm sure it will be cheaper than paraquat**..."

When we launched *No-Till Farmer* in 1972, the burndown herbicide that was key to making this reduced system work was paraquat. While it has lost much of its popularity over the years, particularly to glyphosate, it still plays a role in controlling certain weeds in a no-till program.

Marketed today as Gramoxone, paraquat is among several herbicides that could have been used much more extensively in recent years to help control marestail and to interrupt the continuous glyphosate use cycle. Its decline was due to sharp reductions in the price of glyphosate (Roundup) and the greater versatility of glyphosate with numerous weed sizes and cycles.

"Paraquat is probably most useful when rapid dessication of weeds is essential to allow for tillage or planting," says Ohio State University weed scientist Mark Loux. "For example, a combination of paraquat plus atrazine or metribuzin will result in more rapid death and dessication of chickweed or purple deadnettle, compared to glyphosate, when applied in the spring under cool conditions."

"Our first Roundup article appeared in the **June 1973 issue…**"

GLYPHOSATE CHEMISTRY. The introduction of Roundup as a burndown herbicide dramatically boosted the no-till acreage in North America.

SLOWER BUT MORE EFFECTIVE. When it came to no-till weed control, Roundup translocated to the roots while parauqat burned the leaves.

Originally synthesized in 1882, paraquat's herbicide properties were not recognized until 1955. The non-selective contact herbicide was first manufactured and sold by Great Britain's Imperial Chemical Inc. (ICI) in 1962.

Roundup Expanded No-Till Acreage

Our first glyphosate (Roundup) article in *No-Till Farmer* appeared in the June 1973 issue. The page had the word "Warning" in big type and pointed out that this new contact herbicide was not yet cleared for crop use. Referred to as Mono 2139, Monsanto announced the new postemergence chemistry controlled more than 100 annual and perennial weeds.

More than 2,500 field tests conducted over 7 years demonstrated the benefits of this herbicide in no-till programs. This included control of troublesome perennials with well-established root systems such as Johnsongrass, quackgrass, Canada thistle, Bermuda grass and nutsedge.

Many no-tillers got their first glimpse of Roundup in plots that were part of the 1974 and later National No-Tillage Conferences in Hawaii. Most attendees, such as a Wisconsin no-tiller attending the 1975 event, were convinced Roundup would be priced lower than Paraquat.

It never happened!

University researchers our editors talked with anticipated that the clearance of the new herbicide product (now named Roundup) would occur for no-till corn, soybeans and other crops by 1974 or 1975. However, EPA registration of Roundup as a crop burndown herbicide ahead of planting corn, soybeans, wheat, barley, oats and grain sorghum didn't come until early 1976.

Roundup vs. Paraquat

A 1976 *No-Till Farmer* article pointed out several distinct differences between paraquat and Roundup. First, Roundup is translocated to the roots of weeds while paraquat is not.

Second, growers often see an almost immediate kill of weeds with paraquat — within a few hours. On the other hand, Roundup may result in a slower kill of existing weeds. Part of this delay is because the herbicide needs time to work its way into the roots of the weeds.

While the visible effect of applying Roundup to most annual weeds occurs within 2-4 days, it may take up to 10 days with some perennial weeds. ❂

A PARAQUAT HISTORY LESSON

Despite company consolidations and name changes, the same firm has handled paraquat for over five decades.

July 2017

Up until now, the manufacture and distribution of this burndown herbicide (now called Gramoxone) has pretty much been done by several companies, despite a number of company name changes. And it's undergone still another recent change with the sale of Syngenta Crop Protection to ChemChina.

To meet the antitrust requirements of several governments around the world, ChemChina was forced to divest itself of a few crop production products, including paraquat. The U.S. paraquat product registration and marketing rights were sold to AMVAC Chemical. However, Syngenta retains the ownership and the marketing rights of Gramoxone in the U.S.

British Role in No-Till

When *No-Till Farmer* was started in 1972, Imperial Chemical Inc. (ICI) of the United Kingdom had been manufacturing paraquat for 10 years as a burndown weed control in orchards, coffee plantations, around buildings and railway right-of-ways. The forerunner of ICI was Plant Protection Ltd., which had been formed in 1937 and became part of the ICI ag division in 1964.

Paraquat sales in the U.S. were handled by Chevron Chemical Co. and by Chipman Chemical in Canada. After Chevron discovered the exciting new use of this burndown herbicide in no-till among Kentucky growers, they pursued a label and quickly jumped on the no-till bandwagon. Both firms played key roles in the early acceptance of no-till in North America.

ICI set up a sister company called ICI United States in 1963 that eventually became ICI America in 1971. Having promoted no-till extensively, Chevron was disappointed when they lost the U.S. marketing rights to paraquat, which shifted to ICI America in the 1970s.

Since the first National No-Tillage Conference in 1993, Syngenta and its predecessor companies have co-sponsored each of these mid-winter events. While the company is pretty much the same, it's been a co-sponsor under several different names, including ICI Agricultural Products, Zeneca Ag Products, Syngenta Crop Protection and now ChemChina.

Other Company Mergers

Today, we've seen other mega-mergers that include Bayer/Monsanto, Dow/Dupont and Syngenta/ChemChina. But during the past 45 years, several companies that were major early-day no-till players joined the Syngenta family.

These are names your fathers and grandfathers will likely recognize, such as Ciba-Geigy, Novartis, Velsicol, Sandoz, Ciba and Zeneca. Other chemical firms that became part of the Syngenta family over the years include Zoecon, Hooker, Stauffer, Chipman, AstroZeneca Ag Products, Chevron, Abbott Labs and Diamond Shamrock to name a few.

While paraquat was later branded as Gramoxone, it still plays a key role in no-till programs. Even though it has lost much of its popularity over the years to Roundup, Ohio State University weed scientist Mark Loux says this burndown herbicide could have been used much more in recent years to control marestail and small annual weeds and to interrupt the continuous glyphosate use cycle. ✺

GLYPHOSATE CONCERNS GROWING

September 2015

Over the past four decades, much of the no-till acreage increase that has taken place across North America has been due to the introduction of Roundup herbicide and glyphosate-resistant corn and soybeans. Yet the continued use of glyphosate has led to serious concerns about the future of this highly effective weed control product.

A 2015 *No-Till Farmer* survey indicated 98% of readers rely on glyphosate to some extent for weed control. So, the continued lack of adopting alternate weed control chemistries is definitely a worry.

Huge Savings Possible

Managing glyphosate-resistance is more cost effective than ignoring the problem. And there's going to be even more economic incentive to do so as the number of resistant weeds continues to grow.

Staffers at USDA's Economic Research Service (ERS) found 93% of the U.S. soybean acreage was planted with herbicide-tolerant varieties in 2013, along with 85% of the corn. While ERS economists indicate the acres treated with glyphosate alone declined during a recent 6-year period, it's because soybean growers have applied other herbicides as more glyphosate-resistant weeds show up.

EARLY-DAY DATA. North Carolina State University Roundup data from the early 1980s showed that it also provided late season grass control.

Growers reported a decline in the effectiveness of glyphosate in controlling weeds on 44% of the soybean acres in 2012, which impacted yields, production costs and profits. As an example, soybean growers who reported reduced glyphosate effectiveness lost $22 per acre compared with growers not having weed-resistance concerns.

To overcome weed resistance, the most common response was to switch to herbicides other than glyphosate. This was done on over 84% of the corn acres and 71% of the soybean acres where glyphosate-resistant weeds were a major concern due to reduced glyphosate effectiveness.

Unfortunately, growers increased glyphosate rates to deal with the density of glyphosate-resistant weeds on 25% of the corn acres and39%of the soybean acres where glyphosate had become less effective.

More Changes Needed

There's no doubt growers must overcome weed resistance concerns by adding more herbicide chemistries and doing a better job of managing weed programs. Otherwise, no-tillers are going to be in a heap of trouble when it comes to controlling super weeds in the coming years. ❋

MANY CHOICES. No-tillers like Indiana grower Stan Smock had to consider numerous factors when selecting the proper burndown herbicide.

ARGENTINA NO-TILLERS PLANT 'BROWN-BAG' ROUNDUP READY BEANS

March 2002

When it comes to no-tilling Roundup Ready soybeans, your competitors in Argentina have a distinct advantage due to lower seed prices. The inability to enforce intellectual property rights has given that nation's bean producers a comparative advantage.

After it became clear in 1998 that Roundup Ready soybean seed was selling for less in Argentina than in the U.S., Congress directed the federal government's General Accounting Office to analyze the situation.

"This study clearly showed soybean seed was cheaper in Argentina than in the U.S., while seed corn was priced the same," says University of Illinois ag economist Peter Goldsmith. "This pricing phenomena reflects differences between corn and soybeans and a firm's ability to protect its intellectual property."

Can't Protect Rights

Roundup Ready soybean seed is cheaper in Argentina because there's no system to protect the intellectual property rights that would allow suppliers to charge an exorbitant price for seed. Some 80% of Argentina's soybean seed is saved by farmers or distributors who "brown-bag" the seed and sell it without any company logo.

Goldsmith says there are major trade disputes between North and South America in regard to intellectual property. Multinational companies have little influence in South American countries, as they are dependent for sales on distributors who may be brown-bagging their seed. In some instances, Roundup Ready soybean seed is sold as a loss leader to retain other parts of the ag chemical business.

With Argentina suffering from severe financial concerns, the Argentine government doesn't have the incentive or dollars to address intellectual property problems.

As a result, you will likely see U.S. soybean exports slipping away to lower-cost nations. With cheap seed and technology, other countries produce yields that may be higher and more profitable than what is grown here in the states.

What Can Be Done?

Goldsmith says the U.S. government could demand better enforcement of World Trade Organization guidelines in regard to intellectual property rights. But South American nations are reluctant to do so as long as U.S. farm price-support programs remain in effect.

One thing is for sure. The Argentine situation regarding Roundup Ready soybean seed pricing certainly doesn't let you compete on an equal footing for bean sales around the world. ✳

Frank Lessiter

$21 BILLION LOSS. If herbicides were banned, weed control costs would increase by $7.7 billion and there would be $13.3 billion in yield losses.

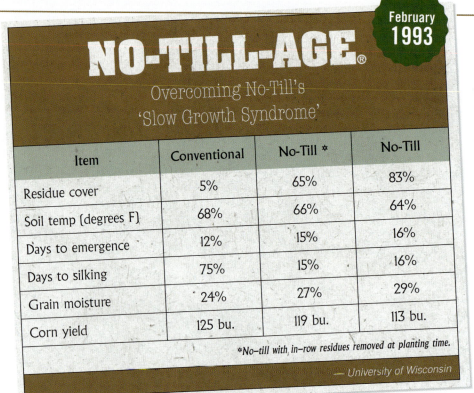

NO-TILL-AGE®

Overcoming No-Till's 'Slow Growth Syndrome'

Item	Conventional	No-Till *	No-Till
Residue cover	5%	65%	83%
Soil temp (degrees F)	68%	66%	64%
Days to emergence	12%	15%	16%
Days to silking	75%	15%	16%
Grain moisture	24%	27%	29%
Corn yield	125 bu.	119 bu.	113 bu.

*No-till with in-row residues removed at planting time.

— University of Wisconsin

NOT ALL GLYPHOSATES THE SAME

With so many brands of glyphosate, no-tillers often wonder how these different products compare.

"No matter how many herbicides are introduced, the question always comes back to whether glyphosate product A performs as well as glyphosate product B," says University of Nebraska weed scientist Brady Kappler.

To get answers, field studies with Roundup Ready soybeans and herbicide-tolerant corn were conducted over 3 years at five Nebraska locations. The 10 glyphosate products were also evaluated in wheat stubble at two locations.

While the herbicides were evaluated at 16 and 32 ounces per acre in 2001 and 2002, treatment rates were reduced to 12 and 24 ounces per acre in 2003. AMS was added to all treatments at a rate of 17 pounds per 100 gallons. All plots were evaluated for grass and broadleaf weed control at 10, 15 and 25-30 days after treatment.

Different Formulations Matter

These herbicides included Roundup UltraMax, Roundup Ultra, Roundup WeatherMAX, Roundup Ul-traDry, Touchdown, Clearout 41 Plus, Glyfos Xtra, Cornerstone, Glypbomax and Glypbomax Plus.

Most products represent the isopropylamine salt of glyphosate, although Touchdown is formulated as a diammonium salt of glyphosate. Roundup UltraDry is formulated as the mon-ammonium salt of glyphosate and Roundup WeatherMAX is formulated as a potassium salt of glyphosate.

"With a difficult to control weed species, such as barnyard grass, or in a more demanding climate such as western Nebraska, differences are easier to find although they will be random and varied," says Kappler. "In most climates, there seems to be little or no differences between brands. Rate, environmental factors and cost most likely play a larger role in the decision process than the brand name."

As a general rule, the lower the price, the less service you receive. "If you don't want to take any risks, then you may want to pay more for your glyphosate to have service and support to fall back on in case of problems," he says. "However, if you are not concerned about service, then the lower cost glyphosates may be the way to go."

Surfactant Usage Critical

Kappler says the label will explain if "you must add surfactant," "you may add surfactant" or "no surfactant is needed." While the "no" and "must" statements are easy to understand, the "may" statement may be confusing.

"May" means the glyphosate product contains some surfactant, but that it might not be enough in some situations. If you are spraying tall weeds or are in very dry conditions, adding 1 quart of surfactant per 100 gallons of solution may be beneficial. Adding AMS will help combat hard-water problems and improve glyphosate efficacy, says Kappler. ❋

MONSANTO DUMPS ROUNDUP READY WHEAT

NO-TILL WHEAT. While many growers years ago were against the introduction of Roundup Ready wheat, some now wish it had taken place.

July 2004

While numerous farm and environmental groups celebrated Monsanto's decision to shelve its Roundup Ready spring wheat project, other growers and organizations are not happy with the decision.

In mid-May of 2004, Monsanto postponed its introduction of Roundup Ready spring wheat until a time when other wheat biotechnology traits are being introduced. Instead of being introduced as early as 2005, it may now be 2012 or later before Roundup Ready spring wheat is released.

Wheat Industry Said No

Numerous food processors, such as General Mills, believe herbicide-resistant wheat offers little benefit to consumers. They see considerable opposition to genetically modified foods in Japan and Europe.

In Canada, officials of the Canadian Wheat Board are praising Monsanto's decision to postpone introduction of Roundup Ready spring wheat. That's because foreign customers purchasing 87% of the group's spring wheat sales last year indicated they wouldn't buy genetically-modified wheat.

Monsanto began the development of Roundup Ready wheat in 1997. A half dozen years of field testing demonstrated it performs exceptionally well under the most difficult production environments for spring-planted wheat and offers the potential for 5%-15% higher yields.

Resistance to Biotech

Because of industry resistance and the dwindling size of the spring wheat market, Monsanto pulled the plug. The spring wheat acres in the U.S. and Canada has declined nearly 25% since 1997, and even more in the higher-cost weed control target market for this biotech product.

Allan Skogen says one of the reasons for the dropoff in spring wheat acreage is due to the biotech benefits offered with corn, soybeans and canola. In 2003, the Valley City, N.D., farmer spent more dollars on seed, fertilizer and chemicals for wheat than with corn. It worked out to about $40 per acre for chemicals with wheat and only $18 per acre for corn, including the tech fees.

Yet some growers believe they've lost the potential for 50 cents a bushel in future cost savings and yield increases with Monsanto's postponement of Roundup Ready spring wheat.

Other companies are also developing biotech wheat. As an example, Syngenta is working on adding scab resistance traits to wheat and hopes to have varieties ready for market in 7-10 years.

WHEAT EXPORT CONCERNS. Western Canadian wheat growers felt genetically-modified wheat would reduce sales to Japan and Europe.

Changing Strategies

Monsanto's wheat research investment in fiscal year 2004 was under $5 million, or less than 1% of the company's $500 million annual research and development budget. Company officials indicate shifting resources will allow the company to focus research on other crops in regard to yield improvement, vegetable oil traits, stress tolerance, agronomic pest resistance traits and feed improvement traits.

Products such as Roundup Ready Flex for cotton and an improved soybean oil for food manufacturers are already moving closer to commercialization. Good results in 2003 field trials for drought-tolerant corn were expanded to more trials this year. ✾

143

KNOW WEATHER CONSEQUENCES WITH ROUNDUP READY CORN, ALFALFA

August 2006

While Roundup Ready crops offer numerous advantages, a number of no-tillers learned the hard way last spring what can happen when things don't go as anticipated.

Faced with replanting, some no-tillers had to scramble when it came to figuring out a way to eliminate a failed stand of Roundup Ready com.

"As the acreage of Roundup Ready corn has increased, it's natural that more producers will be faced with the situation where poor weather resulted in an inadequate initial Roundup Ready corn stand," says Mark Loux, Ohio State University weed scientist. "Glyphosate has typically been used to control the first planting of conventional corn in replant situations, but will obviously not control Roundup Ready corn."

A considerable number of Roundup Ready replanted corn acres in Ohio were treated last spring with post-emergence grass soybean herbicides such as Select, Assure II, Fusion, Poast and Poast Plus. While these herbicides have soil residual activity, Loux says there's the potential to injure replanted corn if the radical or coleoptile part of the plant comes in contact with a high enough herbicide concentration.

While they effectively control emerged corn, Loux says labels for these herbicides do not support their use as a preplant treatment with corn. Instead, they're only labeled for control of volunteer corn in a soybean crop — not in a corn replant situation.

"Our assessment is that there are really only three effective and legal options to kill an existing stand of Roundup Ready corn in a replant situation," he says. "These are tillage, Gramoxone or Liberty."

Taking Out Older Alfalfa

With glyphosate-resistant alfalfa, no-tillers need to think about how they will take out an older alfalfa stand down the road. This was among the critical questions asked by university weed scientists attending a 2006 forage meeting sponsored by Monsanto.

To take out a Roundup Ready alfalfa stand, Monsanto researchers recommend tankmixing 1 pint of 2,4-D with 1/2 pint of dicamba per acre in the fall (preferred) or spring on alfalfa that is actively growing and at least 6 inches tall. To control volunteer Roundup Ready alfalfa in corn after planting, apply 8 ounces per acre of Stinger.

With no-till, corn is the recommended rotational crop after Roundup Ready alfalfa due to better herbicide options for controlling volunteer plants. Since taking out an alfalfa stand won't get every plant, you could find yourself in a world of hurt with no-till soybeans. ✳

Frank Lessiter

AERIAL APPLICATION. While some no-tillers relied on timely aerial applications of Roundup, other growers felt it didn't offer as much precision accuracy as desired with no-tilled corn and soybeans.

Frank Lessiter

TACKLING WEATHER DELAYS. Aerial pesticide application can help overcome unfavorable weather when timely application is a necessity.

> ## "Soil: Loosen it and lose it."
> ### — *Piers Woodriff, Somerset, Va.*

PARAQUAT BAN COMING IN EUROPE?

October 2007

Following an unprecedented court action in Sweden, paraquat might soon be banned throughout the European Union. The major active ingredient in Gramoxone, this chemical has been used extensively by a number of North American growers since the late 1960s.

OK, But Not OK

Banned since 1983 in Sweden, paraquat was added to the European Community's list of authorized substances in 2003 after an intensive review by the group's pesticide safety group. But in 2004, Sweden — supported by Austria, Denmark and Finland — asked the European Court to remove paraquat from the list of acceptable pesticides. These countries argued that the review procedure was faulty and failed to protect the environment, human and animal health claims that the court largely upheld in a recent ruling.

While Syngenta officials were surprised by the court's decision, another surprise was that continued use was not allowed while the ruling is under appeal.

In Europe, more than 500,000 farmers rely on paraquat because of its fast contact action, rapid deactivation in the soil and the ability to combat soil erosion by preserving weed roots that bind with the soil.

North American Concerns

"We were certainly surprised and disappointed by the decision made by the European court," says Les Glasgow, technical brand manager for non-selective herbicides at Syngenta Crop Protection in Greensboro, N.C.

"But as far as this decision is concerned, it is not a safety issue. Rather it is a procedural decision in which the Swedish regulators are charging that the European Commission did not follow the correct procedure in allowing re-registration of paraquat."

Glasgow expects the court's decision will not have any impact on paraquat sales in more than 100 countries around the world.

NO MORE FOAM MARKERS. The development of GPS technology made utilizing foam markers for sprayer accuracy a product of the past.

Frank Lessiter

"We don't intend to change our paraquat strategies," he says. "It is safe and is governed by stringent government criteria, such as by the EPA here in the states. We are monitoring the decision and haven't seen or expect to see any significant impact here in the states." ❁

NOZZLE ACCURACY COUNTS. It's important for no-tillers to test whether each nozzle of the sprayer is applying the correct amount of pesticide.

Frank Lessiter

November 2010

WITH ORIGINAL ROUNDUP READY PATENTS EXPIRING, IS PLANTING BIN-RUN SAVED SEED IN YOUR FUTURE?

Few growers have thought about legally transferring Roundup Ready soybeans out of their grain bins to dump in the planter or drill when seeding no-till fields. But with the patents expiring on the older Roundup

GLYPHOSATE RESISTANCE WORRIES. Each year, more no-tillers are having to find new ways to deal with serious weed resistance concerns.

Ready soybean technology in 2014, it might be a possibility.

Monsanto has agreed to maintain export approvals for this older Roundup Ready technology through 2021, even though the patent's expiration in 2014 will turn it into an unprotected generic trait.

The company also agreed not to collect royalties on this older technology after 2014. Other seed companies will not be obligated to destroy or return seed with the expiration of these trait licenses.

Monsanto says it will not use variety patents against U.S. farmers who save varieties containing the expired Roundup Ready trait for use on their own farms. However, this may not be the case with other seed companies.

No-Tilling Bin-Run Beans?

When *No-Till Farmer* readers were surveyed about whether they would plant bin-run Roundup Ready soybean seed after the patent expires, 33% indicated they would. Another 43% were not sure, while 24% prefer getting the latest traits with newer bean varieties.

Some 86% indicated the reasons for using bin-run Roundup Ready soybean seed was that it would be a cheaper seed alternative; 54% said it would be due to the elimination of tech or licensing fees; and 31% said it's because they don't believe anticipated yield increases with the newer Roundup Ready 2 Yield technology will pay off.

Some 22% of growers would plant bin-run Roundup Ready seed as a protest against Monsanto for its poor treatment of growers and overly aggressive pricing. If they plant off-patent, bin-run Roundup Ready soybeans to trim costs, 21% would boost seeding rates.

We also asked growers what yield increases they're expecting from the new Roundup Ready 2 Yield technology. Over half of the growers expect a yield boost of less than 5 bushels per acre. Another 17% anticipate increases of 6-10 bushels per acre.

Some 35% of these growers expect Monsanto to drop the price of the Roundup Ready 2 Yield soybean technology to compete with alternative herbicide-resistant technologies.

Not A Done Deal

Even with the original Roundup Ready soybean technology becoming generic, some no-tillers believe seed companies will find a way to discourage saving bin-run seed. But if a variety is certified under the Plant Variety Protection Act, you can save seed to use for next year's crop. However, it's illegal to sell the seed to other growers.

Since Monsanto expects to focus on the new Roundup Ready 2 Yield technology by 2012 and likely won't offer the original Roundup Ready soybeans through the 23 seed companies they own, a few growers maintain it may be difficult for growers to save older varieties for seed.

Payback Time Coming?

One seed producer believes growers will choose bin-run Roundup Ready soybeans, especially if newer varieties don't deliver a substantial yield boost.

"I expect farmers to plant bin-run Roundup Ready soybean seed to stick it to Monsanto for charging high tech fees, excessive seed costs and having an arrogant attitude toward farmer needs," he says. "Growers are mad about the way they've been treated by Monsanto and a number will plant bin-run Roundup Ready soybeans just to get even."

Monsanto's major troubles began 2 years ago when it aggressively raised seed prices. Rather than introducing new technologies at lower-cost as in the past, they went for a higher premium — a move that definitely hurt 2010 seed sales.

There could be repercussions for other companies that market Roundup Ready, LibertyLink or offer the new DuPont and Dow AgroSciences soybean technologies. Seed and herbicide sales might both take a hit if growers decide to plant more bin-run beans. ✻

NO-TILL LOOSE ENDS...

'MAYBE THAT GUY WASN'T SO CRAZY'

November 1972

In the early 1970s, Jim Smith worked at the John Umstead Hospital in Butner, N.C. He managed 11,000 acres and the production of milk, eggs and meat in feeding the mental hospital's 3,000 patients.

Smith was no-tilling corn into sod one day in April when a neighboring farmer stopped and asked what he was doing. He told the grower he was no-tilling corn.

The farmer didn't say anything further to Smith, but drove to the hospital's office and told the general manager that one of the mental patients was running a planter through an alfalfa sod field and needed to be locked up. The general manager later related the story to Smith.

Wait, there's more to the story.

Sometime in late July, the farmer told the hospital general manager the corn no-tilled into the sod was looking darned good and maybe it would be OK to let the patient out of lockup. ✻

Herbicide Oldies
Making a Comeback

"Like me, maybe you'll realize that the **more things change in weed control,** the more they stay the same."

Back in 1972, the most popular herbicides being used by no-till farmers were paraquat, atrazine, 2,4-D, Princep and Banvel. What's interesting is that these herbicides are still being used to some extent by many no-tillers decades later.

One major concern that hasn't changed over the years is the debate

October 2005

over the safe use of atrazine between farmers, scientists, government regulators and environmentalists.

Atrazine Worries Continue

Even though no-tillers say atrazine is a vital part of their weed control program and is used on 75% of Iowa's corn crop, environmentalists question

the pesticide's long-term ecological impact and human health concerns.

Already banned in Wisconsin's sandy soil areas, the herbicide that has been on the market for more than five decades has faced several recertification deadlines. While Environmental Protection Agency scientists once considered the herbicide as a potential cancer risk, it now appears unlikely to cause cancer in humans.

HARVEST AND SPRAY. While combining wheat, John Johnson sprayed double-crop bean herbicides with nozzles positioned under the combine.

Frank Lessiter

NO-TILL-AGE®

April 1992

Number of Growing Season Pesticide Applications
(Data from 506 fields)

Tillage System	Corn	Soybeans
Spring plow	1.4 trips	1.5 trips
Spring disc, chisel	1.6	2.0
Fall chisel	2.3	1.9
Fall plow	2.5	2.1
Ridge till	2.5	2.5
No-till	2.7	3.1

— MAX Program

In fact, Syngenta toxicologists say more than 200 atrazine studies have looked into possible human health problems and found it to be safe.

New Dicamba Uses Coming

As biotechnology continues to expand, new weed control concerns keep cropping up. One example is finding a way to remove volunteer, herbicide-tolerant soybeans from no-tilled corn fields, the exact opposite of trying to figure out how to keep corn out of soybean fields.

University of Minnesota weed scientist Jeff Gunsolus suggests using a dicamba product (originally known as Banvel) such as Hornet, Stinger or Widematch to eliminate Roundup Ready beans from no-till corn fields.

HUNDREDS OF HERBICIDE JUGS. Handling of pesticides is much safer today than in early no-till days.

ARE MORE HERBICIDES NEEDED WITH NO-TILL?

Among the questions that came up during a recent Shell Chemical-sponsored workshop for agronomists and weed scientists on the use of Bladex was whether more herbicides are needed with no-till than with conventional tillage.

As you might guess, there were no simple answers.

George Kapusta, Southern Illinois University weed scientist, says weed control during the first 4-5 weeks is the same with no-till as with conventional tillage with the same amount of herbicides and plentiful rainfall.

"No-till weed breaks come in mid-July and that's when problems come up," he adds. "You may also find some shift in weed species with no-till."

Bob Schieferstein, a Shell Chemical Co. field development and technical service rep, says Iowa studies show farmers should increase herbicide rates by 25% when no-tilling.

By going with split herbicide applications, University of Wisconsin weed scientist Doug Buhler feels farmers won't have to increase herbicide rates when no-tilling compared with weed control in conventional tillage.

In the eastern states, opinions vary. "As a general rule, we jump herbicide rates by about 15% with no-till," says George Bayer, weed scientist with   in Syracuse, N.Y.

Frank Webb, University of Delaware weed scientist, says their studies show better weed control under no-till than with a chisel plow system when the same amounts of herbicides are applied.

Nate Hartwig, Pennsylvania State University weed scientist, says farmers in his state are applying 20%-25% more herbicides with no-till.

"No-till growers in our 5-acre corn club are spending $10 more an acre for herbicides," he adds. "But I think it's necessary to use higher herbicide rates with no-till in our area."

Oliver Russ, Kansas State University weed scientist, has been able to get the same weed control with the same amount of herbicides applied in both no-till and conventional tillage. ✾

Frank Lessiter

30-YEAR WAIT. After waiting 3 decades for approval, no-tillers were given the go-ahead in 1992 to use 2,4-D for weed control in soybeans.

DIRECT SPRAY RIG. Ohio no-tillers Tom and Frank Moritz modified a 2-row cultivator to spray paraquat to rescue their no-tilled soybeans.

"Studies show **better weed control under no-till than with a chisel plow system** when the same amounts of herbicides are applied..."

NO-TILL-AGE®

August **1987**

How Herbicide Actions Change with Weather Conditions

(Measured by herbicide absorption into weed leaves and translocation into the weed plant)

Temperature	Relative Per Acre	Herbicide Absorption	Herbicide Translocation
95 degrees	40%	56%	17%
95	100%	70%	30%
65	40%	33%	67%
65	100%	43%	8%

— *Mississippi State University*

With Roundup Ready and Liberty Link herbicide-resistance technology on the market, research is under way to develop crops that are tolerant to dicamba. Based on more than 10 years of research by University of Nebraska biochemists that led to the identification of a gene that can make dicamba-sensitive crops such as soybeans tolerant to the widely used herbicide, Monsanto recently signed an exclusive licensing agreement. This technology will lead to development of soybean and other broadleaf crops that are highly tolerant to treatment with this herbicide.

Finally, After 17 Years

In 2005, a favorable comprehensive assessment of 2,4-D under the EPA re-registration program was released. The agency's scientists concluded that 2,4-D does not present human health risks when label instructions are followed. Discovered 60 years ago, the phenoxy-type herbicide has been a valuable no-till weed control product.

The EPA assessment also concluded that proper 2,4-D usage does not present an unacceptable environmental risk, says Don Page of the industry task force on 2,4-D, who spoke on this situation at the first National No-Tillage Conference in 1993. An economic evaluation conducted by USDA researchers indicated the loss of 2,4-D would cost the nation's economy $1.7 billion annually in higher food production and weed control expenses.

NO-TILL-AGE®

What No-Till Weed Control Cost in 1996
(Herbicide cost per acre)

Crop	$0-$9	$10-$19	$20-19	$30-$39
Corn	—	17%	68%	15%
Soybeans	2%	27%	40%	31%
Grain sorghum	—	33%	67%	—
Small grains	76%	24%	—	—

— Survey of 1996 National No-Tillage Conference attendees

While these herbicides have long-term roots, recent changes offer more flexibility in managing weed control challenges. Like me, maybe you'll realize that the more things change, the more they stay the same. ✿

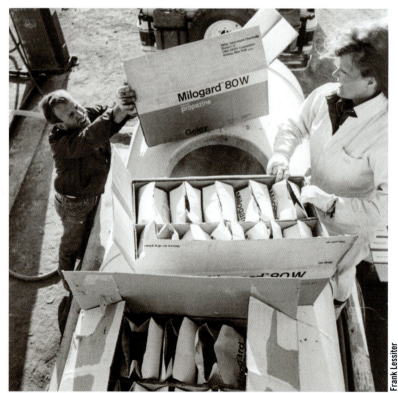

NO-TILL GRAIN SORGHUM. Kansas no-tiller and aerial applicator John Notestine says only a limited number of herbicides were available for no-till grain sorghum in 1970s.

SHIIELDED PLANTS. Direct herbicide spray rigs proved popular with southern growers raising no-till cotton.

25% MORE HERBICIDES? As no-tillers gained weed control experience, more herbicides were not needed.

Frank Lessiter

No-Tillers Well Ahead of the Pack When
It Comes to Covering Up

"Despite the success we've had with no-till, selling cover crops is going to be a tougher task..."

The way some government agencies and ag groups are promoting cover crops these days, you'd think it was a brand-new idea. While cover crops are a hot topic, it's not because of any major research breakthroughs or being brand new on the ag scene.

Instead, cover crops offer a soil health message that's starting to catch on with folks both in and out of agriculture — with numerous benefits that no-tillers have enjoyed for many years. But cover crops are certainly not new.

As a kid growing up in the late 1940s on our family's Michigan Centennial Farm (owned more than 100 years by the same family), I remember Dad seeding red clover after cutting silage corn for our dairy herd.

The *No-Till Farmer* editors are no "Johnnies come lately" to the cover crop craze. In fact, articles on the value of cover crops appeared in some of our very first issues in the early 1970s. You'll see a handful of our early coverage in this chapter.

At the Delmarva No-Tillage Conference that attracted more than 1,000 farmers in December of 1972, we reported on how Jim Parochetti stressed the value of including cover crops in a no-till program. The University of Maryland agronomist said farmers are shortchanging themselves

RYE IS PERFECT COVER CROP. Neal Hering no-tills corn into a rye cover crop that may reach a height of 5 feet. The Westminster, Md., no-tiller's corn yields in 1972 were as high as 180 bushels per acre. He finds cereal rye offers early spring growth, is easy to kill, provides needed organic matter, atomizes raindrops so they don't damage the soil and provides wind erosion protection.

prevents early weed germination, offers numerous no-till management benefits and conserves moisture during dry years.

"A good cover on no-till ground in many cases can prevent early germination of fall panicum," says Parochetti.

Looking back at early copies of *No-Till Farmer,* there was an article in the August 1973 issue entitled "Aerial Seeding Is Working with Cover Crops." That issue also included an ad for paraquat herbicide that asked the question, "Who'd plant rye in a stand of corn?" The answer ... innovative no-tillers looking to trim costs by seeding a fall cover crop.

Many Benefits

What's interesting is how many no-till pioneers saw the benefits of seeding cover crops well ahead of growers using other tillage practices.

if they no-till corn into stalks or other crop residue instead of seeding cover crops.

He pointed out how cover crops provide erosion control, offer effective weed control as ground shading

"Farmers should be able to grow their own cover crop seed on set-aside acres..."

COVER CROPS EARN A 266% RETURN WITH CONTINUOUS NO-TILL

July 2015

Cover crops definitely involve extra expenses, but members of the family-owned Rulon Enterprises in Arcadia, Ind., have found they more than pay their way with 6,000 acres of no-tilled corn and soybeans.

In the fall of 2014, cover crops were seeded on 3,527 acres. At a cost of $26 per acre for seed and planting, the total cost was almost $92,000.

In looking at the cover crop payback, they've penciled out the benefits at $69.17 per acre. This includes the need for less fertilizer, higher corn and soybean yields, fewer disease concerns, drought protection, a reduction in erosion, increased Conservation Stewardship Program government payments and improved soil quality.

They calculate a 266% return on their investment. "You can slice every benefit category in half and we still calculate a 50% return on our investment in cover crops," says Ken Rulon. ❀

"83% of no-tillers seed cover crops vs. only 2% of all farmers..."

These innovative no-tillers relied on cover crops to improve soil health, reduce soil, wind and water erosion, trim fertilizer costs, control weeds, reduce compaction and improve other economic and environmental benefits.

Recent refinements include interseeding covers into growing grain, seeding covers with the combine, rolling cover crops to avoid using burndown herbicides, expanding the use of mixes to meet specific needs and many other new innovations in recent years.

In our 10th annual No-Till Benchmark Study conducted in the winter of 2018 among *No-Till Farmer* readers, 83% were using cover crops. What is truly amazing is the fact that these no-tillers seeded an average of 451 acres, an astounding 40% of their acres to cover crops.

These results are even more dramatic when compared with data from growers using all types of tillage systems in 13 states that drain into the Mississippi River Basin that stretches from Minnesota to Louisiana. Based on data from 2017, these growers used cover crops to protect less than 2% of their cash crop acres.

With lower commodity prices and growers closely watching costs, it's certainly not cheap to add a cover crop. Results from a recent Conservation Technology Information Center survey indicates the average cost of $37 per acre is a major barrier to adopting cover crops.

Cover crop use is nowhere close to being a common farming practice. The word is getting out, but there is still a long road ahead — except among no-tillers who are leading the pack.

A Tough Sell

While no-till caught on faster than hybrid seed corn did a few generations back, Bill Richards believes it's going to be tougher to sell the benefits of cover crops to farmers.

The veteran no-tiller and former NRCS Chief from Circleville, Ohio, says growers see production costs drop quickly with no-till. Like others, Richards recognizes that seeding cover crops can be expensive and maintains it's not as easy to see the immediate savings, as is the case with no-till.

"There's an awful lot of anecdotal evidence as to what they can do and many farmers are doing a fantastic job with cover crops," he says. "Even my sons are a hard sell, because cover crops cost us about $50 an acre with seed and planting expenses, which is significant for us."

Richards says cover crop seed costs have increased dramatically. "In fact, my sons have told me to quit going to meetings and stop telling people about cover crops. They say all I'm doing is raising cover crop seed costs," he says.

One suggestion Richards offers to trim seeding costs would be to allow farmers to grow their own cover crop seed on USDA set-aside acres. It wouldn't cost the government anything and he maintains that if farmers could produce their own seed, many more cover crop acres would be planted. ❋

COVER CROP EXPLOSION TAKES NO-TILL BACK TO ITS ROOTS

Like fashions from yesteryear, cover crops have come full circle and are once again becoming popular among many no-tillers.

Just a few generations ago, cover crops were standard procedure on many family farms. Life was different then, as small family farms had a variety of animals and planted a diverse selection of crops to support themselves and their livestock. Many produced much of their own fertilizer by including legumes in their rotations and spreading manure.

As farm consolidation took hold, new herbicide developments and larger machinery led to ever-expanding acreages. Rotations soon included fewer crop options.

"Fertilizer and herbicides were simply more cost-effective at the time," says Steve Groff, a Holtwood, Pa., no-tiller and cover-crop innovator. "And with heavy use starting in the 1940s through the 1980s, we gradually lost the biological components in our soils to synthetic fertilizer use."

But after years of applying herbicides, other pesticides and fertilizers, farmers took note that their soil was changing — not in a good way.

"As we lost diversity in the rotation, producers started noticing poorer soils and lower yields," says Dale Mutch, Michigan State University cover crop specialist. "The products weren't giving enough back to the soil and soils started losing organic matter. This turned more farmers toward no-till and, finally, back to cover crops."

Riding the Market

While waning yields may have been enough to get some producers looking into cover crops, increasing input costs also brought more no-tillers back to the table.

"Growers are saying they have good soil now and can grow 220-bushel corn, so why bother with cover crops?" Groff says. "But when nitrogen prices go up, they start thinking harder about cover crops."

With a corn-and-soybean rotation, most no-tillers in the Upper Midwest are limited to winter annuals like cereal rye and winter wheat.

"Wheat harvested in July opens a whole new cover-crop option for no-tillers to use," says Mutch. "They have the whole month of August to plant oilseed radish, crimson clover, red clover, annual ryegrass, Austrian winter peas and more.

"Fertilizers will feed the crop, but not the soil biology. Cover crops stimulate microbes and return organic matter to the soil."

Groff says active soil life has helped him handle weather extremes.

"We recently got 12 inches of rain in a week," he says. "My fields aren't washing away, and I expect to get back in them sooner than conventional-tilling farmers because I've invested in no-till and cover crops. My soil structure is more stable and I've earned those rewards by investing in my soil."

Getting into the Game

Groff first became interested in cover crops when he gave cereal rye a try for erosion control. After hearing about hairy vetch and crimson clover that could

MORE PROFITABLE. Shorter-season corn hybrids allow Steve Groff to seed his cover crops much earlier.

Steve Groff

MAKING MIXES MATTER. "We're not harvesting the cover crop, so why not plant as many things together as makes sense?" says Steve Groff, a long-time no-tiller and tillage radish advocate from Holtwood, Pa. "The diverse plants and root systems create a smorgasbord for the soil."

NITROGEN PRODUCERS. David Brandt likes to see nodules the size of a man's thumbnail on the roots of hairy vetch and Austrian winter peas. That's a sign he'll reach the targeted nitrogen goals for the next crop.

FULL-PRESS COVERAGE. David Brandt uses a White planter to no-till cover crops into wheat stubble. He's used cover crops for four decades to improve weed control, reduce reliance on purchased fertilizers and improve no-till crop yields.

In 1978, the Carroll, Ohio, no-tiller started planting hairy vetch to produce nitrogen for his upcoming corn acres and cereal rye to reduce weed competition in his soybeans.

"The hairy vetch grew so well that it provided 70% of the nitrogen needed for the corn, and the cereal rye residue choked out the weeds in the soybeans," Brandt recalls.

Brandt quickly made a full commitment to covers, changing his crop rotation to corn, soybeans and wheat. He's been experimenting with mixes that include pearl millet, sorghum-sudangrass, cowpeas, soybeans, sunn hemp, Ethiopian cabbage and annual ryegrass.

supply cheap nitrogen, he started using covers on his no-tilled vegetable acres.

Groff says the one of the toughest thing for no-tillers to realize about adopting cover crop is that it's just one piece of the puzzle.

"They have to tweak their system. Some no-tillers change rotations and others get more innovative," Groff says. "There are people who spread cover crop seed with aerial applications and those that spread into still-standing corn with high-clearance seed spreaders.

"We're not harvesting the cover crop, so why not plant as many things together as makes sense in mixes? The diverse plants and root systems create a smorgasbord for the soil."

40 Years of Covers

After switching to no-till in 1972, David Brandt noticed his corn and soybean rotation was losing ground in the yield department. Weeds were a major factor.

FULL CIRCLE. Cover crops, such as buckwheat (shown here), have re-emerged as tools no-tillers can use to boost fertility, fix nitrogen, break up compaction and boost the organic matter of their soils.

Cropping Program	Corn Yield per Acre
Corn after cover crop	180 bu.
Corn with no cover crop	77
Soybeans after cover crop	131
Soybeans with no cover crop	54
Wheat after cover crop	63
Wheat with no cover crop	63

— 2016 Cover Crop Survey conducted by the Sustainable Agriculture Research and Education program and Conservation Technology Information Center

ANNUAL RYEGRASS LIFTS CORN YIELDS

Corn no-tilled behind annual ryegrass in a continuous corn system thrived in a dry year in southern Illinois. It beat conventionally tilled corn by better than 50 bushels per acre.

In University of Illinois trials no-tilled continuous corn grown behind annual ryegrass averaged 121 bushels per acre vs. 52.5 bushels with corn grown with conventional tillage. ❋

Steve Groff

EARLY HARVEST OPTION. No-tilling 99-day corn is very competitive with a 112-day hybrid for Pennsylvania's Steve Groff. The shorter season hybrid offers higher corn prices and opportunities to better utilize cover crops.

"For the past 5 years, I've mixed Austrian winter peas with oilseed radishes to grow and store more nitrogen in the soil," Brandt says. "Now I'm starting to work with even more cover crop mixes."

More Rotation Benefits

When Brandt added wheat back into his rotation, he started with just a few acres. He planted cover crops after wheat before no-tilling corn or soybeans the next spring. The results made the rotational change worth the effort as he got a 5%-10% yield boost with both corn and soybeans.

A mix of oilseed radishes and Austrian winter peas reduced problems with soybean cyst nematodes and slugs in subsequent crops.

"I'm trying to build better soil health so my cash crops are healthier, and I want to do that while reducing the inputs I buy," Brandt says. "I think cover crops are the solution."

Wheat-Free Options

Rotating to wheat makes cover cropping easier, but it isn't a necessity. In the Midwest, Groff notes, no-tillers can opt to plant shorter-season soybean varieties or corn hybrids.

"A 99-day corn hybrid vs. a 112-day corn hybrid gets the corn off the field a week or two earlier and provides added days to allow a cover crop to become established and grow," he says. Planting shorter-season varieties may actually benefit no-tillers by spreading out pollination and harvest — both functions that can be negatively impacted by weather.

Moving to the Next Level

Mutch says it's no surprise that no-tillers are at the forefront of the return to cover crops. The two practices build upon each other, piling up the benefits.

"No-tillers are innovators and risk takers. They're not afraid to try something different," he says. "They see cover crops can take their no-till farming system to the next level, so they go for it." ❋

WILD, CRAZY WAYS TO SEED COVER CROPS

It's easy to find no-tillers who are successfully seeding cover crops in more than a half dozen ways.

Besides seeding with a drill or planter, no-tillers are using high-clearance sprayers for interseeding cover crops in late summer into growing cash crops, mounting seed hoppers on combines to seed during corn or bean harvest, aerial seeding, using rotary fertilizer spreaders and even mixing cover crop seed with liquid manure. Plus, there are likely a few other seeding options that have slipped my mind.

Here are a few examples of unusual cover crop seeding options featured over the years in *No-Till Farmer*.

1. Back in the mid 1990s, Nate Andre of Wauseon, Ohio, mounted two Great Plains Mtg. seeding units along the sides of a combine to seed a rye cover crop while harvesting soybeans or corn. Power for the seeding units came from an electric windshield motor. **2.** Featuring low soil disturbance, a modified tanker places liquid hog manure in the root zone while seeding cover crops on the Virgil Gutshall farm at Blain, Pa. A Gandy Orbit-Air seeder is mounted above the toolbar for seeding cover crops between 30-inch rows when corn is in an early growing stage. **3.** Jeff Moyer, the farm manager at the Rodale Institute in Emmaus, Pa., is among a number of growers who have done considerable work with rolling down a living cover crop prior to no-tilling corn or soybeans. **4.** Kevin Staats, who no-tills 1,200 acres of corn and soybeans near Danville, Ohio, seeds cereal rye in mid-October with a boom and small air seeder mounted on the back of his combine. **5.** Michigan State University ag engineers developed a liquid manure-slurry cover crop seeding system that produces excellent stands and saves time and money. **6.** At Loysville, Pa., Charles Martin modified a high clearance sprayer with a boom that can be raised or lowered for seeding cover crops into standing corn or soybeans.

NO-TILLING SOYBEANS IN RYE. Southern Illinois no-tiller Terry Dahmer drills soybeans with good results into rye that is growing 5 feet or taller.

COVER CROPS PAY. Based on University of Delaware data from the early 1980s, cover crops can trim no-till corn costs by 75 cents per bushel.

13 COSTLY MISTAKES TO AVOID WITH COVER CROPS

September 1979

One of the more neglected aspects of no-tillage is the proper use of cover crops. Unfortunately, many farmers simply take cover crops for granted and unnecessary mistakes often result.

Back in 1979, the *No-Till Farmer* editors asked several university agronomists for ideas on the "most common mistakes no-tillers make with cover crops." Here's their list of what you should avoid.

1 **Failure to seed.** This results in costly soil erosion, moisture loss and poor weed control in the following no-till crop, says Morris Bitzer, University of Kentucky agronomist. As farmers gain more experience with no-tillage, most realize the importance and benefits of seeding cover crops.

2 **Seeding too late.** Poor planning results in a cash grain crop remaining in the field too long. By using earlier-maturing soybean or corn varieties, a crop can be harvested in time to get a cover crop seeded.

3 **Poor cover stands.** Bitzer suggests using higher seeding rates with later cover crop seeding. Apply nitrogen to help establish a fall-seeded crop.

"Poor stands would also include the failure to adjust the pH for a legume cover crop," says Bill Lewis, North Carolina State University weed scientist. "Fall moisture may also be more critical for legume establishment. The time when legumes can be established is less than for a small grain cover crop. Plus, legumes are more sensitive to low moisture conditions after they've germinated."

4 **Wrong cover crop selection.** There's no easy answer to decide which cover crop is best for your conditions. This depends on the time of seeding and the crop that will be seeded next spring.

"Cover crops can reduce erosion where crop residues are removed or sparse," says L.S. Robertson, a Michigan State University soil scientist. "Oats planted in late summer provide a blanket to the soil while they're growing as well as after they are frozen in the winter. Cereal rye is the old standby for those who do not fall plow and is very feasible with the use of modern-day herbicides."

5 **Failure to realize some cover crops are better erosion fighters.** "As we go into the federal government's farm program, we're concerned about a comparison of legume cover crops vs. small grain cover crops for erosion control," explains Lewis. "A legume cover crop does not provide as long-lasting a mulch as rye or wheat."

6 **Planting covers in fields heavily infested with difficult to control perennial weeds.** Lewis and Gene Krenzer, a crops specialist at North Carolina State University, recommend conventionally planting cover crops and using preplant incorporated herbicides in fields with Johnsongrass or nutsedge.

"If at all possible, apply Roundup in late summer prior to seeding a winter cover crop," says

> "Growing a cover crop only for cover **would be a big mistake...**"

Lewis. "In the case of soybeans, a recirculating sprayer could be used to control weeds that have grown above the soybean canopy."

7 ***Killing a cover crop before no-tilling corn.*** "The experienced no-tiller will seed the next cash crop ahead of spraying down the cover crop," advises Bitzer. "This results in more uniform killing of the cover crop."

8 ***Allowing the cover crop to get too tall.*** The planting date for corn or soybeans in the spring will determine the best cover crop to use. Legumes are more appropriate for late planting while small grains are recommended when no-tilling soybeans or corn early in the spring.

9 ***Mice damage to no-tilled corn.*** This can be a problem when a cover crop gets too much growth prior to no-tilling a cash crop. If this occurs, the cover crop should be cut at a high height or removed for hay or silage before no-tilling corn.

10 ***Failure to realize nitrogen management for corn may be critical.*** Under heavy mulch conditions, it's best to apply nitrogen in two applications, says Krenzer. Apply one third of the nitrogen at planting time and the other two thirds when the corn is several inches tall.

"Right now, we cannot confidently predict the amount of nitrogen supplied by a legume cover crop for the following corn crop," adds Lewis. "So, this may contribute to the difficulty in managing the amount of needed nitrogen."

11 ***Failure to select proper herbicides, rates or volume to kill a cover crop.*** Krenzer says killing off

a legume cover crop may be improved with an application of Banvel or 2, 4-D at least 3 days before applying paraquat. Use at least 40 gallons per acre of solution along with the proper spray nozzles and pressure to force the herbicides through the cover crop canopy.

12 ***Failure to control insects and recognize problems.*** Use a soil-applied insecticide. Monitor for cutworms and armyworms and know which soil-applied insecticides to use to control these pests.

13 ***Cover cropping with no specific use in mind.*** Farmers in the Southern Great Plains can't afford to grow a cover crop unless it has a practical use.

"Growing a cover crop for the purpose of cover only would be a big mistake," says Allen Wiese, Texas A&M University weed scientist at Amarillo, Texas. "However, this is not true in eastern Texas."

In this area, water from rain or irrigation is expensive, scarce and must be used to produce a return. In the northwest Panhandle of Texas, seeding winter annual small grains serve the dual purpose of providing cover that is necessary to prevent wind erosion and furnishing grazing for cattle.

"The biggest mistake farmers make is destroying the residue left after harvest," says Wiese. "Unless we have a crop failure, there's enough stubble to control erosion and in many cases increase soil moisture storage.

"This cover is part of the crop and is free. But it is wasted when plowed under. Yet this is not a no-till mistake — the mistake is the tillage." ✳

ROLLING COVER CROPS. NRCS staffers in New York apply glyphosate before later rolling the cover crop and no-tilling corn in a single pass.

CRIMPED AND CONTROLLED. A roller crimps hairy vetch stems every 7 inches, closing down a plant's vascular system and leading to its demise.

NO-TILL-AGE®

Cover Crops Compared for No-Tilled Corn

Cover crop	Nitrogen in Top Growth
Hairy vetch	293 bu.
Bigflower vetch	201
Common vetch	131
Crimson clover	190
Arrowleaf clover	114
Austrian peas	125
Subterranean clover	58
Red clover	56
Alfalfa	61
Sweet clover	53
Barley	58

— University of Maryland

MANY COVER CROP CHOICES

Our *No-Till Farmer* data shows subscribers have had experience with about 80 different cover crops.

African cabbage	Kale
Alfalfa	Kura clover
Alsike clover	Lentils
Amaranth	Lupins
Annual fescue	Medic
Annual rye grass	Millet
Arrowhead clover	Mung beans
Austrian peas	Mustard
Barley	Oats
Beets	Oilseed radishes
Berseem clover	Okra
Bigflower vetch	Pearl millet
Birdsfoot trefoil	Peas
Black oats	Persian clover
Browntop millet	Phacelia
Buckwheat	Proso millet
Canola	Radishes
Carrots	Rapeseed
Cereal rye	Red clover
Chard	Safflower
Chickpea	Sainfoin
Chicory	Sorghum-sudangrass
Clover	Soybeans
Corn	Spinach
Common vetch	Sudangrass
Cowpeas	Sunflowers
Crimson clover	Squash
Egyptian clover	Sudangrass
Ethiopian cabbage	Sunflower
Field peas	Sunn hemp
Flax	Sweet clover
Forage collards	Teff
Forage peas	Tillage radishes
Forage radishes	Turnips
Forage turnips	Triticale
Foxtail millet	Vetch
Grain sorghum	Wheat
Hairy vetch	Winter camelina
Japanese millet	White clover

NO-TILL LOOSE ENDS...

AVOID GOING 'WHOLE HOG'

In the winter of 1973, I fielded a phone call from an Illinois farmer who had never no-tilled a single acre. He was looking for planter attachment recommendations on going total no-till with 1,500 acres of corn and soybeans that year.

Even though I told him it was a huge mistake to shift all of his acres to no-till time, he went ahead and did it. Several years later, he told me going 100% no-till in a single year had been a good move.

I had a similar conversation a few years later with a Wisconsin farmer who went from moldboard plowing to no-tilling 1,200 acres of corn and soybeans in a single year.

While I believe in the benefits of no-till, never in my life have I advocated that someone go "whole hog" on thousands of acres. Instead, I've always suggested new no-till enthusiasts start out with only a few no-till acres, work out the kinks and expand slowly from there.

Anything else is much too scary for me! ✻

> ## "If soil degradation continues at its present rate around the world, **farming will be extinct in 60 years.**"
> — *United Nations Report*

September 1980

8 REASONS TO 'COVER' YOUR FIELDS

Cover crops certainly aren't new — they've been used for more decades than many growers like to remember. Yet cover crops are growing in importance since they play a key role in today's no-tillage programs.

To give you some insight into the value of cover crops in a no-till program, here are eight ideas from various areas.

1 When Ralph Rasnic has fertilizer applied on his corn stalks in the fall, he has a custom applicator mix rye with the nutrients. Along with the rye cover, the Riner, Va., farmer applies 40 pounds per acre of nitrogen along with potash and phosphorous based on soil test results. To work the rye and fertilizer into the ground, Rasnic runs a disc over 12-inch tall corn stubble.

2 Not only can a cover crop help hold valuable soil in place all winter, but it can also trim nitrogen bills by up to 90% and boost crop yields.

At the University of Delaware, ag engineer Tom Williams and agronomist Bill Mitchell have found vetch can result in considerable energy savings. By seeding vetch, they reduced nitrogen purchases for no-till corn from 110 down to only 20 pounds per acre.

With vetch providing most of the needed nitrogen, the total energy requirement for no-tillage dropped to the equivalent of only 12.9 gallons of gas per acre compared with 32.2 gallons with conventional tillage.

3 Results of a 3-year cover crop nitrogen study in Delaware show the importance of combining cover crops with no-till. Several cover crops were compared with a control plot that had no cover crop seeded, but did receive 100 pounds of applied nitrogen.

Seeding oats as a cover crop increased yields with all levels of nitrogen. Since this cover is normally killed by winter temperatures, it must be planted early enough to develop good ground cover.

Even more striking result-wise was an oats and vetch mixture, which produced an extra 19 bushels of corn per acre.

4 To protect corn fields over the winter, Stanley Dunn seeds rye after silage is harvested. Then the Charlestown, W.Va., farmer takes off about 8 tons of rye silage per acre in mid-May before no-tilling corn.

5 Some 10 years of work at Clemson University has shown a vetch and rye mulch resulted in 3.11 inches less water runoff and 2.38 tons per acre less soil erosion loss each year without any drop in yield compared with corn grown on plowed ground.

6 In a Georgia study, no-tilling corn and soybeans into a rye cover boosted corn yields by 52 bushels per acre and soybean yields by 13 bushels per acre. Erosion control and improved utilization of moisture were major factors leading to the higher yields.

7 University of Florida studies with crops seeded into an in-row subsoiled rye mulch led to greater soybean and grain sorghum yields, taller plants, greater crop population and fewer weeds.

8 When LeWayne and Allen Bartel seed wheat, they intend to use it as a cover crop ahead of full-season, no-tilled soybeans the following spring. But the Buhler, Kan., brothers have the option of cutting the wheat for grain.

"We wait until the wheat heads out around the end of May," says LeWayne. "Then we kill it back with the same herbicide tankmix we use for double-crop beans before no-tilling our full-season beans."

July
1980

NO-TILL SAVINGS NO LONGER 'CHICKEN FEED'

In three of the 1980 summer issues, the editors took a close look at the no-till benefits in terms of trimming fuel, labor and machinery costs. The data documented what was needed to grow 400 acres of corn with no-till and conventional tillage.

By switching to no-till from conventional tillage with 400 acres of corn, a grower would:
- ◆ Save 4.21 gallons per acre of diesel fuel, a total of 1,684 gallons with 400 acres.
- ◆ Slash farm labor needs by 44 minutes per acre, saving 293.5 hours when cropping 400 acres.
- ◆ Reduce the no-till machinery investment by $128.05 per acre, dropping to $88,117 compared with needing $127,637 of machinery to grow 400 acres of corn with conventional tillage.

Fuel, labor and machinery data shows a $134.94 per acre savings from shifting from conventional tillage to no-till. Switching from conventional tillage to minimum tillage saved less than 10% of the no-till benefits. ❈

NO-TILL LOOSE ENDS...

April
1984

NO-TILL IN ANY LANGUAGE!

A big booster of no-till, Ned Trovillion designed a unique symbol that publicizes the many advantages of no-tilling and not plowing — in a number of languages. Designed to look like an international highway sign, the symbol shows a horse-drawn plow in a red circle with a slash through it.

"I thought the sign would be a good way to help promote no-tillage," says the Soil Conservation Service district conservationist in Johnson County, Ill. "In any language, the signs means more no-till and less erosion." ❈

ENGLISH SPANISH PORTUGUESE FRENCH ITALIAN
GERMAN DUTCH POLISH DANISH

More Earthworms
'Liven Up' No-Till Fields

"The plow is one of the most ancient and most valuable of man's inventions; but long before he existed the land was plowed, and still continues to be plowed, **by earthworms…"** — *Charles Darwin, 1881*

"Early one evening, after a rain shower, my son and I walked across one of our corn fields where I had just disced and planted in the residue," recalls Ed Hawley, of Stockton, Ill. "The earthworm holes were as thick as pegboard in that field. You would have to see it to believe it.

"Then we walked into a neighboring conventionally-tilled field which had suffered erosion from the storm. There wasn't an earthworm or a worm hole to be seen."

Hawley's observation is not unusual among farmers who've been working with no-till.

Less Plowing, More Worms

"Ground worms do our plowing for us," says Toby Johnson of Baldwinsville, N.Y. "Since we've stopped disturbing the topsoil with a plow, the worm population has become substantially greater; it's surprisingly mellow

TODAY'S 'PLOWING' SYSTEM. Earthworms have replaced the moldboard plow and other minimum tillage tools as the primary means of working the ground in no-till fields. In 1943, Edward Faulkner noted that one of the distinctive features of badly eroded soil was a lack of beneficial soil life.

"Increasing earthworm activities increases the speed with which water soaks into the ground..."

near the surface. There's a good firm seedbed with plenty of water because no-till, unlike plowing, does not disturb the capillary action of the water."

Earthworms Important

Earthworms are important because their existence indicates the presence of other organisms, according to two soil scientists, Art Peterson at the University of Wisconsin, and Charles Rieck of the University of Kentucky. And they say reduced tillage concepts fosters the existence of all such forms of life.

Probably the biggest single benefit of large populations of earthworms is the increased porosity of the soil. "Any means of increasing earthworm activities increases the speed with which water soaks into the ground," says Peterson.

Water enters the soil through the network of tiny channels. The channels result from cracks, root growth and the activity of soil organisms, of which earthworms are the largest. Every hole made by an earthworm that emerges onto the surface becomes a channel through which water and air enter the soil.

"Reduced tillage practices aid water intake in two ways," Peterson continues. "For one thing, a rough soil surface keeps the tiny channels open to the surface. When the surface soil is pulverized to make an extremely fine seedbed, these channels are sealed off.

"Leaving residue on the surface keeps the upper few inches of the soil cool and moist, so the worms will continue burrowing right up into and through the top inch of soil. If the surface gets dry they may not come within 3-4 inches of the surface."

Peterson compared worm populations under various tillage systems by counting the number of castings, which are the little mounds of dirt left around the hole when a worm emerges onto the surface.

"In one test we compared the number of castings present in a conventionally-tilled field with one where corn stalk residue had been gathered up, chopped and blown back on top of the plowed ground," he says. "There were 10 times as many castings in the area with the chopped stalk residue on top."

He expects the same ratio to hold true when he compares no-tilled ground to conventional tillage.

Earthworms, of course, are not solely responsible for the channels in the soil. "The amount of soil moisture directly attributable to the activity of earthworms probably will not affect yields in a normal crop year. But in an exceptionally dry year it could cause a tremendous increase in yield," Peterson says.

Earthworm activity also improves soil structure and aeration, Peterson says. They aid in decomposing organic matter, although most of this is done by much smaller microorganisms such as bacteria that are undetectable to the naked eye. There are about 2 billion microorganisms found in a thimbleful of good black soil.

Practices that can have an impact on earthworm populations include tillage, soil pH, soil type, soil depth, soil drainage, infertility, amount of organic matter, precipitation, temperature and pesticide use.

Indicator of Soil Life

Rieck says earthworms provide a visible indicator of the microorganism content of the soil. He says biological activity is much greater in no-tilled fields than in conventionally-tilled

THE FARMER'S EARTHWORM HANDBOOK, OUR MOST POPULAR NO-TILL BOOK!

First published in 1995 and reprinted a number of times, this 114-page book on "managing your underground money-makers" is the all-time best seller in the *No-Till Farmer* bookstore.

Some 19 chapters delve into all aspects of on-farm earthworm management. Topics range from seeding fields with night crawlers, increasing worm populations as well as earthworm friendly fertilization and pesticide application, manure management and their role in sustainable agriculture. The book also looks at the worm's impact on tillage systems, cover crops, crop rotations and outlines a dozen major benefits of taking steps to increase earthworm numbers in your no-tilled yields.

Written by former *No-Till Farmer* writer Dave Ernst, it shows how earthworms can turn over and improve as much as 8,000 pounds of valuable soil on an acre of land in only 2 week's time.

The book is a down-to-earth compilation of valuable information on the underground critters that make no-till successful — and is just as valuable today as it was when written in 1995.

CHECK THE EARTHWORM ARITHMETIC

From The Farmer's Earthworm Handbook...

◆ 25 earthworms per square foot equals 1 ton of worms per acre.
◆ The work schedule for earthworms in your no-till fields is 100 days a year.
◆ One ton of worms produces 100 tons of castings, which is equal nutrient-wise to spreading two-thirds of an inch of manure per acre.
◆ The improvement in drainage worms can make with the macropores in the soil is equal to installing 4,000 feet of 6-inch tile per acre.
◆ The nutrients that worms typically add yearly to an acre include 4 pounds nitrogen, 30 pounds phosphorus, 72 pounds potash, 90 pounds magnesium and 500 pounds calcium. This added nutrient value is worth $34.15 per acre.

— *Bill Becker, crop consultant, Springfield, Ill.*

"There are about 2 billion microorganisms in a thimbleful of good black soil..."

fields. Since earthworms are a higher form of life, they're most visible.

"Actually, no-tillage may be stabilizing levels of microbial life rather than increasing them," Rieck continues. "With conventional tillage we destroy these microorganisms by stirring the soil. No-tillage is getting us back on the right track.

"There are likely to be 10%-25% more bacterial organisms and 15%-30% more fungal organisms in no-tilled soils than in conventionally-tilled ground, depending on the time of year."

Insecticide Effects Temporary

"It's regrettable that rootworm insecticides often kill some earthworms," Peterson says. "But this effect is limited to only the corn row and is not particularly detrimental to the worm population as a whole.

This is also a temporary effect, as the earthworms will be back by the middle of the growing season.

"The presence of earthworms indicates healthy conditions for all the other microorganisms that actually do most of the work of decomposition. Killing all these would amount to soil sterilization, and that is virtually impossible to do with anything short of hours in an autoclave — a steam oven like they have in hospitals." ✳

MANY FRIENDS LIVE UNDERGROUND

Crop scouting usually happens in broad daylight. But it's best to scout after dark if you want to estimate how many earthworms you have in your no-tilled fields.

The last time he counted, Steve Groff found an average of 2-3 worms per square foot in his no-tilled fields. The long-term no-tiller from Holtwood, Pa., says the longer you no-till, the more worms you'll have — since you're not destroying their homes.

Besides the obvious advantage of not chopping up the worms and destroying their burrows with tillage, no-till boosts worm populations in another way.

November 2005

"If you keep a lot of residue on the soil surface, they will multiply," says Groff. "Earthworms multiply when they reach a certain size rather than a certain age. If you can grow them quicker, they're going to multiply faster. So a key is keeping residue on the ground for them to feed on."

In addition, worms chew up the residue and recycle it later for no-tilled crops. But what really matters is the fact that worms are doing Groff's tilling.

"They do a better job than steel tillage tools," he says. "They treat the soil better. Worms are my allies, as they're out there working for me, and they never send me a bill for what they do." ✳

REACHING A HAPPY MEDIUM. Large populations of earthworms often have difficulty surviving in coarse-textured or wet, high clay soils. But most soil types, such as silt loam soils, fall between the extremes and produce large populations under no-till cropping conditions.

SHOULD NO-TILLERS PLANT WORMS IN THEIR FIELDS?

February 2002

While *No-Till Farmer* readers often ask if they should consider seeding fields with more worms, the answer is usually no.

It's not something Purdue University soil physicist Eileen Kladivko recommends. She seeded night crawlers in 14 fields in research projects over the years and it only worked in a few instances.

She says it comes at a one-time cost of a couple hundred dollars an acre. Most inquiries about seeding earthworms come from no-tillers or growers switching to no-till who become more interested in soil health and how earthworms can improve soil quality.

University of Maryland researcher Ray Weil says some high quality soils that have been no-tilled for more than 20 years likely contain 2,000 pounds per acre of earthworms. By comparison, heavy-tilled soils might contain only 200 pounds of earthworms per acre.

What's the value of the worms in your no-tilled fields?

Several garden catalogs have advertised the sale of earthworms at $29.95 per pound. If you had 2,000 pounds of earthworms in a no-till field, their sale value would be $59,900 compared to only $5,900 per acre for a farmer who still moldboard plows.

For a grower with 900 acres of no-tilled crops with 2,000 pounds per acre of worms at those garden catalog prices, these worms would be worth an amazing $5.4 million! ✺

MANAGING NO-TILL'S OWN LIVESTOCK

November 2001

When you think of livestock, most likely you picture herds of cattle or hogs. But not if you're no-tiller Bob Neidigh of Bremen, Ind., whose underground livestock includes earthworms to increase soil nutrients, drainage and yields.

Instead of raising cattle or hogs, the no-tiller and retired mechanical engineer concentrates on managing earthworms, which include benefits such as higher soybean yields and excellent drainage.

Neidigh primarily follows a wheat and soybean rotation while seeding a winter cover crop in between. But what seems to set him apart from many farmers is his philosophy on earthworms.

"The earthworm is nature's plow and soil builder," he says.

"Cover crops caused my worm populations to explode..."

Frank Lessiter

TAKE TIME TO DIG. Illinois no-tiller and strip-tiller Jim Kinsella says soil pits let you see how soil microorganisms are helping boost your yields.

Winter Cover Critical

To Neidigh, a good winter cover crop is just as important as his main crops because they stop erosion, hold water, build topsoil and feed the earthworms. But the true benefit from cover crops is demonstrated with Neidigh's earthworm populations.

Cover crops caused his worm population to explode. Where 50,000 worms per acre are average, Neidigh has counted as many as 2.3 million worms an acre — or 40 times the average.

His calculations show this many earthworms are capable of producing 79 tons of castings per acre each year. Since 1997, Neidigh has had worm castings tested in a lab to determine their nutrient makeup. The tests showed the nutrients in the earthworm castings included 9 pounds urea, 71 pounds phosphorus, 169 pounds potassium, 211 pounds magnesium and 1,176 pounds of calcium.

There are additional nutrients from worms, he says. Nitrogen from worm attrition equals 5,460 pounds per acre, nitrogen from worm urine makes up 4,306 pounds per acre and worm drainage is the equivalent of installing 162 miles of 6-inch tile placed on 5-foot centers.

"Earthworms produce 60% of their body weight in urine each day and urine is high in urea," he explains. While earthworms only live about 1 year, Neidigh says 2.3 million worms per acre produce about 8 tons of protein each per year, which are broken down by soil bacteria to produce nitrogen.

Over the years, Neidigh's soil organic matter has climbed from 1.7% to 3.8%.

"Phosphorus has grown steadily and potassium has climbed from the 150 pound range to well over

"The earthworm is nature's plow and soil builder..."

300 pounds an acre," he says. "Magnesium has also doubled along with a 50% rise in calcium, zinc, iron and copper."

The farm's cation exchange ratio has also shown a steady rise from 8 to over 12.

No-Till Benefits

Neidigh is 100% no-till and has stayed on course by seeding cover crops and leaving all residue in his fields, which has contributed to his high worm populations. The top 2 to 3 inches of Neidigh's no-tilled soil is loose and spongy. "It soaks up water like a sponge," he says.

A few years ago, heavy spring rains flooded several of Neidigh's fields.

"Many of my neighbors flooded out and had to re-plant the low spots. My low spots also flooded, but the water went down so rapidly that I suffered no damage," he says.

Neidigh's fields feature very heavy soils underlain with yellow clay. "To work the ground to a fine powder, as many farmers do with conventional tillage, you can get good germination if conditions are ideal. But if they get repeat rains like we often have after planting, the ground seals off and the water can't move down into the ground," he explains.

To understand this concept, Neidigh suggests thinking about having a tractor stuck in a mud hole with the wheels spinning.

"The soil particles under the wheels are forced together tightly and the open wheel tracks will hold water

for weeks," he explains. "The same thing happens in conventionally-tilled fields, but to a lesser degree."

His soybean yields average 63 bushels an acre. Neidigh attributes his yields to a combination of biological additives, healthy soil, high soil bacteria and earthworm populations fed by the cover crops.

In accordance with his earthworm philosophy, Neidigh says, "The least amount of soil disturbance is best. Let nature take its course and make the earthworm your most effective plow.

"On our farm, earthworms are tunneling, chewing, digesting and producing nitrogen, phosphorus, potassium, magnesium and calcium 24 hours a day. We consider this pretty cheap labor and an excellent way to increase both no-till yields and profits." ✳

EATING OUT TONIGHT. Pennsylvania no-tiller Steve Groff finds night crawlers like to feast on the residue of the forage radishes he seeds as part of his extensive cover crop program. The faster that worms grow, the sooner no-tillers will see an expansion of worm numbers in their non-tilled fields.

"Earthworm numbers were reduced 70% by 5 years of tillage after sod..."

LESS TILLAGE, MORE WORMS. In Nebraska fields that were continuously no-tilled for 5-20 years, seven times more earthworms and 10 times more earthworm cocoons were found that in nearby tilled fields. With more intensive tillage, earthworm populations and benefits drop dramatically.

EARTHWORMS ARE NO-TILL'S GREATEST ALLIES

January 2005

Earthworms are nature's tillers and their presence is a key component in successful no-tilling. But how many do you have? And how many worms do you want?

"It takes about 500 years to make just an inch of topsoil," says Paul Hay, a University of Nebraska extension educator at Beatrice, Neb. "We're burning through our soils a lot faster than that, so we need to make every effort to preserve them.

"Just because you have the best organic matter, moisture or soil condition doesn't guarantee many earthworms. They may have died for some reason and they take time to regenerate."

Like any good researcher, Hay looked at existing studies before developing his own earthworm trials.

Forages Vs. Grains

"University of Wisconsin agronomists looked at earthworm counts that were remarkably different," Hay explains. "Forage crops in rotations had a lot more earthworms than cash grain crops. Even so, no-till cash grain ground was a lot better than tilled ground in terms of earthworm numbers in all rotations. Reducing tillage increased the number of earthworms, as did manure application."

"University of Illinois researchers learned that a continuously plowed field averaged only 10 earthworms per square meter. When you no-tilled that ground, the number doubled to 20 earthworms."

Other factors that affect earthworm counts included having soybeans and other legumes in the rotation, which yielded 60 earthworms per cubic meter in plowed fields and 140 worms in no-tilled fields. Clover, however, boosted populations to 400 worms per cubic meter. Add heavy manure applications and earthworm numbers increased to 1,300.

"They get pretty thick when you have good soils and moisture content with moderate weather," says Hay. "In a pea and wheat rotation, Oregon trials found five times as many earthworms that can increase infiltration."

Huge Worm Losses

"Earthworms in Ohio tests were reduced 70% by 5 years of tillage after sod. No-till fields rated a 30-fold increase over tilled fields. This study is replicated throughout the world."

To see for himself, Hay counted the earthworms in several southeastern Nebraska fields. Average rainfall is 28 inches per year, with very dry winters and hot, dry summers. Soils are silty, clay-loams.

He ended up digging in 24 fields that had been continuously no-tilled for 5-20 years. The average count was 5.8 worms per cubic foot with 1 worm cocoon per cubic foot. In 12 tilled fields, the numbers were reduced to only 0.8 worms per cubic foot and 0.1 cocoons where they were not reproducing as fast.

Hay's research did not find any direct correlation between decreased earthworm populations and the application of anhydrous ammonia.

"I had the highest earthworm count in a field that was not using anhydrous," he explains. "But the third-, fourth- and fifth highest counts were in fields with anhydrous applications. I didn't find a clear association between worm count and fertilizer source."

Improved Soil Structure

Another aspect of Hay's research dealt with the impact of soil structure. On average, the no-tilled fields ranged from good to excellent, while the tilled fields ranged from below medium to poor.

"The minute you start tillage, the soil structure is affected, as earthworms are at the top of the feeding scale in our soils," he says. "Bacteria, protozoa and nematodes are all very important microbiology that we're still studying. Earthworms are on the top of the feeding chain. It's the largest animal that's present in the ecosphere, so it's important that you look at it very carefully in evaluating your soil's overall health." ✳

NO-TILL-AGE®

January 1997

Where Earthworms are Found

Based on 1996 spring and fall samplings at 11 sites.

Cropping Systems	Earthworms Per Acre
No-till	277,254
Minimum tillage	81,441
Conventional tillage	55,934
Forest land	288,958

— Essex, Ontario, Regional Conservation Authority

EARTHWORM ARITHMETIC

No-tiller Bob Neidigh devised a mathematical formula to compare the power of earthworms to tractor horsepower.

Earthworm population = **2,352,240** worms per acre.
Weight of worms = **15,616** pounds per acre.
Castings produced = **79.46** tons per acre per year.

Worm power = **86.8** horsepower per acre or 8,680 horsepower per 100 acres.
Worm power in moving soil = **37,932** tons per acre per year.

Compare **37,932** tons per acre of worm power per year to a tractor which moves **1,200** tons of soil per acre while plowing. ✳

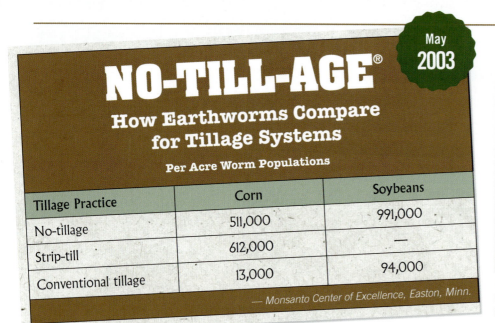

NO-TILL-AGE®

How Earthworms Compare for Tillage Systems

Per Acre Worm Populations

Tillage Practice	Corn	Soybeans
No-tillage	511,000	991,000
Strip-till	612,000	—
Conventional tillage	13,000	94,000

— *Monsanto Center of Excellence, Easton, Minn.*

Mike Wurmnest

LEAF EATERS. Earthworms have been known to pull no-till corn leaves down into soil channels for feeding purposes.

IT'S TIME FOR EARTHWORM TRIVIA

Q: How many earthworms do you need in no-tilled fields?

A: To gain the benefits of a good earthworm population with no-till, you want 8 to 10 worms living 6 to 8 inches deep in every square foot of soil.

Q: How do you determine how many earthworms you have?

A: Earthworms go deeper when the soil gets hot and dry. Take soil samples at half a dozen test sites per field, 6 to 8 inches deep and 12 inches square. Break the soil apart and see how many worms you can find in each sample. The magic number is 8-10 worms per square foot.

Q: How do you get more earthworms to inhabit your no-tilled soils?

A: Encourage more earthworms by planting cover crops, adding manure and compost, reducing the amount of tillage and disturbance of the soil and keeping the soil covered with a layer of mulch such as shreddings. Worms like warm (not hot) and moist (not saturated) conditions.

Q: What are the ideal living conditions for an earthworm?

A: They require an environment with lots of crop residue and a calcium-rich soil. They like shaded conditions such as provided by cover crops. They can tolerate a range of temperatures from freezing to 100 degrees F. They live in almost all soil types, except very coarse soils (sands) and very acidic soils.

Q: What about using anhydrous ammonia and nitrogen fertilizer with no-till? Does it hurt the worms?

A: Purdue University agronomist Eileen Kladivko says that while anhydrous ammonia kills worms that it comes in direct contact with, this likely has a very small impact on a field-scale basis. Anhydrous ammonia is typically injected in row middles (every 30 inches), so this chemical will only kill a few worms found directly in the injection shank zone.

Research at Purdue University suggests this is a small percent of the population (less than 10%) and that the worm populations where anhydrous ammonia is used are not necessarily low.

Some studies have shown that while earthworm and soil organism populations are reduced in the immediate vicinity of the band of anhydrous ammonia, with time, the populations of soil organisms often increase to levels greater than before the application was made. This is a response to stimulation and enrichment by the fertilizer itself.

Why No-Tillers Turn Out
Top Corn Yields

"They're good managers, pay attention to detail and recognize it takes a complete package to turn out bin-busting yields…"

Opponents of no-till have argued for a number of years that yields simply won't be as good from no-tilled fields compared to where extensive tillage is used. Yet over the years innovative no-tillers have proven this old wives' tale is wrong.

Besides the fact that most no-tillers turn out yields at less cost per bushel than neighbors using more tillage, no-tillers also seem to capture more than their share of top honors in local, county, state and national yield contests.

In this chapter, we'll look back at why no-till fares so well in the national corn yield contest, analyze some flawed research in regard to no-till and explain why growers often harvest higher yields than university scientists get in their research plots. ❁

TOP CORN YIELD CONTEST RESULTS FAVOR NO-TILL

December 2017

In 2017, David Hula harvested the highest yield ever recorded in the national corn yield contest. The Charles City, Va., grower turned in a world record yield of 542 bushels per acre from a no-tilled irrigated field in the National Corn Growers Association contest.

In 2016, Randy Dowdy turned out the top yield in the same contest with 521 bushels per acre from an irrigated no-till field on his Valdosta, Ga., farm.

Both the Hula and Dowdy families have been consistent yield contest winners for a number of years.

Looking at the 2017 contest data, Hula's 542 bushels per acre yield was 135 bushels higher than the top yields in irrigated categories where other tillage systems were used. His yield was 172 bushels higher than the top yields in the dryland divisions where other tillage systems were used.

Dowdy's yield was 164 bushels per acre higher than the top yields in dryland contest categories where other tillage systems were used. His yield was 46 bushels per acre higher than for the irrigated category that featured other tillage systems.

When asked to list three critical factors that lead to higher yields, Hula recommends planting when soil conditions are just right, getting the most out of starter fertilizer and providing essential nutrients when the corn needs them the most during the growing season.

Hula says paying attention to detail and thinking outside the box are essential in finding new ways to turn out more profitable no-till yields. "I don't want to do anything to my crop that's going to take some yield away, because we can't build or create yield," he says. "Instead we're just protecting the potential yield."

Dowdy's basics for producing higher yields include planting when soil conditions are optimum, scouting fields regularly to stay ahead of potential pest or nutrient concerns and constantly evaluating new crop production ideas.

So why do no-tillers tend to end up with the top yields in many contests? Why aren't growers relying on more extensive tillage also setting yield records? That's a question we asked several crop educators and suppliers, and here's what they had to say.

PAY ATTENTION TO DETAILS. Virginia no-tiller David Hula credits thinking outside the box when evaluating new ideas with helping him produce a world record corn crop of 542 bushels per acre in 2017.

Managing Differently

Long-term no-tillers have a major advantage since they farm healthier soils that have improved dramatically over the years. Thanks to their use of best management practices, healthier no-till soils tend to turn out the highest yields.

A number of these no-tillers rely on irrigation to overcome moisture shortages throughout the growing

"Healthier no-till soils tend to turn out the highest yields..."

"I don't want to do anything **to my crop that's going to take yield away...**"

CRITICAL MANAGEMENT STRATEGIES. Crop rotations, cover crops and paying attention to detail are the keys to turning out high-profit yields.

These high-yield no-till managers understand the soil and production capabilities of corn and recognize why intensive tillage can limit yields. They stay on top of full-season nutrient needs through constant soil, stalk and leaf analysis and make additional fertilizer applications as needed.

Precision Pays

Among growers turning out top yields, equipment gets serious attention, as they recognize a fine-tuned planter can lead to a near perfect seed drop at the right depth. It's also likely these growers plant at much slower speeds to end up with almost perfect stands.

These no-till innovators constantly seek the latest cropping ideas from suppliers, crop consultants, university educators and fellow no-tillers. By combining the "best of the best" crop management ideas with the many benefits of no-tilling, they turn out top yields year after year.

Basically, they're good managers, pay attention to detail and recognize that it takes a complete package to turn out top no-till yields. ❋

season. They recognize how surface residue reduces evaporation and keeps the soil moist. They understand how increased soil biological activity starts with eliminating tillage and leads to more efficient use of nutrients throughout the growing season.

GAIN BENEFITS FROM NO-TILL RESEARCH

Most innovative no-tillers are looking for a much better and more profitable way to produce high-yielding crops. By searching for results that are both practical and success-oriented, they can take full advantage of the broad picture of the associated benefits without worrying about sophisticated statistics to separate small yield differences.

When it comes to making significant changes in cropping programs, no-till innovators have learned that combining both practical and science-based information offers the best of both worlds.

For the best no-till results, maybe researchers need to switch to including more large-scale field data in their research projects.

A case in point would be the highly successful onfarm model developed by South Dakota State University researcher Dwayne Beck at the Dakota Lakes Research Farm at Pierre, S. Dak. Not only does Beck restrict projects to practical field-scale no-till research, but he's been able to make the total farming operation, which is funded by local growers, profitable at the same time.

Thanks to many of the latest technology breakthroughs in the past decade or so, it's possible for no-till researchers to enjoy the best of both worlds. They can provide valuable data not only from small-scale research plots, but also test their ideas with valid, large-scale replicated field tests while working with a few innovative no-till farmers.

Accurately calibrated yield monitors and today's related "big data" collection opportunities give a researcher the chance to work with long-time successful no-till farmers to set up replicated strips that can provide several acres of valid yield results. ❋

December 2014

A FLAWED LOOK AT NO-TILL

Some college professors still don't get it when it comes to seeing the benefits of no-till. While thousands of growers around the world are cashing in with no-till, there are still folks who pour cold water on the idea of being able to get good yields without tilling the soil.

A case in point is found in a recent analysis by a team led by researchers at the University of California-Davis. In a study published in *Nature* magazine, these researchers examined results from over 5,000 side-by-side tillage system observations that were part of 610 peer-reviewed university studies.

First, The Good News

The authors admitted that no-till can improve long-term productivity, profits and food security, particularly with what's taking place today with climate change. They recognized no-till is less time-consuming and can be more cost-effective than conventional tillage, but still concluded that no-till often leads to significant yield declines.

To put it bluntly, their research methodology was flawed. In fact, they even admitted that they neglected to consider the impact that residue retention and crop rotation had on no-till.

With no-till leaving the majority of residue on the soil surface, I don't understand why these researchers don't believe residue retention is a critical part of an overall no-till package. In fact, the researchers indicated that if they had looked at the value of residue retention and rotation, their no-till results would have been different.

Even though there are an estimated 275 million acres of no-till in the world today, somehow these so-called experts are not convinced it's a good thing.

In regions with a moist climate and sufficient rainfall, such as in Northern Europe, the researchers calculated no-till led to a 6%-9% drop in yields when compared with conventional tillage. Yet when no-till, crop rotation and residue retention were combined in dry climates around the world, the authors found yields were significantly better than conventional tillage due to increased soil moisture.

A recent study by the International Food Policy Research Institute looked at the benefits of no-till in an entirely different way. Some 35 years from now, this group's staff calculated no-till adoption could increase corn yields in Sub-Saharan Africa countries by as much as 31%. Widespread use of no-till with corn, rice and wheat would mean 10% fewer Africans would go to bed hungry each night.

They're Wrong!

Unfortunately, the researchers in the California study painted a bleak picture for further promotion of no-till around the world. But as *No-Till Farmer* subscribers know, it's already working in almost all areas, under various climate conditions, with many soil types, a wide variety of crops and numerous environmental conditions.

Regardless of what this flawed international research project shows, no-till will play a key role in feeding the world's ever-expanding population over the next 35 years. ❁

NO-TILL EQUIPMENT ADJUSTMENTS, MODIFICATIONS. Before manufacturers started designing machinery specifically for no-till, there was always a crowd at field days evaluating how successful no-tillers were making planter and drill adjustments or checking out the latest farm shop ideas.

February 2015

GROWER TO GROWER. Onfarm test plots often result in higher yields and returns than what researchers obtain from small-scale university trials. By only evaluating a single idea at a time, it's difficult for scientists to develop a total no-till package that can dramatically boost yields and profits.

29 REASONS WHY GROWERS HARVEST HIGHER NO-TILL YIELDS THAN SCIENTISTS NORMALLY FIND IN RESEARCH PLOTS

The biased University of California-Davis study (see page 178) got our editors to wondering why successful no-tillers obtain better results than the yields often reported from numerous small-scale research studies conducted by university folks and some seed and fertilizer suppliers.

To provide the answers, the *No-Till Farmer* editors approached a few university educators, consultants and no-till growers. You'll be amazed at their candid responses.

1 Researchers are all over the board when it comes to defining reduced tillage practices, which make comparisons extremely difficult. Terms such as no-till, reduced-till, mulch-till or conservation tillage are used loosely. Many times the scientists don't bother to explain how much or what kind of soil disturbance actually occurred in their research trails.

2 Some small-scale plots measure only 10 by 30 feet to remove statistical inconsistencies. Yet, tractor and planter tire traffic can cover up to 40% of this area, and reduced traffic is among the items that makes no-till shine. Even when compaction occurs, the scientists often don't want to use this data since it may mess with their research study's protocol.

3 Small-scale plots don't allow for the use of real-world-sized equipment. As an example, research done in an Illinois farm situation found 1,000-foot long plots had a difference of only 4 bushels per acre with corn compared with a huge 29-bushel variability difference in 50-foot plots.

4 Developing a long-term no-till "systems" approach is not what most scientists evaluate. Instead, they prefer to research procedures with only a few variables. On the other hand, growers have much more experience in developing successful no-till systems.

5 The peer review of no-till scientific research papers makes total systems analysis difficult. To run statistics that are acceptable to peer reviewers, everything needs to be kept constant, except for one or two variables. This means planting time, timing of weed control, herbicide rates, form and timing of fertilizer and recommended plant populations are often wrong for the no-till plots, yet still correct for the intensively tilled plots. Since planting in comparison plots must be done on the same date, no-till crops are often planted when the soil is still wet, which leads to sidewall compaction and lower yields.

6 Most research studies are funded with 2- or 3-year financial grants, while 4 or more years are normally needed for no-tilled soils to recover from many years of intensive tillage.

7 Rotating tillage practices doesn't allow the soil to reap the full benefits of no-tilling.

8 Some no-till plots are evaluated only during the first year after intensive tillage, with no consideration given for inputs such as cover crops or crop rotations. The full benefits of a mature no-till system can't be captured in a single year research study.

9 Since small plots are also often used for other agronomic research, studies are sometimes done in plots that have been no-tilled for as little as 1 year. When this occurs, there's likely to be a yield decline for the first few years until the rhizosphere population in the soil re-establishes itself.

UNIVERSITY RESEARCH TOURS. Among the longest running mid-summer events is the University of Tennesse's Milan No-Till Field Day.

10 Many short-term no-till studies don't recognize valuable soil property changes. Long-term plots often indicate no-till yield benefits that weren't recognized during the first several years.

11 While it takes 3-5 years to fully establish a no-till system, researchers often don't want to make any significant changes in the research protocol during the later years to get valid results. On the other hand, a no-till farmer after 5 years may have modified his planter, added cover crops, changed fertilizer placement or made other refinements that boosted yields.

12 Having earned a graduate degree doesn't necessarily make someone a good farmer.

13 While no-till growers believe in the educational benefits of university research, they don't want to wait 3 years to decide if an "experimental treatment" is good or bad.

14 Onfarm research is good, but has its own issues. Sometimes, farmers no longer want to continue a comparison study once a particular idea is found to be superior.

15 Some graduate students working on a doctorate degree may end up being clones of their major professor, unfortunately developing the same poor thinking process or biases against no-tilling.

16 A serious issue among university scientists is the need to publish scientific papers. If the research actually advances the knowledge base, that's sometimes secondary to the need to "publish or perish" to protect their academic jobs.

17 Some educators in the 1970s and 1980s didn't have the necessary skill levels to effectively evaluate no-till treatments. Did the plots completely control competition from weeds? Did the plots have adequate corn stands? On soils that do not crack, was there adequate mulch cover? Did the reviewers have enough knowledge to read the articles critically and recognize such concerns?

One crop consultant stated that early in his career he came across a pair of plots at a major university that compared no-till and conventionally tilled soybeans. Since Treflan herbicide (which should be incorporated) was used in both plots, the no-till plot was a weedy mess and yielded significantly less as the herbicide wasn't incorporated. Yet the resulting research conclusion in the early 1980s was that no-till soybeans yielded significantly less than conventionally-tilled soybeans.

18 Results from one trial showed a corn yield boost with no-till on soils that do not crack and had adequate mulch cover. In the same trial, there was no response with no-till on soils that crack regardless of available in-season moisture. Did the peer reviewers know enough to consider the various soil factors?

19 When a farmer growing continuous cotton with intensive tillage switches "cold turkey" to no-till, a yield drop is likely. Adding corn, wheat and cover crops to the no-till rotation can bring the soil back to life. Such tillage and crop rotational changes re-aggregate the

NO-TILL EQUIPMENT CONCERNS. Among the concerns growers see with small plot work is the failure to properly adjust planters and drills.

soil, allow a buildup of earthworms and other organisms to re-establish needed soil porosity and enable the establishment of needed mycorrhizae numbers.

Once these processes have advanced, no-till should perform as well as, or possibly better, than conventional tillage. But educators need to recognize the no-till yield boost is largely due to the improved uptake of water and nutrients, which result from better soil quality and increased mycorrhizae activity.

20 Farmers look for good data on which to base decisions, but don't always find meaningful university data. The art of farming requires "farmer's intuition" along with in-the-field observations.

21 Demands on a university or government scientist often make timely attention to the details of farming extremely difficult. Researchers may be teaching classes or attending conferences on days when their plots desperately need weed or insect control. Unless they have an excellent farm manager, who can handle the fieldwork, these pest control needs may not be met in a timely manner.

22 Some researchers have proclaimed in peer-reviewed papers that no-till will not work under certain soil and climate conditions. Yet nearby farmers are profitably no-tilling the very same soils.

23 Budget restraints often prevent researchers from investing in adequate no-till equipment. Many university and federal government research facilities are using antiquated planters and drills that are not set up correctly or are missing attachments that many no-till farmers are using successfully.

NO-TILL STRATEGIES. The art of being a successful no-tiller boils down to "farmer intuition" along with making accurate field observations.

NO-TILL-AGE®

What's Most Important in Making Tillage Decisions?

February 1990

Practice	% Farmers
Soil erosion	33%
Weed control	22%
Machinery/fuel costs	20%
Labor requirements	19%
Government rules, programs	6%

— Garst Seed Co.

24 A no-till farmer takes a hands-on approach and tackles his own equipment modifications, calibrations, adjustments, planting date, seed depth, etc. By comparison, some scientists rely on technicians to handle the "hands-on" work based on their "not so detailed" verbal instructions.

25 Offers from manufacturers to help with equipment setup or to provide the latest attachments are often rebuffed by researchers. Accepting this help apparently is viewed by some as an admission that scientists are not knowledgeable when it comes to no-till equipment usage. University farm managers and field technicians often do not welcome industry help, as they feel it may threaten their standing with researchers.

26 Since communication between researchers and field personnel is often poor, wrong instructions can be given. Unless the researchers do the actual fieldwork themselves, it falls on the shoulders of technicians to make sure a trial is done correctly.

27 Technicians may be discouraged from providing research trial suggestions. Some are still unwilling to do this "no-till thing."

28 Since the typical university farm may be evaluating multiple tillage practices and a wide range of agronomic studies at the same time, the farm staff can't be an expert on all cropping and tillage ideas.

29 A researcher usually has other responsibilities so he or she can't make sure no-till planting, spraying or other operations are always done at the proper time. ✺

NO-TILL-AGE®

Cashing in on Value-Added Farm Products

Based on a 40,000 pound semi-load (714 bushels) of flaxseed

Added Value	Truckload Value	Per Bushel Value
Flaxseed sold at elevator	$2,500	$3.50
Flaxseed (with freight) sold in Brazil or Japan	$90,000	$126.05
Packaged on-farm in 1-pound bags	$280,000	$392.16
Packaged on-farm in 25-gram bags with added sweetener	$800,000	$1,120.45
Flaxseed nutrition bars	$5.6 million	$7,843.14
Medical capsules	$20 million	$28,011.20

— Rick Heintzman, Onaka, S.D.

HOW PLOT LENGTHS IMPACT YIELDS

100-Foot Rows...

Plot	Corn Yield Per Acre
1	193 bu.
2	197
3	196
4	197

4 bushels of variability

50-Foot Rows...

Plot	Corn Yield Per Acre
1	190 bu.
2	165
3	194
4	192

29 bushels of variability

— Calmer Agronomic Research Farm, Alpha, Ill.

NO-TILL LOOSE ENDS...

EX-PLOWING FANATIC BECOMES NO-TILLER

While Piers Woodriff now no-tills, it took him a long time to get to this point in his cropping program.

"I was once a plowing fanatic — plowing 14 inches deep with four 16-inch moldboard plow bottoms," says the Somerset, Va., grower. "I still remember the day when a shrewd old corn grower asked me whether a big sponge or a little sponge held the most water, which got me thinking about reducing water runoff with no-till."

April 1983

But the experience that actually convinced Woodriff to shift to no-till took place during a rainstorm when he was walking around a conventionally-tilled field in boots, raincoat and with an umbrella in a newly planted field of corn.

"Some 5 or 10 inches of rain had already filled up my 2 feet of subsoiled loose dirt," he says. "Then came another hard inch of rain. There was nowhere for the water to go except down the hill — along with my expensive lime, fertilizer, organic matter and soil.

"That's when I started to think more seriously about no-tilling." ❋

Double-Cropping Puts
More Bucks in the Bank

"Soybean yields in a double-crop wheat and bean rotation would have to drop to 20 bushels per acre before full-season soybeans were as profitable…"

Double cropping is certainly not a new idea. For many years, growers using conventional tillage in many areas planted soybeans after wheat, corn and beans or after harvesting early-maturing vegetables such as peas.

But there's no doubt that replacing intensive tillage, which dries out the soil and reduces yields, with no-till has dramatically increased double cropping opportunities. This is particularly true with soybeans no-tilled immediately after wheat harvest south of a line that runs from Kansas City to Baltimore, Md.

Nationwide, around 6% of U.S. soybeans are double-cropped each year.

Three Crops in Two Years

Veteran no-tillers John and Alexander Young have been double-cropping soybeans after wheat on their Herndon, Ky., farm — where the birth of no-till took place in 1962 — for more than 5 decades. The biggest reason they've double cropped is to protect profits.

1. Over 70% of the U.S. double-cropped soybeans were no-tilled in 2004. **2.** Innovative no-tillers seeded soybeans with their combines as they harvested wheat or barley. **3.** These double-crop beans were seeded into 4,400 pounds per acre of wheat residue. **4.** 31 combines in this Brazilian field harvest wheat followed immediately by 12 no-till drills double-cropping beans. **5.** Growers find no-tilled wheat plus double-cropped soybeans earn more dollars than full-season beans.

In 2013, the Youngs were averaging about 150 bushels of dryland corn per acre. However, double-cropping wheat and soybeans earns as much net income as harvesting 200-bushel corn.

The 2,000 acres they double crop — out of 4,000 acres — are only uncovered for 4 months during a 2-year cycle. Double-cropping also introduces a diversity of soil organisms and reduces the possibility for pests and diseases when compared with no-tilling continuous corn.

Among their keys for success is their self-propelled sprayer to control pests and apply fertilizer in a timely manner so high-yielding wheat can be grown in the double-crop system. They spray wheat five or six times, corn three times and soybeans twice. They consider streamer bars a must for nitrogen applications.

Based on University of Kentucky research, no-tilling double-crop soybeans after June 25 results in the loss of 1 bushel of beans per day. If a farmer harvests 100 acres of wheat per day and soybeans earn $12-$15 per bushel, there's a loss of at least $1,200 for every day they're behind planting.

As quickly as the Youngs harvest wheat, they no-till soybeans — usually beginning around June 10.

The Youngs start drying wheat when moisture levels are 20%-25%.

"You can spend lots of money drying wheat, but make more than that up with earlier bean planting," John says.

Managing residue is another necessity, as it's essential that wheat residue be evenly spread ahead of the no-till planter. One year, a chopper stopped working on the Youngs' combine and it left a 100-yard strip of windrowed residue in the field before they discovering the problem. Figuring it would be fine, they no-tilled soybeans through the windrow of straw. Unfortunately, the resi-

6. Many no-tillers double-crop soybeans on the same day as wheat harvest to avoid losing soil moisture. **7.** Harvesting wheat at 18%-20% moisture can advance the soybean planting schedule by 3-5 days. **8.** Ohio agronomists suggest seeding 4 beans per foot in 7-inch double-cropped rows. **9.** Homemade seed platform offers extra down pressure weight for early-summer dry weather seeding. **10.** A wide choice of double-cropping opportunities allow no-tillers to grow three cash crops in 2 years.

due was so thick that row cleaners couldn't clear it and the coulters couldn't cut through it, leading to almost non-existent bean stands.

Doubling Up

Because of no-till's time and dollar savings, the opportunity for double cropping soybeans has moved further north. No-tillers located in the southern areas of Iowa, Illinois, Indiana and Ohio now have a better chance of double cropping because they have more days before the first killing fall frosts.

Factors that no-tillers need to consider include the potential for rainfall in late July and August, as well as adequate soil moisture to support the soybeans.

Ohio State University soybean specialist Jim Beuerlein suggests cutting wheat at 18%-20% moisture to advance no-tilling beans by 3-5 days. To extend bean-growing season, he recommends harvesting wheat and no-tilling soybeans the same day.

For fields near Interstate 70 that can be no-tilled in 7 1/2-inch rows on July 4, he's found beans with a relative variety rating of 3.4-3.8 is suitable.

Because late-planted soybeans flower about 30 days after emergence and end up with less growth, Beuerlein suggests increasing the seeding rate to 4 seeds per foot of row when no-tilled in 7-inch rows. An earlier plant canopy can help control weeds and make much more efficient use of available soil moisture.

Banking on Benefits

Thanks to no-till, farmers in the southeastern US. have increased the double-crop soybean acreage by more than 50% since 1970. Among no-tillers in this area, feed salesman Jimmy Howard can still afford to no-till corn, soybeans and wheat on 450 acres while holding down a full-time job.

DOUBLING UP WORK. In the early 1970s, Allis-Chalmers pioneered the idea of harvesting small grains and no-tilling double-crop soybeans right behind the combine to extend growing season.

The Mooresville, N.C., grower started no-tilling when he took over his father's cattle, poultry and hay operation. Howard no-tills double-crops soybeans into wheat stubble, which is his most profitable rotation.

Howard harvests wheat during the first half of June and im-mediately no-tills soybeans so fields don't dry out. He saves a costly field trip by skipping a burndown herbicide application before planting and only making a single later herbicide application.

He normally plants Group 5 soybeans, explaining that Groups 3 and 4 have the potential for higher yields, but are also a higher risk due to the area's high humidity. In recent years, he's planted late Group 4 beans — around 4.9 — for earlier harvest. He usually ends up no-tilling wheat behind corn, but prefers to plant the small grains after soybeans.

In addition to getting three crops in 2 years with no-till, Howard estimates wheat yields have increased by 10 bushels per acre, soybeans by 5 bushels per acre and corn by 5-10 bushels per acre.

"We have a longer growing season. As long as we plant beans before July 1, they mature before early frost and we can get that extra crop," Howard says.

Part of Howard's double-cropping success is configuring his planter to deal with residue. He plants soybeans in 15-inch rows using an 11-row Kinze 3000 planter outfitted with row cleaners to push wheat straw out of the row area. He uses cast-iron closing wheels because they work better than rubber tires or spiked wheels under most soil conditions. Staggered closing wheels avoid pinching the soil and causing sidewall compaction. By running one wheel out front, he moves the sidewall over past the center while the other one moves the soil back and crumbles it up. He also relies on 13-wave coulters to work up a loose band of soil. ❋

NO-TILL-AGE®

Check Out New Double-Cropping Options

Double-Crop Combination	Net Return per Acre
Wheat, amaranth	$892
Canola, amaranth	$656
Wheat, sunflowers	$419
Wheat, soybeans	$347
Wheat, buckwheat	$313
Canola, sunflowers	$284
Canola, soybeans	$184
Canola, buckwheat	$113
Wheat, pearl millet	$105

— University of Missouri

'FRANK, YOU'RE THE FIRST EVER!'

Deere & Co.

A MISSING PHOTO. The Deere public relations team bragged that they could fill any equipment photo requested by farm editors. But they drew a blank when we asked for a shot of a combine harvesting wheat and a no-till planter double-cropping soybeans in the same field.

Will McCracken served for many years as the John Deere ag division public relations manager. He prided himself on being able to meet any editor's specific request from the extensive John Deere photo library.

In the winter of 1973, I requested a photo of a John Deere combine harvesting wheat along with a John Deere planter no-tilling double-crop soybeans together in the same field. Since Allis-Chalmers had provided photos like this, I was sure Deere could do the same.

A few days later, McCracken called and told me this was the first time he'd ever had an editor ask for a product photo that Deere did not have.

McCracken assured me he would have the shot by the middle of the summer. And a few months later, he met my request by sending along a photo with a John Deere rig double-cropping soybeans behind wheat. Unfortunately, it was a lousy photo, and didn't come close to the Allis-Chalmers photos in telling the double-crop story.

For the first time, my 1973 no-till double-crop photo request had stumped Deere's public relations folks. They thought they had all the bases covered when it came to ag equipment product photography … until we asked for this specific no-till image. ✺

12 IDEAS TO BOOST DOUBLE CROP BEAN YIELDS

Here's a rundown on a number of proven ideas that may help you do a better job of double-cropping soybeans.

1 *Watch planting date.* Selecting an early maturing wheat variety can let you harvest 5-7 days earlier than with late varieties.

2 *Aerial seeding.* This may work if beans are seeded into standing wheat or barley about a month prior to combining grain.

May 1979

Following a 1-inch rain, 3 bushels of soybeans were aerially-seeded on May 12-16 at the University of Tennessee Milan field station. These beans yielded 32.6 bushels compared with 43.4 bushels for 20-inch row double-cropped beans harvested on the same date.

Aerial seeding doesn't work when weeds are a serious problem, particularly if Johnsongrass, Bermuda grass or other perennial grasses are present.

3 *Harvest wheat earlier.* Early harvest of wheat coupled with drying is worth an extra $15-$20 an acre, says John Barrett, USDA ag engineer stationed at Purdue University. Besides earlier bean planting, wheat quality is improved with drying.

Harvesting wheat at 21% moisture moved bean planting ahead by 6 days, gaining an extra 4.5 bushels of beans per acre.

4 *No-till beans as you harvest.* If the soil is moist when you harvest wheat, enough moisture will likely be conserved to ensure bean emergence without the need for additional rainfall at planting time. However, if the soil is tilled, then rain will be needed for bean emergence.

5 *Use the right variety.* Ohio agronomists say a soybean variety that's moderately early for your growing area will probably work best. However, extremely early maturing varieties do not yield as well as later maturing varieties. But due to a shorter growing season, you run the risk of losses from frost with later maturing varieties.

6 *Circle July 10.* Ed Stroube, Ohio State University weed man, says to skip double-cropping if you can't get beans in the ground prior to July 10. By this time, potential yields will have been reduced to the point where the practice probably will not be profitable.

7 *Don't cut too much straw.* Strong demand for straw has contributed to some weed control problems with double-crop beans.

With barley and wheat straw selling for $30 or more a ton in many areas, growers often cut small grain close to the ground to get as much straw to sell as possible. Without realizing it, this may be costing them money in terms of reduced double-crop soybean yields.

In University of Delaware tests, 5- to 6-inch barley stubble was compared with 10- and 15-inch grain stubble heights. The taller stubble boosted bean yields by around 5 bushels per acre, caused beans to grow taller and contributed to rapid shading of the ground. High stubble also caused soybean pods to develop well above the ground, making bean harvest easier and reducing field losses.

8 *Row spacing is critical.* Delaware agronomists recommend planting no-till double-crop beans in 20-inch or narrower rows. This gives a good leaf canopy early for weed control. With 20-inch rows, you need five plants per foot of row and four plants with 15-inch rows.

9 *Two biggest sprayer concerns.* Avoid brass nozzles and worn tips, says Don Daum, Pennsylvania State University ag engineer. He recommends no-tillers use flat fan stainless steel tip nozzles positioned on 20-inch centers.

10 *"Knock-down" option.* University of Illinois agronomists have found combining Dow General herbicide with diesel fuel can work as a "knock-down" herbicide instead of using Roundup or paraquat for no-till soybeans planted in wheat stubble.

11 **New herbicides coming.** Pay careful attention to what new compounds and tankmixes can do for double-crop weed control.

12 *Consider coulter width.* Russell Hughes of Hillsboro, Ill., uses a 1-inch fluted coulter when double-cropping beans. "I can drive faster using a 1-inch wide coulter than most people using the 2-inch wide coulter," he says. "The 2-inch coulter throws out more dirt, which means the press wheel doesn't have enough dirt to cover the beans."

CRUNCHING DOUBLE-CROPPING NUMBERS

In Kentucky, no-tillers normally seed more full-season soybeans than double-crop beans since the yield reduction with double-crop beans is about 20%.

Back in 2012, University of Kentucky farm management specialist Greg Halich worked out budgets for full-season and double-crop soybeans for southwestern Kentucky growers. Including an adjustment for new crop basis, these budgets were based on $4.40 corn, $10.40 soybeans and $7.70 wheat.

The budgets accounted for additional costs associated with double-cropping, including fuel, machinery repairs and depreciation. Cash rent was set at $175 per acre, diesel fuel at $2.75 per gallon for 15-mile, one-way trucking, 40 cents per unit of nitrogen, 45 cents per unit of phosphorus and 40 cents per unit of potassium.

Around Hopkinsville, Ky., where double-cropping is popular among no-tillers, Halich assumed 70-bushel wheat yields, 35-bushel, double-cropped soybean yields and 44-bushel, full-season soybean yields.

With these budget figures, a wheat and soybean double-cropping scenario was the best moneymaker. There was a net profit of $230 with wheat and double-cropped soybeans, compared to only $79 per acre with full-season beans.

Put another way, the double-cropped soybean yield would have to drop to only 20 bushels per acre before full-season beans were as profitable.

Halich also worked out a budget for various crop combinations for northwestern Kentucky, where little wheat is typically seeded, using 65-bushel wheat yields, 38-bushel, double-cropped bean yields and 50-bushel-per-acre yields for full-season soybeans.

For this area in 2012, there was a $228 per acre net profit for wheat and double-crop soybeans compared with $140 with full-season beans. ❋

NO-TILL-AGE®

July 2005

Full-Season vs. Double-Cropped Soybeans
(Based On 2004 U.S. Acreage)

Soybeans	Acreage	% No-Tilled
Acres	76,013,582	39%
Full-season acres	71,414,741	36%
Double-cropped acres	4,598,841	72%

—*2004 National Crop Residue Management Survey, Conservation Technology Information Center, West Lafayette, Ind.*

NRCS

Frank Lessiter

MONEY-MAKER. In 2012, Kentucky ag economists determined that wheat and double-cropped no-till soybeans earned an extra $88 per acre compared with only no-tilling full-season soybeans.

189

Interseed Second Crop Before Harvesting First One

"Interseeding a cash crop or a cover crop into a growing crop is among the hottest farming ideas in some time…"

Although double-cropping offers no-tillers the chance to harvest two crops in a year from the same field, a yield penalty of as much as 20% on late-planted soybeans can't be ignored. There's also the risk of dry conditions that take place in late spring or early summer impacting no-tilled soybean yields.

Other terms used to describe this interseeding system include interplanting and modified relay intercropping (MRI). Regardless of the system's name, the goal is to have a well-established soybean plant that's already 6-8 inches tall when wheat or barley is harvested.

Even in the 1970s, no-tillers were looking at interseeding soybeans into growing wheat to boost yields, eliminate equipment issues with heavy straw and reduce the risk of hairpinning residue in the seed furrow.

Interseeding allowed growers to often turn out bean yields equal to full-season soybeans while harvesting two crops a year from the same field. It reduced soil erosion, led to less compaction and eliminated the need to burn straw or use herbicides to kill existing vegetation.

RELAY CROPPING. In the early 1990s, soybeans were no-tilled into wheat growing in 7 1/2-inch rows when development of the grain crop was in the medium to hard dough stage. Seeding soybeans normally takes place about 10 days prior to wheat harvest, say Clemson University agronomists.

Searching for Answers

In 1993, Clemson agronomist Jim Palmer and ag engineer Clarence Hood began searching for the best way to adapt equipment for interseeding. At the Edisto Research and Education Center in Blackville, S.C., their plots had 11 rows of winter wheat spaced 13 inches apart while leaving a pair of 24-inch wide non-seeded zones for wheel traffic. Eight rows of soybeans were interseeded between the wheat rows 3 weeks ahead of grain harvest.

Used with both crops, a modified grain drill was mounted on the three-point hitch of a 60-horsepower tractor. It had 76-inch center wheel spacings to match the rows and tractor wheel tracks.

At the Blackville location, the researchers found interseeding had no impact on wheat yields. However, 2 years of trials at Dillon, S.C., and Griffin, Ga., showed wheat yields dropped as much as 25%. Even so, the interseeded soybeans yielded significantly more than conventional double-cropped soybeans under both irrigated and dryland conditions.

Modified Relay Intercropping

In Ohio, a similar system where soybeans are no-tilled into standing wheat up to 30 days ahead of small grain harvest is called MRI. Long-

MODIFIED DRILL. Mounted on a three-point hitch toolbar, Clemson University's 8-row modified grain drill could effectively no-till either soybeans and wheat with only a few minor adjustments.

DOUBLE, TRIPLE DUTY FOR HIGH CLEARANCE UNITS. No-tillers not only use high-clearance rigs to apply herbicides insecticides and fungicides, but also to apply fertilizer and seed cover crops.

TIMING IS CRITICAL. Beans seeded 20 days before wheat harvest led to the best yields.

"Interseeding annual rye grass into V4-V7 stage corn, **led to a 5-bushel bump in soybean yields…**"

term Ohio State University research in Wooster and other Ohio locations has shown MRI wheat will yield about 90% of conventional wheat that hasn't lodged. Ohio agronomist Jim Beuerlein found yields were similar for 7- and 10-inch wheat rows. The difference between seeding wheat in 7- and 14-inch rows was 3 bushels per acre.

Soybean production in the MRI system was shown to be more speculative than wheat due to the need for adequate rainfall in July and August. Results from the Ohio research plots averaged 34 bushels of soybeans per acre while wheat averaged 71 bushels per acre.

A lack of light often has a profound effect on the amount of growth with intercropped soybeans. Soybeans planted too early into growing wheat often become tall and spindly while weak plants normally don't grow well. In the Ohio trials, MRI soybeans interseeded 20 days prior to wheat harvest offered the most consistent yields.

HIGH CLEARANCE SEEDING. Georgia researchers added Tye planter units to a high-clearance sprayer to interseed twin rows of double-crop soybeans into growing corn in the early 1980s.

John Woodruff

SEEDING COVER CROPS INTO STANDING CORN. The Sussex Conservation District in Georgetown, Del., purchased a Miller Nitro air seeder to help area growers evaluate the interseeding of cover crops. In 2015, the unit was used to interseed 4,000 acres of cover crops from late July to mid-October.

Interseeding Cover Crops

Interseeding has also become an invaluable tool to seed cover crops into standing crops, especially with shorter growing seasons in northern climates.

In a 2014-15 cover crop survey conducted by the Sustainable Agriculture Research & Education group, 21% of cover crop users named establishment as their number 1 challenge. This was followed closely by the amount of time and labor that was required for seeding and managing the cover crop.

Thanks to accurate placement with GPS guidance and improved seeding technology, no-tillers are seeding annual ryegrass, cereal rye, clover and other covers into growing soybeans and corn. While seeding has typically been done with drills, highboy seeders with drop tubes mounted on self-propelled sprayers are used to seed covers into corn as tall as 6 feet.

In 2012, jaws dropped at a field day in Pennsylvania when an unmanned machine dubbed the "Rowbot" seeded cover crops as it traveled between rows of 30-inch corn rows. Use of the 22-inch wide machine removed crop height as an obstacle to getting covers seeded

STRONG START. For Ontario no-tiller Gerard Grubb, interseeding is the only way to get cover crops established at his northern location. Annual ryegrass increases soil tilth and soil structure.

by traveling along the ground.

Robotics aside, no-tillers are finding plenty of success interseeding with more basic equipment.

Spinning on Seed

Loran Steinlage started interseeding covers on his no-till and strip-till ground in West Union, Iowa, with a fertilizer spinner while the corn was still small and growing. Since then, he's modified a half dozen toolbars

for interseeding. He's now using Dawn Biologic's DuoSeed Cover Crop Inter-Seeder row units to interseed annual ryegrass and clovers into V4-V6 stage corn.

No-tiller Allen Dean of Bryan, Ohio, tried interseeding cover crops with a helicopter in late summer, thinking it would do a decent job of distributing the seed. When that wasn't the case, he had an airplane fly on later-season cover crops for

"An unmanned robot seeded **cover crops** between 30-inch rows of corn…"

several years, but still wasn't happy with the results.

"In our area there are a lot of homes, odd-shaped fields and trees," he says. "We knew that if we had a ground rig, we'd be able to do a better job of distributing and getting the seed exactly where we wanted it in the field."

Dean's self-propelled sprayer has a front-mounted boom that was modified into an inter-row air seeder that leads to near perfect cover crop establishment.

In Indiana, Ag Conservation Solutions no-till consultant Dan Towery is conducting interseeding cover crop trials into corn at the V4-V7 stage. While they haven't yet gathered enough data to provide good yield comparisons, a group in Quebec that is trying the same idea is reporting a 5-bushel bump in the following soybean crop's yields, as well as a 4-8-bushel per acre increase for corn.

Ontario no-tiller Gerard Grubb has noticed an increase in soil tilth and improved soil structure from interseeding cover crops. With his northern location and short growing season, interseeding is the only way he's able to establish an adequate cover crop.

Early vs. Late

No-tillers like Steinlage and Grubb prefer to interseed cover crops when corn plants are in the V4-V7 stage. Towery says the key for success is to seed cool-season cover crops that will grow several inches tall while the corn is still small. Once the corn shades the ground, cover

TWO TASKS IN ONE TRIP. Many no-tillers have shown a keen interest in combine modifications that enable growers to seed cover crop mixtures in a timely manner while harvesting corn.

crop growth slows considerably and becomes what Towery refers to as semi-dormant. It's just sitting there, not using a lot of water and not using much nitrogen.

However, a concern with seeding this early is whether the cover crop will steal water or nutrients from the no-till corn crop. The point of inter-seeding into corn plants at the V4 stage is to provide enough sunlight to establish the cover crop. Crop competition shouldn't be a problem unless no-tillers seed covers prior to the V2 stage.

If a no-tiller wants to use a cover

crop such as cereal rye that may not survive early shading conditions, later interseeding may be a better option.

That's the case for Mike Werling, of Decatur, Ind., who credits seeding cereal rye with being a great interplant option for no-till soybeans while creating a large amount of biomass to help suppress weeds. Based on his own experiences, he prefers to seed covers when corn is in the black layer stage since that's when corn starts to drop its leaves, which provides needed sunlight to the cover growing crops. ❁

BEFORE HARVEST, NO-TILL SECOND CROP

INTERSEEDING COVER CROPS INTO V4 STAGE CORN. By interseeding cover crops at this stage of development, there's enough sunlight to establish a cover crop. Once the corn shades these plants, the cover crops move into a semi-dormant stage and don't steal nitrogen or water from the corn.

May 1976

Interseeding soybeans, grain sorghum or cover crops into standing small grain is one of the hottest farming ideas to come along in some time. A number of farmers have made the idea work and more farmers are giving interseeding a try.

For these innovators, interseeding means seeding a crop such as soybeans or grain sorghum into standing wheat or barley 4-6 weeks ahead of small grain harvest. This gives the second crop a week head start, which can be crucial in years when an early frost hits.

The second crop may be seeded by airplane, helicopter, no-till planter, or with a no-till grain drill. While the tractor or drill tires may trample some small grain, yield hits are minimal.

Boost in Net Income

Our *No-Till Farmer* editors surveyed a number of farmers who had experience with interseeding during 1975 or 1974. Some reported interseeding successes while others reported failures.

> "Interseeding success boils down to **time** of seeding, population and weed control..."

When Jack Umholts of Earl Park, Ind., interseeded 150 acres of soybeans into wheat, he ended up with 63 bushels of wheat and 37.5-bushel soybeans per acre. Gross income per acre was $471, with a net income of $383.16.

Richard Schmidt of Bremen, Ind., flew soybeans into wheat 6 weeks before harvest. He figures this was worth an extra $75 per acre in income. If he had waited until the wheat was combined before no-tilling soybeans, the beans would not have ripened before an early frost.

Richard Quinton of Heyworth, Ill., harvested 62.5 bushels of wheat and 17.4 bushels of interseeded soybeans per acre last year. The soybeans had been aerially seeded on May 30. He cut the wheat stubble high since the beans were already a foot tall at wheat harvest.

"Last year we planted soybeans on May 2, May 15 and May 24," says Umholts. "I'd prefer to start after May 20 if ground conditions are right. Planting too early means cutting off 2-3 leaves off the soybean plants at wheat harvest, which hurts yields. The beans also do

"Frost hazards, yield reductions due to late planting and a higher probability of rain in June have encouraged soybean interseeding…"

much better on our low organic matter soils since the wheat plants seem to stand much better."

Theodore Sinker of Highland, Kan., plans to go back to conventional double-cropping. His interseeded soybeans yielded only 12 bushels per acre after 60-bushel wheat.

Warren Buhler of Rushville, Ind., says switching to barley will enable him to no-till beans in mid-June without interseeding.

Buhler interseeded soybeans 5 weeks ahead of wheat harvest for 4 years. He used an 8-row, 20-inch no-till planter with the second and seventh planting units removed to leave space for the combine wheels.

To avoid knocking down

NUMEROUS COVER CROP SEEDING OPTIONS. Planters, drills, robots, airplanes, spinners and helicopters are among the options no-tillers are relying on to seed cover crops in a timely manner.

wheat, he switched out the tractor tires for old-style 1-inch wide steel lugged skeleton wheels, which led to much less wheat damage. Two weeks after planting, you couldn't see where Buhler had driven through the wheat.

Farmers were also asked to share their keys for successful interseeding. Lloyd Younger of Bethany, Ill., says it boils down to time of seeding, population per acre and weed control. Joe Rush of Severance, Kan., says there must be ample moisture so the beans germinate before the soil dries out.

Umholts recommends using a helicopter for seeding, as an airplane distorts the air pattern and leads to poor coverage. With the helicopter, he flies on 1 bushel per acre of soybeans east and west and in another north and south trip seeds another bushel to get excellent stands.

Interseeding Grain Sorghum

Mark Larson interseeded a medium-to-early maturity grain sorghum hybrid on July 10. The goal was to harvest three crops from the land in a year's time if the weather cooperated — settling for two crops if the weather turns extremely dry.

He got a 50-bushel wheat yield, cut 2,000 pounds of grain sorghum per acre and harvested the stover for his cows.

This year he will interseed an earlier grain sorghum hybrid on July 1. If frosts nip the grain sorghum before it's completely ripe, he'll cut the crop as stover.

Why Farmers Interseed?

Frost hazards, yield reductions due to late planting and a higher probability of rain in June than in July encouraged many people to interseed soybeans into small grain. Yet University of Illinois agronomist W.O. Scott maintains most problems associated with interplanting have not been solved.

When seeding by air or broadcasting with ground equipment, seed is left on top of the soil. A timely rain or very heavy dew is a must for germination.

Some farmers have drilled no-till soybeans into standing wheat. While this reduces the need for rain, it doesn't eliminate the need for moisture.

Another question concerns the best time to interseed, as one of the obvious advantages of intercropping over double-cropping is early planting.

"Soybeans can be interseeded too early," says Scott. "If they are blooming when the wheat is combined, they'll probably be so tall that the bean tops will be clipped off by the combine."

NO-TILL-AGE®

How Cover, Mulch and Manure Impact Earthworms

Field Condition	Weight of Worms per Acre
Fallow	170 lbs.
Cover	250
Mulch	850
Manure, mulch and cover	1,700

— Paul Reed Hepperly, Maryville, Tenn.

Soybean yields as high as 45 bushels per acre have been reported when planted immediately after corn. Yet this double-crop scheme appears to have limited potential due to ultra-late no-till soybean planting after corn.

Never Give Up

In this area, early maturing corn can't be harvested before mid-July and soybeans should be planted before August 1. So there's only a 2-week period for harvesting corn, preparing land and planting soybeans. And there is usually lots of rain in southern Georgia in late July.

Yet John Woodruff, a University of Georgia agronomist stationed at Tifton, Ga., feels relay planting of soybeans prior to corn harvest might overcome these double-cropping limitations.

He reasoned that no-tilling soybeans with a high-clearance tractor could overcome the "time bottleneck" for getting beans in the ground during a more favorable period. Woodruff also reasoned it could be beneficial to leave corn residue on the ground to reduce high soil surface temperatures and conserve soil moisture on extremely hot summer days.

To overcome soil compaction, the plan was to design a planting unit, which could no-till soybeans close to the old corn rows. Since area farmers use in-row subsoiling to disrupt soil compacted layers when planting corn, planting close to the corn plants would hopefully allow the soybeans to utilize the original subsoil furrow.

Using a high-clearance tractor from Powell Manufacturing, the researchers designed a high-clearance unit to no-till twin soybean rows along each former corn row. A Tye drill was modified with planting units sitting on 2-inch stilts and a fluted coulter was used to open each seed

INTERSEED BEANS INTO GROWING CORN

John Woodruff

AHEAD OF ITS TIME. In the 1980s, Georgia agronomists built this interseeding rig to extend their double-crop soybean growing season.

To avoid a "time bottleneck" in harvesting corn and planting double-crop soybeans, researchers in south Georgia are planting the second crop before harvesting the first. Down in south Georgia, double-cropping irrigated soybeans behind corn is among the hottest farming trends to come along in some time.

> "Relay intercropped soybeans no-tilled 7 days later yielded the best..."

furrow and to slice through surface trash. Seed was placed in the furrow with a double disc opener, then covered with a narrow 2-inch Tye packer wheel. A small hydraulic motor was used to drive the Tye drill box and deliver seed to the openers.

Pioneer 3369A corn was planted on March 6 in 36-inch rows. Soybeans were no-tilled into the standing corn on July 13 and 20 and beans were seeded at a rate of 75 lbs. per acre.

Later Planting Better

The soybeans no-tilled on July 20 yielded much better than beans interseeded on July 13. The earlier planted beans averaged 9.2 bushels per acre while July 20 beans averaged 18 bushels per acre.

In the July 20 trials, yields averaged 19.9 bushels per acre where nothing was done to the corn stalks, 18.2 bushels per acre where corn stalks were topped and 15.9 bushels when the stalks were removed.

NO-TILL-AGE®

September 2004

What's 2 Inches of Extra Water Worth with No-Till?
(Based on availability during the growing season)

Crop	Increased Yield
Corn	20.8 bu.
Wheat	12.6
Soybeans	10.7
Cotton seed	7.1
Sunflowers	13.4

— *Central Great Plains Research Station, Akron, Colo.*

"We had trouble getting and keeping a soybean stand with relay planting, especially on the earlier planting day," concludes Woodruff. "Sencor herbicide injury was much greater than normal, which may have resulted from soil compaction or weak plants as a result of stem etiolation."

NO-TILL LOOSE ENDS...

HOW 'SWEET' NO-TILL REALLY IS

Since 1973, Whiteford Packing Co. has been no-tilling sweet corn, soybeans, cover crops and sweet peas.

April 1983

"Over 800 acres of yellow and white sweet corn varieties were no-tilled last year with a John Deere 7000 8-row planter with outstanding results, considering our dry growing season," says Bill Hanna, Jr.

The Whiteford, Md., vegetable grower no-tilled soybeans with a Haybuster 1068 drill after baling clover and timothy hay or seeding immediately behind the pea harvesters. Yields of double-cropped beans averaged 35-50 bushels per acre. Hanna has now shifted from using V-press wheels to a single press wheel on the drill.

Sweet peas were no-tilled into soybean stubble with comparable results to a conventionally-tilled crop.

Hanna cites 13 reasons why he likes to no-till vegetables:

1 Conserves soil and reduces erosion by up to 90%, saving 5 to 8 tons topsoil per year.
2 Better utilizes available moisture.
3 Improves labor efficiency by 250%
4 Reduces fuel usage by 40%
5 Saves time.
6 Increases yields.
7 Makes double-cropping easier and allows use of erodible land that can't be effectively farmed in any other way.
8 Enjoys better harvesting conditions.
9 Keeps nutrients at target levels by reducing costly leaching.
10 Increases organic matter, thus boosting nutrient holding capacity.
11 Saves horsepower, hardware and machinery investment.
12 Prevents or restricts lodging.
13 Saves mental and physical stress.

Innovative No-Tillers Rekindled Interest in Old Cropping Standbys

"No-tillers are always looking for fresh ideas that can offer **a competitive edge with their cropping operation**..."

As growers in the 1970s and 1980s gained experience with no-till, some innovators found new ways to combine no-till with other cropping options. The goal was to find new ways to conserve valuable moisture, especially in areas with limited rainfall, and pump up both yields and farm income.

Even today, innovators are still looking for ideas that can give them an edge with their specific farming resources, geographic location, management abilities, cropping history, equipment and weather conditions.

In this chapter, we'll look at a half dozen cropping ideas that no-tillers have adopted for their own farms as reported on in *No-Till Farmer*. These include stories on how readers made no-till work with skip rows, strip intercropping, polymer seed coatings, companion cropping, twin rows and ecofallow programs.

Boosting Yields with Skip Rows

No-tillers in the semi-arid Great Plains face unique challenges in raising corn with often meager moisture. One option is skip-row planting, which involves leaving some rows un-

SKIP-ROWS BOOST DROUGHT TOLERANCE. Skip-rows prevent corn plants from using all of the available moisture too early in the growing season. The skip row corn at left is compared with traditional corn at right in these 2012 photos, which proved to be an extremely dry year in western Nebraska.

planted, while increasing the overall plant population.

The idea behind the skip-row system is to prevent plants from using all available soil moisture too early in the growing season. Because water in the soil between widely-spaced rows can't be reached by the crop until later in the season, more water is available in July and August.

Since no-till residue retains more moisture, skip-row planting allows the roots to reach additional water later in the season.

Planting Goof Pays Off

A malfunctioning planter in 2003 led Jack Maranville to discover the benefits of skipping rows in his corn fields.

"I'd done five rounds before noticing something was wrong with the planter and it had skipped a row," says Maranville, who grows corn and sun-

flowers on 10,000 dryland acres near Matheson in east-central Colorado.

"But when I was walking that field just before harvest, I noticed that on each side of the skip, every corn plant had an ear on it. I joked with my brother that maybe I should have planted the whole field like that," he recalls.

Over that winter, he heard University of Nebraska agronomists were researching skip-row corn and he investigated the idea. The Nebraska researchers convinced him to make a serious attempt at skip-row corn on his farm, which rests about 6,000 feet above sea level.

Maranville had already moved into no-tilling, but found it didn't solve one of his major soil problems.

"Some people say when you've been no-tilling for a few years, compaction just goes away," he says. "Well, it doesn't in our area. There's pretty much agreement that there's a hard pan out there, and we still had it."

Maranville believed he had to break up the underlying hard pan. He also wanted to increase efficiency by placing fertilizer 7 inches deep.

He had read about strip-tilling but

NO-TILL-AGE®

July 1989

How One Farmer Compares Tillage Systems
(Total costs per acre)

Tillage System	Corn	Soybeans
Ridge-till	$304.11	$226.05
Conventional tillage	$342.73	$251.96
No-till	$305.67	$235.45

— *Gayle Goold, Fairbury, Ill.*

didn't believe it was appropriate for his dryland fields. So when a neighbor tried strip-tilling, Maranville paid close attention and later tried it himself.

The first time he strip-tilled corn, Maranville says his 230-horsepower tracked Caterpillar tractor had all it could handle pulling 8 rows at 5 mph. The next year he seeded sunflowers into corn residue by splitting the old rows — and found the job required 40 less tractor horsepower.

The results convinced Maranville about the value of strip-tilling. The knives bring up moisture, producing a nice planting bed and,with the hard pan broken, sunflower roots go straight down.

"I call strip-tilling the best of both worlds," he says. "It gives us 9 inches of worked ground, and 21 inches of no-till."

The Next Step

With strip-tilling established and the ongoing drought putting a premium on soil moisture, Maranville turned his attention to skip-rows in 2004.

"We planted two rows and skipped two," he says. "We had been using a 15,000-16,000 corn population, and now we're at a 12,000 population. Within the rows, it's actually a rate that would equal 24,000 plants per acre if the skip-rows weren't there."

Skip-rows maximize limited soil moisture, but Maranville worried about what would happen if the season turned wet.

"Lo and behold, the first year we did this, it rained," he says. Some 14 inches of rain fell between the middle of June and the first of September, slightly more than what's expected as normal for the full year.

"In fields that we strip-tilled, there was never any standing water," he says. "If it rained an inch and a half, we

NO-TILL-AGE®

July 1983

Trips, Fuel Needs for Tillage Systems

Tillage System	Field Trips	Diesel Fuel per Acre
No-till	3-5	1-3 gal.
Minimum till	5-8	3-5
Conventional till	8-11	5-8

— No-Tillage Farming/Minimum Tillage Farming Book

might see a little water out there right after it rained, but we'd come back 20 minutes later and it had disappeared. That really sticks out in my mind."

Skip-Row Results

Pointing to his goals for skip-rows, he says, "We thought that if we could gain the consistency to raise 40-60 bushels of corn per acre every year under these extremely dry conditions, we could make corn work. Last year, we had some 100-bushel corn in the skip-rows. If the temperatures had been warmer, we'd would've had some 120-bushel corn."

Adding corn to his rotations offers numerous carryover benefits. He harvested an extra 300-400 pounds per acre of sunflowers behind corn than behind any other crop. Skip-rows also save 25% on seed costs and about that much on nitrogen.

"Maybe this is just a High Plains phenomenon, especially in dryland corn," he says, "but I think it has great possibilities wherever there is drought sensitivity. It keeps the corn from using all the soil water too soon."

"We planted about 1,200 acres of corn in skip-rows last year, and it didn't let us down." ❋

September 2005

RESEARCH SUPPORTS SKIP-ROW NO-TILL

Skip-row farming in the Great Plains could provide higher yields in dry years based on studies done in a wheat, corn and fallow rotation, says Dale Fjell.

The Kansas State University agronomist says this idea works particularly well with Roundup Ready corn. Early studies indicate that skip-row corn won't harm yields in a good year, but that could help in a drought year.

"The idea is to plant the same number of plants per acre, just in one-third fewer rows," says Fjell.

Kansas researchers are also looking at no-tilling skip-row soybeans. However, Fjell cautions that this should not be confused with skip-row practices used in some Midwestern soybean fields where growers skip rows to save on seed costs and to leave unplanted rows for tractor and sprayer wheels to follow. ❋

GETTING FRIENDLY RESULTS FROM COMPANION CROPPING

THE RIGHT MIX. Nine or more species make up the companion-crop mix on the Robin and Kelly Griffeth farm, but cowpeas and buckwheat are primary components because they attract beneficial insects and harvest sunshine.

Another variation of intercropping adopted by no-tillers is companion cropping. It's where certain crops are planted in near proximity to each other with the belief that shared cultural benefits can be beneficial. These include better yields, pest control, pollination, soil health and increasing crop biodiversity.

When it comes to no-till, few do companion cropping better than Robin and Kelly Griffeth of Jewell, Kan. The father and son team double-crop sunflowers with multiple crop species to reduce inputs and enhance plant health.

With 2 years of companion cropping under their belts in 2013, the north-central Kansas farmers follow their hard red winter wheat crop with sunflowers intermixed with up to nine companion species selected for specific attributes.

Continuous no-tillers since 1995, they raise 3,700 acres of corn, soybeans, wheat, grain sorghum and sunflowers. They were already believers in cover crops when an unplanned "experiment" revealed the companion-crop potential.

> **"A 10-crop seed mix was planted after late June or early July wheat harvest..."**

"It was a fluke," Robin says. "I was planting buckwheat in July 2011 and had about 4 acres left with not enough seed to finish. I hate leaving blank spots in fields, so I mixed up some leftover sunflowers, winter peas, five clovers, canola, chickling vetch, safflower and a couple of different radishes to bulk up the volume enough to finish."

To his surprise, the sunflowers in the companion mix yielded well.

Designer Seed Mix

Kelly came up with a different companion mix the following year — common vetch, chickling vetch, spring forage peas, winter peas, cowpeas, crimson clover, oilseed radishes, purple-top turnips, buckwheat and sunflowers.

At a normal rate of 21,600 seeds per acre, sunflowers made up about 10% of the mix. Buckwheat had the largest number of seeds per acre at 64,800, with the other species ranging from 12,500-17,000 — for a total of 206,924 seeds per acre. The 10-crop seed mix was planted as soon as possible after late June or early July wheat harvest.

NEW PROFIT OPPORTUNITY. A slow-starter that's extremely aggressive in the reproductive phase, a sunflower crop rises above an intense companion mix that enhances plant health and contributes to higher oil premiums.

Using their White 6531 31-row planter equipped with Keeton seed firmers, Thompson closing wheels and 15-inch rows, they no-tilled sunflowers in 30-inch rows. The companion mix was planted in the centers as a row filler, nitrogen producer and beneficial-insect attractant.

In the past 2 years, they've planted nearly 1,000 acres of double-crop sunflowers and companion species. They also drilled 54 acres in 7 1/2-inch rows with a Great Plains CTA-4000HD rig. Seeds are no-tilled 1 1/2 inches deep, the optimum depth for sunflowers.

This year, Robin and Kelly are going to mix all of the seed together — including the sunflowers — in the planter boxes. They don't expect a perfect stand, but believe using a soybean plate and optimum air pressure will produce good results.

With moisture competition in a 25-inch annual-rainfall area, Robin finds sunflower roots extend much deeper in search of moisture than the roots of the companion species. At harvest, the Griffeths use Lucke Manufacturing sunflower pans mounted on a rigid Case IH head.

"Most of the taller plants are warm season and have been killed by frost by the time we harvest sunflowers," Robin says. "If a species is going to cause a harvest problem, it should be left out of the mix. The beauty of the flowers we plant is they get 6-7 feet tall."

Reduced Input Costs

For the Griffeths, companion planting eliminates the need for an insecticide and a broadleaf herbicide. At a cost of $15-$20 per acre each for insecticide and herbicide applications, Kelly says the companion-crop seed mix is basically paid for. The entire seed mix, including sunflower seed, costs $65 per acre — about the same cost as double-crop soybean seed.

The profit impact comes at harvest. "Yields are comparable to sunflowers without a companion crop — they're certainly not less," he says. "The real profit advantage is grain quality. The sunflowers have better plant health with far less insect damage and that gives us better oil content. With the companion crops, the oil premium has been the highest we've ever received."

In addition to higher grain sorghum and corn yields following sunflowers, the Griffeths add to their income by renting companion-cropped fields for winter grazing.

The Griffeths plan to extend their companion-crop concept to no-tilling corn in 30-inch rows with a companion mix of lupins, cowpeas and sunn hemp. It will have to be on double-crop corn, however, since crop insurance will not cover the practice.

"Making a profit is our primary purpose," Robin says. "We take care of our cash crops first, and anything we try has to pencil out. We'd love to try companion crops on full-season crops but, from a risk management standpoint, you don't want to make an insurable crop uninsurable."

POLYMER COATINGS PUSH SOYBEAN INTERCROPPING FARTHER NORTH

RUNNING START. Coated seed corn planted early (above) emerged as soon as conditions were favorable, compared with uncoated corn. At right, coated beans got off to an earlier start in wheat.

Since polymer-coated soybeans won't emerge for about 25 days, no-tillers can interplant the coated soybean seed in early May, when the wheat crop is still young and less susceptible to damage by tractor and planter wheels. By interplanting soybeans into growing wheat, no-tillers add 3-4 weeks to the bean growing season.

In on-farm trials, wheat averaged 58 bushels per acre and soybean yields ranged from 20-30 bushels.

Purdue University figures were used to calculate variable input costs based on $5.40 soybeans and $2.62 wheat. The relay intercropping returns averaged $32 per acre more for intercropped soybeans than full-season soybeans and $66 per acre more for intercropped wheat compared to single-crop wheat.

Wide-Row Concerns

Doug Cunningham, a farmer from Alvin, Ill., wasn't convinced relay intercropping would work. "When they recommended growing wheat in 15-inch rows instead of much narrower rows, that was definitely a concern in regard to yield," he says. "While I was worried about both weeds and yield, neither has been a problem."

Cunningham harvested 64 bushels of wheat and 21 bushels of soybeans per acre. While first thinking those yields were not very good, it penciled out as well as raising a 150-bushel corn crop. His goal is to raise 70-bushel wheat and 40-45-bushel soybeans with relay cropping.

By adding wheat to a no-till corn and soybean rotation, Britt Weiser maintains soil erosion could be reduced by 40-60%. If double-cropping of wheat with soybeans is found to be economically beneficial in Illinois, the USDA agronomist says this system should be promoted as a best management practice for reducing soil erosion and sedimentation.

Editor's Note: *When Landec was sold to Monsanto in 2006, the company entered into a 5-year polymer license with Monsanto for the use of the Intellicoat polymer seed coating technology. In recent years, the seed coating technology appears to have disappeared from the farming scene.* ✳

Around 2002, a soybean seed coating product was developed with the potential to allow northern no-tillers to cash in on the benefits of relay intercropping, which is also known as interseeding or interplanting.

Following 3 years of on-farm evaluations, Landec Ag of Monticello, Ind., introduced the Intellicoat polymer for soybeans. *(See page 92 for more on this seed.)*

Reacting to specific growing conditions, this polymer seed coating technology takes on moisture over time and regulates the timing of soybean germination. The soybeans are typically no-tilled at 200,000 seeds per acre 45-50 days prior to projected wheat harvest. Since the seed coating delays germination, the soybean plants are still relatively small and suffer little damage as wheat is harvested.

Claude Butt, a Landec Ag senior agronomist, says once the appropriate number of heat units have accumulated around the polymer-coated soybeans, germination begins. During application of the polymer coating, germination can be adjusted for a pre-determined number of days and also be treated with fungicide.

June 2010

REDUCING EROSION, HIKING YIELDS WITH STRIP INTERCROPPING

The concept of no-tilling two crops in alternating narrow strips is called strip intercropping. Even though the advent of herbicide-tolerant crops during the 1990s made managing these strips easier, the system works best for no-tillers with smaller acreages.

No-tillers who strip intercrop with corn do so to boost yields in outside rows. However, Craig Fleishman strip intercrops to reduce erosion from water moving down the gently rolling hills in his operation at Minburn, Iowa.

The veteran no-tiller and ridge-tiller started strip intercropping in 2000, alternating 12 rows of corn with 12 rows of soybeans. The strips break up the wind, rain and the resulting erosion that create gullies while the 12 rows of corn slow the flow of water moving down between the hills.

Reducing Soil Erosion

At two locations, Fleishman divides the ground into three blocks, stripping two blocks and no-tilling corn in the entire third block. He moves the block of corn each year.

"Neither farm is classified as highly erodible, but we tend to get some gully washouts," he says. "It's not practical to put a grass waterway in every spot where that could happen. The corn stalks in the strips break up the drainage pattern."

Roundup Ready and LibertyLink corn and soybeans make weed control in strip intercropping easier, but Fleishman has also strip intercropped successfully with non-GMO soybeans.

"You have to be on your toes to do it accurately," he says. "Roundup Ready, GPS and auto-steer make it a lot easier to strip intercrop than for guys who were doing it back in the 1980s."

He says strip intercropping is not widely practiced as no-tillers want to get over the ground faster. With 24-row strips, you're likely to lose much of the benefit of having higher corn yields with more outside rows.

Four rows of corn alternating with four rows of soybeans would provide the maximum number of edges in a field, but Fleishman finds 12-row-wide strips work well.

Strips Take Time

Strip intercropping offers dual benefits by boosting yields and preventing erosion, maintains Jim House. The Port Stanley, Ont., Canada, no-tiller began strip intercropping corn, soybeans and winter wheat 17 years ago.

Since wheat lowered the bean yields, he moved to stripping corn with soybeans. Three 21-inch corn rows were spaced in 7.5-foot wide strips. Both sets of outside corn rows consist of twin rows spaced 7.5 inches apart. Soybeans were planted in 15-inch rows in 15-foot wide strips.

CATCHING SUNLIGHT. Long-time no-tiller Jim House of Port Stanley, Ont., strip-intercropped 7 1/2-foot wide corn strips next to 15-foot-wide soybean strips.

House no-tilled a 32,000 corn population per acre in the center rows and 50,000 in the outside twin rows to maximize exposure to the sun. The narrower strips exposed more corn to the sunlight than wider strips to get the beneficial "edge effect."

Higher Corn Yields

House says corn in the strips consistently yielded 100 bushels per acre above nearby corn fields. Yields in some of House's strips ran as high as 400 bushels. From 1999 to 2008, corn grown in the strips averaged 275-300 bushels, thanks to more exposure to the sun along with higher plant populations.

Because the outside rows of corn shaded the soybeans, the yields were slightly lower than in a full field of beans.

House believes in strip intercropping, but says it's not for everyone, particularly no-tillers with thousands of acres to cover in a timely fashion. ✻

DOUBLING UP ON NO-TILL WITH TWIN ROWS

Improvements in planter technology in the '90s helped usher along a more revolutionary idea in raising no-tilled or strip-tilled corn — twin rows.

In the typical twin-row system, each set contains two rows normally planted 8 inches apart that are no-tilled on 30-inch centers, leaving 22 inches between the twin rows. Staggering seed placement in the opposing rows gives each plant the maximum possible space.

Advocates believe twin rows offers advantages over traditional row spacing, including more room for each plant, better use of sunlight and a quicker canopy that provides superior ground shading, better weed control, improved water conservation and reduced soil temperatures during the warmest months of the growing season.

Twin row advantages lead to a larger root system for improved uptake of moisture and nutrients, as well as better stalk size and improved standability in corn. Allowing for higher plant populations, twin rows offer farmers significantly higher yields than traditional 30-inch rows and thus more profit, advocates say.

Twin Win

The quest for bigger yields and healthier plants led Ohio no-tiller David McNeilan to twin-row corn and soybeans to take advantage of higher populations and a quicker crop canopy.

McNeilan wanted to give his corn a little more room to breathe — and tap into more nutrients, moisture and sunlight. But he also wanted more plants per acre to chase higher yield goals.

A no-tiller for 15 years on 1,000 acres near Celina, Ohio, McNeilan has no-tilled twin-row corn and soybeans for several years.

"Higher populations are necessary if we're going to continue to increase yields, but you can only put so many plants in a row before they start competing with each other," McNeilan says. "Twin rows seemed like the best solution for us."

TWIN-ROWED CORN. Instead of typically no-tilling a row of corn on 30-inch centers, two staggered rows are planted 8 inches apart.

His planter is set up on 30-inch centers with an 8-inch spacing between the twin rows. That leaves 22 inches of spacing between each set of paired rows.

At 38,000 corn plants per acre, the spacing within each row is about 11 inches compared to about 5.5 inches with 30-inch rows. Seed placement in the twin rows is staggered, forming a triangular pattern designed to maximize room for root development and sunlight utilization, while minimizing evaporative moisture loss and weed growth.

When dry weather compromised yields last year, McNeilan says his twin-row yields held up better than most. He had some yields in the 120-175-bushel range, when much of the corn in his county made only 40-50 bushels per acre. He's not sure it can all be attributed to the twin rows, but the plants had more root mass and the quicker canopy may have helped conserve moisture at the right time.

More Twin-Row Benefits

Even though some dry summers have eaten into yield prospects, McNeilan has observed important differences.

"The corn has thicker stalks and larger root balls," he says. "It's hard to say what that translates into in terms of yields, but it tells me those plants are healthy and doing well." Thicker, sturdier stalks should also lead to fewer standability issues under poor weather conditions.

The twin rows pretty much harvest like single rows on 30-inch spacings. The combine snouts just crowd the two rows in a little bit from the outside.

Although McNeilan expects to see the biggest yield impact on corn, growing twin-row soybeans lets him use the same no-till planter, although he hasn't significantly increased his soybean planting population of 175,000-185,000 seeds per acre. However, he believes twin rows help increase air movement between the soybean plants and may help avoid disease problems. ✺

STRIKING A COMPROMISE WITH ECOFALLOW

Yet another cropping system to emerge from the reduced tillage movement in the U.S. is ecofallow — representing the period between winter wheat harvest and no-tilling a crop the following spring into undisturbed stubble.

The concept evolved several decades ago from a winter wheat, grain sorghum and summerfallow rotation where a crop was only grown in two out of three years in the semi-arid Great Central Plains. By not growing a crop one year, the idea was to store moisture in the soil profile for use during the next two years.

In the year when no crop was grown, limited tillage was originally used to control weeds. The concept took on a new life with the advent of no-till and chemical weed control, which made it possible to capture larger amounts of moisture and control weeds. Folks soon started referring to ecofallow as chemical fallow or chemfallow.

A study by the University of Nebraska of the long-term effects shows growing two crops in this system, such as grain sorghum and corn, is important in preventing the buildup of plant diseases that commonly occurred with continuous wheat under reduced tillage.

Mark Watson, a no-tiller from Alliance, Neb., who chaired the Panhandle No-Till Partnership, says producers moved away over the past few decades from a wheat and summer fallow production system where the only crop grown is a cool-season grass. The next step was to move to ecofallow where there's a cool-season grass such as winter wheat followed by a warm-season grass such as corn or proso millet.

They went from harvesting a wheat crop every other year, to ecofallow with crops grown in 2 out of 3 years and finally to continuous no-till where a crop is grown every year.

Ecofallow Benefits

Gail Wicks, a University of Nebraska agronomist at the West Central Research and Extension Center in North Plate, Neb., points to four key developments that contributed to the success of ecofallow:

◆ Improved management practices conserve an average of 2 inches of soil water each year, a primary reason behind increased yields.

◆ The need to control weeds within 4 weeks of harvest to preserve already-saved water.

TWO CROPS IN THREE YEARS. With chemical ecofallow, Great Plains growers normally have enough stubble to conserve around 2 inches of moisture per year and shade out weeds.

◆ Learning that weed control can be enhanced by selecting competitive wheat varieties.

◆ Refining chemical application to boost weed control while reducing the environmental impact.

With continuous no-till on dryland acres, Great Plains growers follow crop rotations similar to those that irrigated producers have used over the years.

With increased knowledge emerging about the environmental factors that affect soil biological activity and soil moisture, many growers in the semi-arid Great Plains are abandoning fallow altogether in favor of continuous no-till with cover crops.

The goal is to keep a living root in the ground year round to improve soil organic matter, boost water infiltration and retain more soil moisture than with fallowed fields. ✸

CONVENTIONAL TILLAGE. Some 200 tons of topsoil per acre were lost from this conventionally-tilled summer fallowed field following 6 inches of rain that fell in October of 2016. This November 8, 2016, photo shows the results after highway crews scraped topsoil off a road near Colfax, Wash.

Guy Swanson

Soil Losses Can Top
$24,000 per Acre

"Multiply a loss of **$75 of wheat per acre by 20 years and** excessive tillage is very costly…"

I've made a number of trips over the years to visit with innovative no-tillers in the Palouse area that encompasses southeast Washington, northeast Oregon and north central Idaho. This is an area with highly fertile hills, limited precipitation, steep slopes and a tremendous potential for costly erosion. It's also a hotbed for fresh no-till ideas.

Despite these circumstances, the area's climate and highly fertile

Guy Swanson

NO-TILLING AFTER EXCESSIVE RAINS. Three miles away from the scene shown above, Ross Jordan bands fertilizer on November 8 into winter wheat no-tilled in September.

NO-TILL-AGE®

Corn Residue, Soil Losses Compared

Tillage System	Amount of Residue	Soil Loss	Erosion reduction from Plowing
Plow, disc twice, plant	6.3%	7.8 tons	—
Chisel plow, disc, plant	34.6%	2.1	73.7%
Disc twice, plant	20.6%	2.2	71.9%
Rotary-till, plant	26.4%	1.9	75.9%
Till-plant	33.6%	1.1	86.4%
No-till	38.9%	0.7	91.9%

— University of Nebraska

deep soils allow no-tillers to turn out wheat yields as high as 125 bushels per acre while farming these steep slopes.

In the southeastern Washington part of the Palouse, 2016 was the wettest October on record. Some conventionally-tilled summer fallow fields saw losses of as much as 200 tons of valuable black topsoil per acre after a 6-inch rainstorm. Much of this dirt flowed into the Snake and Columbia rivers, reports No-Till Innovator Guy Swanson of Exactrix Global Systems in Spokane, Wash.

What's Your Soil Worth?

Topsoil selling at a Wisconsin garden store is normally priced today at around 40 cents per pound. At a soil loss rate of 200 tons per acre, that's $160,000 in lost income.

"With excessive rains and conventional tillage, losses amount to $121 per ton of lost topsoil..."

The costly soil loss figures used here are adapted from information compiled in 2012 by Iowa State University ag economist Michael Duffy who looked at soil erosion costs on both a ton and per acre basis. In our analysis, we used an average soil erosion loss of 3 tons per acre in modifying the Iowa data to come up with the cost of losing each ton of eroded soil.

◆ $2.10 due to a loss of 2.32 pounds of nitrogen and 1 pound of phosphorus.

◆ $1.64 in water cleanup costs.

◆ $113.00 in landowner costs.

◆ $4.27 in reduced cash rent values.

These figures add up to $121 per ton of lost soil due to erosion. Multiplying it by the 200 tons per acre for soil lost in areas of the Palouse after this excessive 2016 rain, it adds up to an astounding $24,200 per acre.

More Rain, Higher No-Till Yields

Among area no-tillers, the excessive rains in the fall of 2016 represented extra income potential, as the average rainfall in this area of southeastern Washington is only 19 inches per year. These extensive fall rains in 2016 provided a new charge of water flowing through root and worm channels. The following year's crops should be excellent in this area where wheat yields of 125 bushels per acre are common.

The value of losing 200 tons of soil per acre likely leads to a loss of around 15 bushels of wheat per acre. With wheat at $5 per bushel, that amounts to $75 per acre in lost yields. Multiply that by 20 years and the yield loss to excessive tillage is extremely costly. ❋

"With excessive rains, no-tillers can expect to increase wheat yields..."

SAVE 19 TONS OF SOIL PER ACRE

What can happen if sloping ground isn't managed correctly? David Mussulman, an area conservationist for the Soil Conservation Service for a dozen 12 northern counties in Illinois, provided one explanation.

November 1972

As an example, he cites a farmer with a field of Tama silt loam that has a 300-foot long 6% slope. The grower is not contouring, but also is not farming straight up and down the slopes either.

5 or 24 Ton Soil Loss?

Mussulman says the universal soil loss equation shows that in an uneroded condition that soil has a tolerance — an allowable soil loss — of 5 tons per acre. That's how much soil can be lost and the farmer will still be able to maintain high production.

However, continuous corn grown with conventional tillage, including spring plowing, on that field will lead to a loss of 24 tons of soil per acre each year. No rotation more intensive than one year of row crops, one year of small grains and one year of meadow could keep soil losses within the allowable limits when using conventional tillage.

The key to minimizing soil loss is keeping crop residue on the surface. Referring again to the universal loss equation, Mussulman says that if the farmer in this example adopts a conservation tillage system that leaves 2 or 3 tons of residue on the surface at planting time, he can control erosion and grow continuous corn without any difficulty.

Contouring that moderate slope with conventional tillage would slice the soil loss in half. But if the farmer combines conservation tillage with contouring, he could grow continuous corn and keep the soil loss well below the allowable 5 tons.

20-Ton Losses Common

In Jo Daviess county, district conservationist Jerry Misek and extension adviser George Swallow vigorously promote conservation tillage in this area of northwestern Illinois. Where most of the farming is done on slopes, even small erodible channels cut into the soil by the erosive action of flowing water can result in a 20-ton per acre soil loss. With 185,000 acres of cropland and only about 25,000 acres in conservation tillage, there's still a long way to go in the county.

While conservation tillage is being practiced throughout the state, Illinois state resource conservationist Harold Poeschl finds it is more intensive in some counties. It is just as essential in central Illinois as in the hillier areas in the north and south. Level land can really suffer from wind erosion, especially if soybean land is fall-plowed and left uncovered. ✳

NO-TILL-AGE®
Soil Savings Compared to Fall Plowing

August 1983

Tillage System	Soil Saved per Acre
No-till	4.7 tons
Offset disc	2.9
Coulter, chisel	2.9

— Allen County, Ohio, Soil and Water Conservation District

NO-TILL LOOSE ENDS...

DOUBLING UP WHEN NO-TILLING SOYBEANS INTO CEREAL RYE

August 1983

When Tim and Dennis Flach no-till beans into fairly-mature cereal rye, they drive north and south across the field. Next the Montrose, Ill., farmers no-till beans again in a second trip across the field in 30-inch rows, this time running east and west.

They no-till beans at a rate of 45 pounds per acre on each trip, totaling 90 pounds per acre.

Frank Lessiter

No-Till Offers Huge Benefits
to Wildlife Interests

July
2001

"Renting farmland for hunting **can be a profitable enterprise and no-tilling** definitely promotes more wildlife..."

— *Ron Swindler*

Each year during pheasant and grouse hunting seasons, Ron Swindler's farm was becoming more and more like a shooting gallery. So the Mott, N.D., no-tiller decided to do something about it.

Swindler is among a growing number of no-tillers who sell hunting rights to their land. In his case, hunters pay a daily rate to hunt pheasants and grouse on his land. Some years there are even ducks to hunt.

Swindler allows hunting on CRP land and wetlands that are adjacent to 4,500 acres of no-tilled cropland, which are prime areas for pheasants.

"I have a pretty good hatch," he says. "The birds roost in the wetland areas and during the morning and evening hours come on to the farmland to feed on the standing stubble left from no-tilling."

100% No-Till for Years

Swindler no-tills spring and winter wheat, peas, flax, canola and sun-

211

SEED COVER CROPS DURING CORN HARVEST. Ray McCormick mounted this Gandy unit on his combine to seed a variety of cover crops over the full width of the corn head while harvesting no-till corn. Adding cover crops dramatically increased the wildlife numbers in his Indiana operation.

"Birds roost in the wetland areas and **then feed on the no-till standing stubble...**"

flowers. He's been no-tilling since 1976. By 1983, 100% of his land was no-tilled.

He switched to no-till because of the cost savings and the inability to find good labor. More importantly, he saw serious difficulties with his soil, which was made up of only 1% organic matter.

Most of Swindler's soil is now at 3% organic matter. Some is even better than the native sod, which contains 3.8% to 4.2% organic matter.

Out of Control

Swindler started charging hunters to use his land in the early 1990s. He had previously posted his land and hunters could use it only with permission.

"It just got out of control," he explains. "We had to do something with the hunters instead of fighting them. Charging for hunting rights eliminated a lot of the problem hunters."

The first month of pheasant hunting season, which runs from the beginning

of September through early January, is usually booked solid. From early November through Thanksgiving, Swindler's land is open for deer hunters.

It's The Thing To Do

Most of his neighbors are also renting land to hunters, and many are capitalizing even more on the opportunity. As an example, Swindler's nephew rents out a hunting camp while neighbors offer bed and breakfast facilities to hunters.

CRP PROVIDES EXCELLENT GAME BIRD NESTING COVER

July 2001

Cropland enrolled in the Conservation Reserve Program (CRP) provides excellent nesting cover for popular game birds, according to a 3-year study in St. Croix County in the northwest part of Wisconsin. And by no-tilling when bringing CRP ground back into production, the valuable wildlife habitat won't be lost.

The CRP program is building not only more, but better, habitat for ducks and pheasants as well as other wildlife, says Jim Evrard, a Wisconsin Department of Nature Resources wildlife biologist who headed the study on 1,000 acres of CRP land.

A series of chains and steel cables were pulled through grassy CRP land, flushing female ducks and pheasants from their nests. The nests, mostly for mallards and blue-winged teal, were measured and marked so researchers could later return and see how successful they were.

To maintain a stable population in Wisconsin, 20% of the nests for these species must be successful. The study exceeded that, with the success rate for the 3 years averaged out to 33%.

Crowing rooster pheasants were also counted and the number of crowing roosters — an index to the pheasant population — was 10 times higher in CRP fields than on surrounding cropland, Evrard says. Other species found in the CRP fields include gray partridge, jackrabbits, deer, mice, hawks, fox, coyotes and the rare short-eared owl. ✳

Swindler doesn't limit hunters on acreage alone. He looks at how many birds are in a particular area and will reduce the number of hunters if bird numbers are low. If an area is over-hunted, he steers hunters elsewhere.

Besides the standing stubble from no-tilling, Swindler does other things to keep the pheasants coming back. He makes grain screenings available in the winter and will put out water during extremely dry weather.

Some neighbors put out food plots, but Swindler doesn't find increased numbers from doing it.

"It really doesn't seem like it draws the birds in," he says.

Swindler is pleased with the results of renting out his no-tilled land and is happy to make it available to hunters.

NO-TILL-AGE®
Controlling Mice in No-Till Corn

January 1987

Treatment	Corn Yield
No treatment	104
Zinc phosphide in furrow	122
Measurol (8 ounces per bushel)	119
Grass cut for hay before no-tilling	119

— University of Tennessee

"I've always been a hunter myself," he says. "My family were all hunters and no-tilling your ground definitely promotes more wildlife."

NO-TILL GOES HAND-IN-HAND WITH WILDLIFE MANAGEMENT

Ray McCormick

NO-TILL, DEER, DUCKS AND WATERFOWL. Instead of enjoying raising livestock in a no-till program, Ray McCormick has instead concentrated on providing a super environment for deer and waterfowl. As a result, he operates a profitable sideline hunting business on his 100% no-tilled ground.

In a perfect world, Ray McCormick could strike a lasting balance between wildlife habitats and his no-till operation. For now, the No-Till Innovator award winner veteran from Vincennes, Ind., is doing the best he can to encourage higher wildlife numbers with programs designed to protect streams and waterways with grasses, trees and buffers to reduce runoff and erosion.

In addition to a sizeable corn and soybean acreage, he runs a profitable deer and wildlife hunting operation, thanks to the extra cover offered with no-till.

"When your farm reflects the care and the interest you have in natural resources, to me, that's the pot of gold at the end of the rainbow," says McCormick. "We can make an investment that will pay off in extra resource protection for future generations."

He finds it rewarding to see how no-till works hand in hand with wildlife habitat management practices.

"We have successfully achieved crop yields well above our county average," says McCormick. "Meanwhile, our mallard duck populations are among the highest in North America."

Wild Livestock

When evaluating the benefits of adding animals to a no-till program, McCormick has done it with wildlife rather than livestock.

"Building up my land to have wildlife is a huge asset to me," he says. "With deer and ducks, you don't need an electric fence to have animals on your farm. The wildlife go where they want to go and run all over the place. Plus, my landlords love to deer hunt."

While some no-tillers put out food plots for wildlife, McCormick seeds his entire acreage to cover crops each fall. He's found corn and cover crops are great food sources to attract both deer and waterfowl.

"One of the joys of being a no-till landlord is seeing the migratory birds that come by the farm..."

EXCELLENT COVER. Before no-tilling corn, a variety of cover crops seeded on 100% of Ray McCormick's ground provides excellent nesting for game birds such as pheasants and offers a better habitat for ducks and other wildlife.

Ray McCormick

"Duck and deer hunting is used as a tool to attract investors and new landlords..."

"We run a trophy management program on our farm for whitetail deer," he says. "The deer love the turnips, clover and the annual ryegrass we seed in our cover crop program."

Boosting Wildlife Numbers

McCormick offers these ideas to no-tillers who want to improve water quality and wildlife habitat in conjunction with no-till and programs such as the Conservation Reserve Program (CRP) and the Wetland Reserve Program (WRP).

◆ Reseed native prairie grasses in strips or stands as nesting spots for birds. The large, fluffy seeds need to be planted with grass drills after existing weeds and grasses have been eliminated with the seeding of warm-season grasses.

◆ Leave 30- to 60-foot buffer strips along stream banks. "Under the CRP continuous sign up, you can get government payments based on the cash rent values of soil productivity in your area," he says.

◆ Plant a fence row of trees and shrubs to limit erosion and create a wildlife habitat. McCormick has even added wildlife corridors that cut through no-till fields.

◆ Use drop boxes to control water levels and create food supplies for waterfowl. In the spring, the sun warms shallow water in flooded fields and causes an explosion of small invertebrates that feed on no-till residue. The waterfowl feed on invertebrates in the shallow water.

◆ Return nonproductive areas in river bottoms that are prone to frequent flooding to original wetlands. He accomplished this through funding programs with the U.S. Fish and Wildlife Service and the Wetland Reserve Program program with the U.S. Department of Agriculture.

"You'd be surprised how these items can build a very positive relationship with landlords and give you the ability to get a foothold with other landlords who are interested in the same values," McCormick says.

McCormick says one of the joys of being a no-till landowner is seeing the migratory birds that come by the farm, whether it's whooping cranes or snow geese that visit the farm or the ducks he hunts. ❋

HIGH-CAPACITY NO-TILL EQUIPMENT. Big equipment lets Ray McCormick no-till a sizable corn and soybean acreage in a short period of time.

Ray McCormick

BOOSTING WILDLIFE NUMBERS WITH NO-TILL

Bill Poole of Ducks Unlimited Canada says no-till leads to an improved wildlife habitat when compared to conventional tillage. Over time, the altered micro-climates found at the soil surface in no-till fields offer an increased insect population, which provides an essential part of the habitat required by waterfowl and other wildlife.

Direct Benefits

Poole says the increased surface cover and reduced field disturbance of no-till increase the amount of land that can be used as nesting habitat.

A study conducted in the 1980s by researcher Wayne Cowan in western Canada projected production of 25 broods of waterfowl per section of land with no-till. This compared to only seven broods per section with conventional tillage.

Poole says the value of spring-seeded fields as habitat for ground-nesting birds depends largely on the drills used in no-till fields.

"Machines with wide row spacings and narrow on-row packers are most wildlife friendly," he says. "But if they are to be widely used, they must also be farm income friendly and provide yields at least equal to other seeders."

Poole says fall-seeded no-till crops provide even greater benefits to ground-nesting birds. A North Dakota study looked at use of no-till winter wheat fields by nesting ducks. While nest densities were low (eight nests per 250 acres in one year and six nests per 250 acres the next year), nest success rates were 26% and 29% respectively.

"While these success rates may sound low, they were substantially above the 15% required to maintain population numbers and higher than rates frequently found in other studies with different types of habitat," Poole says.

He says no-till's greater nesting areas make it more difficult for predators to find the nests and increases the chances for a successful hatch.

Indirect Benefits

Poole says no-till can create conditions, which lead to increased populations of small mammals. While it may not seem obvious, this benefits ground-nesting birds.

"These small animals are also prey for some predators, particularly foxes, that feed on nests," he explains. "As the numbers of small mammals increase because of no-till, they can make up a higher proportion of the food source for predators and reduce the predator pressure on nests.

"Indirectly, a higher population of voles could result in improved populations with ground-nesting birds."

NO-TILL-AGE®

Gray Garden Slugs vs. Baits
(During 6-day period)

December 1996

Bait	Number of Slugs
Deadline granules	114
Deadline bullets	111
Wilbur Ellis 5% metaldehyde	90
First Choice	78
RCO Slug & Snail bait	56
Alco Slug, Snail & Sowbug killer	49
Neilsen 5% carbaryl	17
Wilbur Ellis 5% carbaryl	13

— Ohio State University

July 2001

MORE NO-TILL, MORE DUCKS

An initiative by Ducks Unlimited in parts of South Dakota and North Dakota is a win-win situation for both agriculture and wildlife.

The group's Winter Cereal Initiative is being carried out in the Dakotas, where over 40% of the waterfowl in North America begin life on these breeding grounds.

Clean tillage farming leaves little upland cover for watershed protection and wildlife production, says Blake Vander Vorst of the Great Plains Regional Office of Ducks Unlimited in Bismarck, N.D.

No-Tilling is Critical

Ducks Unlimited staffers believe one promising alternative that will benefit farmers, wildlife and wetlands is the widespread adoption of fall-seeded no-till crops, particularly winter wheat.

Research conducted by Dwayne Beck at the Dakota Lakes Research Farm in Pierre, S.D., indicates no-till rotations that include winter wheat are profitable. Research has also proven winter wheat stand establishment is more successful in a prior crop's standing no-till residue.

Waterfowl research by the Institute for Wetland and Waterfowl Research found nest densities and nest success to be relatively high in fall seeded no-tilled crops compared with spring planting.

Earn $7 per Acre for No-Tilling

Ducks Unlimited is supporting the project to enhance the acreage of no-tilled winter cereals in four counties in the Dakotas as part of a rotational cropping system.

Participating farmers receive incentive payments for up to 4 years of $7 per acre for no-tilling up to 150 acres per year of winter cereals in a 3 or 4-year no-till rotation. The winter wheat needs to be no-tilled into undisturbed residue of the preceding crop.

Tillage will be replaced by herbicides and fungicide applications will enhance the yield and profitability of no-tilled winter wheat, says Vander Vorst. Post-harvest control of volunteer grains and grasses will be critical in prohibiting the impact of wheat streak mosaic disease on winter wheat.

A winter cereal study conducted last year in Canada near Parkbeg, Sask., represented the second year of a study by the Institute for Wetland and Waterfowl Research. The study was designed to evaluate the attractiveness of no-till fall-seeded cereals and spring-seeded cereals to upland nesting ducks, with a focus on pintails.

More Nests with No-Till

A total of 239 nests were found on 2,100 acres of cereal rye and winter wheat seeded in the fall of 1998. In comparison, only six nests were found on 1,950 acres of spring-seeded ground.

The nesting species on fall-seeded crops was 21% pintail, 40% mallard and 20% northern shoveler, with the balance being blue-winged teal, gadwall and lesser scaup. ❀

2,000-ACRE NO-TILL WHEAT FIELD. The Boe Ranch in eastern Washington's Palouse area no-tills this 2,000 acre wheat field that stretches over 2 miles and features a 1,3472-foot difference in elevation.

Guy Swanson

With No-Till, How Big is Big?

"The question we asked was why mess with tillage if we can find a no-till drill that will cut through the stubble..." — *John McNabb*

When folks start talking about the size of some of the world's biggest no-till fields and the wider and wider no-till equipment being used, big numbers get tossed around.

Some of the biggest no-tilled fields, equipment and total no-tilled acres that the *No-Till Farmer* editors

know about are described here from Australia, Canada, Idaho and Washington. But that's just what we know about.

Biggest No-Till Fields

After we asked No-Till Innovator Guy Swanson about the biggest no-tilled fields he's seen in his travels,

he sent along photos of no-tilled fields in eastern Washington and northern Alberta.

Swanson, the head of Exactrix Global Systems in Spokane, Wash., says the Boe Ranch in the Palouse area of eastern Washington no-tills a 2,000-acre wheat field that stretches 2 miles from north to south. There's a 1,472-foot

---

Guy Swanson

8,000-ACRE NO-TILL FIELD. Located in the Peace River area of northern Alberta, this 7-mile long field is no-tilled by Ken Dechant. He uses the rig shown above to deep band anhydrous ammonia.

Seven Mile Field

Swanson also provided photos from Ken Dechant's northern Alberta operation near the town of Manning. The family no-tills wheat, barley and canola in the Peace River area. The best way to get to the farm is to fly into Alberta's Edmonton airport, rent a car and make the 6 hour, 400 mile drive to the Peace River area.

Dechant's "biggest no-till field" covers roughly 8,000 acres and a single equipment pass covers 7 miles from one end to the other. Previously, the no-till field covered 5,000 acres, before newly cleared land and an abandoned road boosted the acreage.

Anhydrous ammonia is deep banded in the fall after harvest with an Exactrix system with single disc Deere and Case IH openers mounted on 60-foot wide toolbars.

The farm stores enough fertilizer to deep band 14,000 acres over a 20-day period in late September and October. Three farm-built trailers

top-to-bottom difference in elevation on some of the field's steep slopes.

Self-leveling combines are essential since conventional machines would be out of level more than half the time while making as many as 30 passes up and down these steep slopes.

Fields in the steep-sloped Palouse area of eastern Washington, western Idaho and eastern Oregon frequently turn out winter wheat yields running over 125 bushels per acre. Thanks to no-till, growers save moisture, hold valuable soils in place and watch rain and snow pack moisture soak into the ground at a rate of up to 2.12 inches of water per foot — going as deep as 10 feet into the soil profile.

"The 212-foot wide air seeder is **pulled with a pair of John Deere 9400T tractors** hooked in tandem..."

212-FOOT WIDE AIR SEEDER. Australian grower Gavin Zell and his son use this huge rig in a 47,000-acre no-till wheat, barley and chickpea operation.

each hold 4,000 gallons of anhydrous ammonia and 2,600 gallons of liquid fertilizer for application during spring seeding.

No-Till 2,500 Acres a Day

With huge fields and limited moisture, many Australian growers are constantly looking for bigger equipment to no-till in a timely manner. That's why David Trevilyan came up with the 120-foot Multiplanter seeder featured in the August 2005 issue of *No-Till Farmer*. The following winter, he shared his seeding ideas with attendees at the National No-Tillage Conference.

An engineer and farmer who had been no-tilling for 20 years in central Queensland, Trevilyan designed this rig when he couldn't find appropriate no-till equipment to handle the family's acreage.

These innovative rigs were designed to no-till through any residue. "This is the machine that will deliver sustainable agriculture into the hands of Australian farmers," says the owner of Multi Farming Systems in Moree, New South Wales.

At one time, Trevilyan held the world record for seeding 2,337 acres in a 24-hour period with a 120-foot air seeder.

In 2005, a western Australian grower showed interest in purchasing a 302-foot wide model that was already on the drawing board. The grower planned to pull the no-till air seeder that would have been longer than a football field with an 850 horsepower tractor, but the unit was never built.

"Our problem is not how big we can make a machine, but finding a tractor big enough to pull it," says Trevilyan. "The problem is not the horsepower because each tine has only a 2-inch spear tip, but this soon adds up for a machine with 273 tines. It takes about 4 horsepower to

"Over a typical **18-hour daily seeding period,** this big rig can **seed 2,500 acres...**"

DOUBLE TRACTOR POWER. The seeder is towed end-wise with a width of only 18 feet. It takes about an hour to get the seeder into transport mode.

pull each shank under average soil conditions."

Seeding in 12-inch rows with 2-inch openers reduce horsepower needs, even when the operator has to run the unit deep to find moisture. With the specialized no-till planter capable of dropping seed at practically any depth, a few growers have run the units as deep as 9 inches to reach moisture.

Despite its width, it's a precision-depth planter. The press wheel acts as a depth gauge for the hydraulic controlled shank and seed tube to provide consistent seed placement. Each row assembly works independently of the frame, enabling it to follow the contour of the land. A no-till coulter can be mounted in front

of the shanks to slice through weeds and crop residue.

Each press wheel is controlled by a four-hinged parallelogram that allows the shank to follow the contour of the land more closely than with a conventional, frame-mounted designs. Pressure can be adjusted on the move.

The hydraulic powered tines penetrate the soil at a 15-degree angle and run parallel to the surface. This provides the ability to reach deep moisture, build a trench for holding water and delivers accurate seed placement and optimal seed-to-soil contact.

His daughter, Kris Trevilyan, says there's ongoing interest in building even wider no-till air seeders. In

2010, the company built a 212-foot frame for Australian grower Gavin Zell. The rig includes two large capacity seed carts pulled by a pair of high horsepower John Deere tractors.

Two Tractors Pull 212-Foot Air Seeder

Zell and his two sons grow 47,000 acres of wheat, barley and chickpeas near Collarenbri in New South Wales. To seed in a timely manner, they run the biggest equipment they can.

Farming in a remote area of Australia, the Zells had difficulty finding skilled tractor drivers and didn't want to add more tractors and seeders. After testing a tandem tractor hookup in 2010, they pull the 212-foot air seeder with a pair of John Deere 9400T tractors.

For road transport, the seeder is towed end-wise with a width of 18 feet. It takes about an hour to switch the seeder into transport mode.

As described in *Farm Show* magazine, a heavy-duty hitch mounted on the drawbar of the front tractor moves up and down to handle movements between the coupled tractors. A large ball joint on the front of the rear tractor connects the two units.

Hydraulic lines running along one side of the rear tractor plug into the hydraulic lines on the front tractor. This allows both hydraulic systems to power the giant air seeder and each tractor provides high

NO-TILL-AGE®

March 1992

Optimum Machinery Investment for 1,000 Acres

Tillage System	Machinery Inventory
No-till after soybeans and corn	$189,600
Ridge till after soybeans and corn	$199,750
No-till after soybeans, chisel after corn	$248,350
Disc after soybeans and corn	$252,650
Disc after soybeans, chisel after corn	$261,350
Chisel after soybeans, moldboard plow after corn	$270,450

— Capital Agricultural Property Services

Guy Swanson

SAY GOODBYE TO SUMMERFALLOW. No-till allowed John McNabb to eliminate summerfallow and farm slopes that are too steep for conventional tillage.

NO-TILL IS A MONEY-MAKER. John McNabb credits no-till with returning an extra $35-$40 per acre on the steep slopes he farms in southern Idaho.

capacity hydraulic power for one of the two seed carts.

Seated in the cab of the front tractor, the driver controls both tractors through electric cables. A push/pull cable connects to the transmission of the rear tractor to shift gears.

A GPS system and auto steering system on the front tractor is linked to the air seeder steering system to make sure the unit accurately follows the tractors and increases the maneuverability of the rig.

With each tractor burning 13 gallons of diesel per hour, the rig runs at 5.6 mph and seeds 2.5 acres per minute. Over a typical 18-hour daily seeding period, the big rig can seed 2,500 acres.

During the 2011 season, the Zells used the big rig on about half of their 47,000 acres. The remainder was seeded with a pair of 80-foot air seeders and a 60-foot unit, all working 24 hours per day.

40,000 No-Tilled Acres

At one time, John McNabb was no-tilling 40,000 acres a year in Idaho and northern Utah. No-tilling since 1978, the Inkom, Idaho, grower has varied his acreage dramatically

SHORTLINE INNOVATIONS. Numerous drill developments from shortline farm machinery manufacturers helped meet the specific needs of seeding small grains in a no-till environment.

July
2005

MORE ACCURATE NO-TILLING. With modified air seeders, western Canadian growers no-tillers have reduced seed and fertilizer costs.

NO-TILLING 24 HOURS A DAY

Because of a small window of opportunity for spring seeding, western Canada no-tillers are investing in large capacity equipment to get drilling and spraying done in a hurry. Many run their equipment 24 hours a day to take full advantage of favorable seeding conditions.

Robert Jurgens, an ag sales representative for Hi-Way Service in Lethbridge, Alberta, works with customers who no-till huge acreages, including one grower who no-tills 17,000 acres.

"He bought a pair of 500-horsepower STX 500 Case IH tractors 2 years ago to pull a pair of air seeders," says Jurgens. "Then he traded them in on two more high horsepower tractors after just one year of use when the originals still had plenty of value. As a result, I was able to sell this farmer four big horsepower tractors in 2 years, then turned around and sold two of the used tractors to other growers."

Several other customers no-till 6,000 to 9,000 acres with just one large horsepower tractor that pulls a 57 1/2- to 63-foot air drill.

"Another customer purchased a single-disc, triple-shoot drill to use on 5,000 acres of canola," says Jurgens. "With accurate placement of 3 pounds per acre of canola seed, he's saving $22,000 a year in seed costs alone compared to using an older drill. Not only does he place the seed exactly where he wants it for top yields, but he saves on cost since less seed is needed with more accurate placement."

Leasing self-propelled sprayers is a sizeable part of Jurgens' business. "I have one farm customer with a sizeable no-till acreage who was paying $55,000 in custom application charges to handle a portion of his acreage — and that didn't include the cost of chemicals," says Jurgens. "By leasing a self-propelled Case IH sprayer, the annual payments were $48,000.

"He uses the rig to spray his entire acreage and will still end up with good sprayer value at the end of the lease. By leasing a sprayer instead of relying on a custom applicator, he handles critical spraying needs on a more timely basis and also does spraying for neighbors that brings in more cash." ❁

from year to year depending on land rental costs and anticipated commodity prices.

Crops that he's no-tilled include hard red spring wheat, hard red winter wheat, soft white winter wheat, barley, alfalfa and mustard.

McNabb credits no-till with earning an extra $35 to $40 per acre over the years. And he started no-tilling pretty much by accident.

One of his spring barley fields was set back by frost and the crop didn't grow very tall. As an experiment, McNabb seeded winter wheat directly into the light stubble with a deep furrow drill. Since the drill continually plugged up with straw, the rest of the field was disced before seeding.

The disced ground dried out, delaying emergence. But where wheat was seeded directly into the stubble, the wheat germinated normally and got off to a good early start. To McNabb's surprise, wheat yields were 10 bushels higher where the ground was direct seeded.

"That really got our attention," says McNabb. "The question we asked ourselves was why mess with tillage if we can find a drill that will cut through the stubble?"

Experimenting for a few years with a split packer deep furrow drill and several air seeders, he wasn't happy with their depth control. Most no-till drills on the market at the time were only 15-feet wide, and McNabb was used to seeding in 30-72-foot widths. As a result, he didn't think they could cover their big acreage in a timely manner with no-till.

While attending a Manitoba/North Dakota Zero-Till Assn. winter conference in Canada, McNabb heard about some

NO-TILL-AGE®

Acres You Can Handle with 400 Spring-Time Hours

May 1986

Tillage System	Acres
No-till	1,250
Till plant	930
Disc	833
Chisel plow	635
Moldboard plow	533

— University of Nebraska

fertilizer research being done with the Yielder drill. While the original Yielder placed fertilizer between 10-inch rows, new work had been done on deep banding fertilizer between 5-inch paired rows.

Impressed with the deep root masses Yielder owners were getting, he talked to the company. While the biggest Yielder at the time was only 20-feet wide, they agreed to build four 25-foot wide Yielder no-till drills for McNabb.

"By 1981, we were direct seeding nearly 40,000 acres with the four Yielder drills," says McNabb. "We got much better depth control than with the air seeders, were able to travel faster and didn't have to refill the seed and fertilizer boxes as often."

While he found many benefits for no-till, the one that has probably earned the most dollars is the ability to deep band fertilizer. While broadcasting encourages shallow rooting, crop roots grow much deeper when seeking moisture after deep banding.

In years of favorable moisture, McNabb finds deep banding offers a 20% better fertilizer response. With drier soils, the fertilizer response is 80-100% better.

No-till has allowed the family to eliminate summerfallow and farm slopes too steep for conventional tillage due to erosion concerns.

Around 2000, a combination of things — threatening long-term drought, inflated input prices, higher land costs and conflicts with government agencies on leased land — caused the family to trim their acreage.

In 2005, they no-tilled 7,500 acres with about half of that ground in the Conservation Reserve Program. In later years, they cut even further back on grain production, but continue to run McNabb Grain, which is one of the largest white wheat handling and cleaning facilities in southwestern Idaho. ❃

Frank Lessiter

Remembering the Very First No–Till
Plots Planted in Hawaii

"It was a full week of **non-stop no-till learning and vacationing in Hawaii** for only $448 per person…"

Before my wife, Pam, and I bought *No-Till Farmer* in 1981, the publication had been owned by another Wisconsin based publisher, Reiman Publications.

In those early years, we held a yearly no-till conference from 1973 to 1980 in Hawaii. This was part of a larger tour hosted by *Farm Wife News* magazine, which in some years drew as many as 1,200 farmers and spouses

for an early winter get-away to the warmth and beauty of Hawaii during an early week in January.

The no-till aspect of the tours drew farmers who were already no-tilling, those who wanted to learn more about the reduced tillage practice and hundreds of farmers who had never seen no-till before.

The week-long program included talks by dozens of no-till farmers and educators along with tours of the

island's unique agriculture and many of Hawaii's tourist attractions. Plus relaxing time spent on the beach and networking with other growers from around the country.

Morning Lectures, Afternoon Beach Time

Held during the morning hours, the week-long educational sessions included a total of 35 hours of no-till, tax management and estate planning

FIRST GLIMPSE OF NO-TILL. Hundreds of U.S. mainland farmers and their spouses got their very first look at no-till from tours of plots in the Hawaii islands that were held from 1973 to 1980. These 1972 corn plots represented the very first no-till ever planted in Hawaii.

"Over 8 years, the Hawaiian no-till conferences **drew 2,600 attendees...**"

FARM-RAISED SHRIMP. Many no-tillers got their first glimpse of growers rearing shrimp in these farm ponds on the island of Kauai. It was another way for this Kauai farm to diversify its products.

Frank Lessiter

fundamentals and farm tours. In the afternoon, attendees were free to soak up the sun on the Waikiki beaches, go shopping, take a nap or network "one-on-one" with fellow no-tillers.

The groups stayed at the Hilton Hawaiian Village Hotel where early-to-rise farmers would take a morning stroll along the Waikiki beach.

Besides all the no-till learning opportunities, each group visited a 15,000 head beef feedlot, a 1,400-cow dairy, the Arizona battleship memorial in Pearl Harbor, the north shore beaches where international surfing championships are held, the Polynesian Culture Center that featured talks with Hawaii Five-O stars and the famous Don Ho luau.

Another special part of the first no-till tour to Hawaii was having Mr. and Mrs. Hendric Schreiber of Freeland, Mich., join us on their honeymoon!

DIFFERENT SOILS, DIFFERENT ENVIRONMENT. Conference attendees paid close attention as a Hawaii grower explained the many pitfalls encountered in producing both cash and specialty crops in the islands.

FIRST NO-TILL ON TWO ISLANDS. The first year's no-till corn plots were at the Trojan seed production farm on the island of Maui. Over the next 7 years, the no-till plots were planted at a farm on the island of Kauai.

Looking back, the week-long trip in 1973 included airfare from the mainland, six nights on the beach at the Sheraton Waikiki Hotel, a flight to an outer island, bus tours of two islands, Pearl Harbor cruise, Hawaiian luau and much more. All for only $448 per person!

The first no-till event in 1973 included a short flight to the island of Maui where no-till plots had been created at the Trojan Seed Company corn breeding station. These 1973 plots represented the very first no-till acres planted in Hawaii. In the following

1973 NO-TILL TIPS ARE STILL GOOD ONES TODAY

ALWAYS A NO-TILL LEARNING EXPERIENCE. Many of the veteran no-tillers that were part of the tours were always quick to suggest planter adjustments and refinements in the no-till operation based on their own experiences back home.

Several tips from the 1973 no-till conference speakers still hold as true today as they were more nearly five decades ago:

◆ "Two men are key to your success with no-till," said Harry Young of Herndon, Ky. "That's the man on the planter and the man on the sprayer, as these two operations have the biggest effect on your net income with no-till."

◆ "Most no-till corn problems in our area are due to poor stands," said Joy Apple of Cygnet. Ohio.

◆ "We've been successfully aerial seeding wheat and cover crops in our no-till program," says Bill Good, a veteran no-till grower and Allis-Chalmers dealer from Marshall, Mich.

◆ "Many first-time no-tillers leave seed on top of the ground instead of getting it down where it belongs in the no-till slot," said University of Maryland agronomist Joe Newcomer. ❋

FIRST LOOK AT ROUNDUP

Most attendees at the 1974 conference got their first look at Monsanto's new Roundup herbicide that would go on to play a key role in the growth of no-till around the world. Monsanto's west coast rep Dean Brown demonstrated how the new glyphosate chemistry compared with other herbicide options for early-season no-till weed control.

What I remember from those discussions at the no-till plots was the excitement among a number of no-tillers. They were convinced the new glyphosate compound would be cheaper than the paraquat they were using as a contact herbicide.

It never happened. ❁

HOW MANY ACRES COULD YOU NO-TILL WITH 732 HOURS OF LABOR?

June 1983

System	Acres
No-till	1,200
Till plant	1,003
Rotary strip tillage	882
Disc tillage	871
Rotary tillage	841
Chisel tillage	697
Moldboard plow	600

— No-Tillage Farming/ Minimum Tillage Farming Book

SEEING NEW CROPS. During the tours of Hawaiian agriculture, many of the farmers got their first look at a number of exotic crops such as taro, mangoes and papayas for the first time.

Frank Lessiter

"They were sure convinced **Roundup would be cheaper** than paraquat..."

years, the no-till plots were planted on the island of Kauai.

In late 1972, Allis-Chalmers air freighted a two-row no-till planter to Maui. Arriving just 6 weeks before the 1973 tour took place, it was enough time to get corn in the ground so farmers attending our first conference could see no-till corn growing under tropical conditions.

Starting in 1974, we split attendees into three groups and they flew to Kauai on Tuesday, Wednesday or Thursday to take in the tourist attractions on the "garden island"

along with spending "educational time" at the no-till plots.

Since I was the no-till guy, I'd fly to Kauai each year with the groups three days in a row. While I'd flown nine times from Wisconsin to Hawaii, I've probably flown 25 times from Honolulu to Kauai.

Since 1974, the population of Kauai has doubled to nearly 70,000 residents. On my first trip to Kauai in 1974, there wasn't a single traffic light on the island.

While these tours offered a great winter vacation for farmers and

LEARNING IN THE FIELD. Like most American farmers, the no-till attendees liked to get out in the field, run the dirt through their hands, discuss tillage options and compare ideas with farmers from other states while learning firsthand about the special aspects of Hawaii's agriculture.

Frank Lessiter

WELL, IT ACTUALLY WAS A WARM RAIN

Most years, the Hawaiian weather was warm and sunny in early January, but I remember a few "not so good" weather situations.

◆ One year it rained every day during the week-long no-till tour that often promised a once in a lifetime vacation and memory for many farm couples. I recall a Minnesota attendee (a glass half full kind of guy) making the best of the situation by saying, "Well, at least it was a warm rain."

◆ Another year it was cold by Hawaiian standards on a day we took one of the groups to Kauai to view the no-till plots and tour the island. The tour included a boat trip up a scenic river to the Fern Grotto where three Hawaiian men played ukuleles and two women danced in grass skirts.

The temperature was only in the upper 50s, which was OK for us Mid-westerners who were happily donned in shorts and short sleeve shirts. But you knew it was cold by Hawaiian standards when the two Hawaiian dancers in grass skirts were also wearing stocking caps and ski parkas.

◆ One year, the trip to Hawaii turned out to be a rough one due to extreme turbulence with the huge 747 plane abruptly dropping 150 feet several times. With so many sick passengers, the flight attendants ran out of vomit bags.

◆ Another year we arrived back at Chicago's O'Hare airport at 7 a.m. on a Sunday morning to minus 15 degree weather along with a brisk wind.

Wearing Bermuda shorts, a short-sleeved colorful Hawaii shirt, flip flops and no coat, a Wisconsin grower walked to the parking lot in wind chill conditions of minus 37 degrees to discover his car wouldn't start. Dressed in clothes suitable for the beach in Hawaii, he waited 90 minutes for an emergency truck to show up and get his car started. Even today, I get shivers thinking about it. ✺

"The grass-skirted Hawaiian girls were **wearing ski caps and parkas...**"

their wives in a tropical location, the no-till educational program and plots helped make these mid-winter tax deductible as a farm expense in the eyes of the Internal Revenue Service.

That first Hawaiian no-till event in 1973 drew 160 farmers and their wives for a week in the sun while learning more about no-till. These Hawaiian no-till events continued through 1980 and attracted 2,600 total attendees.

About a dozen years later, we introduced the National No-Tillage Conference that has drawn more than 1,200 attendees in a single year. ✿

'IT'S OK … MY WIFE DOESN'T FLY'

For the Hawaiian tours, planes usually departed from several U.S. airports on a Sunday morning. One year, we got a phone call at the office at 9 a.m. on Saturday, just one day before attendees were to leave for Hawaii.

The caller was a North Carolina no-tiller asking if it was too late to sign-up for the tour that started tomorrow. Somehow, we found a few seats still available on a Hawaiian flight leaving the next morning from Chicago.

When we asked whether he needed one or two seats on the tour, he replied his wife didn't like to fly. We told him we were sorry to hear that as she would have enjoyed the tour.

Then he replied, "No, the reason I'm going is to get away from her for a week." ✿

SEA-SICK HOLSTEINS ON A BRAND-NEW ALL-WHITE SHIP

During a visit to the large 1,400 head dairy operation on the main island, the farm manager explained how the cows were shipped in from California. A half dozen cows were placed in as many as 40 huge wooden crates (total of 240 cows) secured to the open deck of an ocean-going freighter.

For one trip, the crates were placed on the deck of a brand new all-white freighter traveling from Los Angeles to Honolulu. It was an extremely rough voyage and many of the cows got sick, or as the saying goes, "were scared s _ _ _ less."

When the ship docked in Honolulu, the all-white ship was covered from top to bottom with loose manure. That freight line never carried another cow from the mainland to Hawaii. ✿

'HANG LOOSE, WE'RE ON HAWAIIAN TIME!'

Frank Lessiter

Back in the 1970s, the Hawaiians were never too concerned about being on time. If you asked someone why he or she was 5 or 10 minutes late, they'd answer, "Hey, hang loose."

One time we'd loaded a group of attendees on a bus and were ready to head out on a tour with no driver in sight. He had driven the bus to the hotel, but then disappeared.

Some 35 minutes later, I found him upstairs in the hotel bar strumming on a ukulele with a local band. When I told him we had to go right now, his response was, "Hey, hang loose."

Bill Good was a no-tiller and Al-lis-Chalmers dealer attending with his son, Don, from Marshall, Mich. These two men and their wives never missed any portion of these eight Hawaiian no-till tours.

Bill always kept his watch set on Michigan' eastern standard time, despite the fact that there was a 6-hour time difference between the Midwest and Hawaii.

In all of those trips in the 1970s, Bill was among the few folks who never showed up late for any session — even though he was still on Michigan time. ✿

'I NEED A WELDING TORCH FOR 4 HOURS'

The Kauai farm where we had the no-till plots also raised several specialty crops and farm-raised shrimp for the Hawaiian market.

When John Deere came out with a new planter in the late 1970s, the farm manager flew back to the states to buy one of the first units to come off the assembly line. Even though the new Deere planters were in short supply, he convinced the company to sell him a 6-row unit and pulled it out into a nearby parking lot.

His next step was to rent a welding torch. Right in the Deere parking lot, he used the torch to cut the 6-row unit into three pieces so it could be shipped by air to Hawaii.

When the Deere folks saw what he was doing to their brand new and carefully engineered planter, they were shocked. ✻

NO-TILL-AGE®

October 2015

What Influences the Decision to No-Till?

Influence	% Of Farmers
Reduced soil erosion	91%
Reduced labor costs	80%
Improved soil structure	79%
Cost of diesel fuel	76%
Reduced loss of soil moisture	76%
Reduced soil compaction	72%
Need for less equipment	67%
Reduced soil, chemical runoff	67%
Better habitat for soil organisms	62%
Increased water filtration	57%
Higher yields and profits	46%

— Best Management Practices Survey, Pennsylvania No-Till Alliance.

NO-TILL LOOSE ENDS...

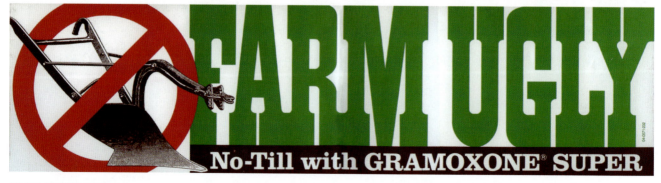

NO-TILL BUMPER STICKER. As part of an ICI Americas "Farm Ugly" no-till marketing program, No-Till Farmer subscribers received this bumper sticker in the May 1989 newsletter.

Early-Day No-Till Innovations
Came Out of Farm Shops

"Major equipment manufacturers had the attitude that **all growers had to do was add coulters to planters** to handle no-till conditions..."

In the early days of no-till, growers weren't satisfied with the equipment choices to make this new reduced tillage practice work. At that time, most drills and planters were designed to work effectively under conventional tillage or minimum tillage conditions.

Unfortunately, most of the major farm equipment manufacturers had

the attitude that all growers had to do was attach a set of coulters to planters and drills to effectively tackle most no-till conditions. Most were also hoping no-till was a fad that would soon go away so they could concentrate on selling higher horsepower tractors and heavy tillage equipment.

However, early-day no-tillers quickly learned existing seeding equipment wasn't capable of making no-till work with different residue and soil conditions.

As a result, farmers spent many hours in their farm shops coming up with their own solutions for tackling no-till's most pressing problems.

233

NO-TILL-AGE®

February 1992

Costs for Till, Plant and Spray Trips

Tillage System	Corn	Soybeans
No-till	$18.57	$23.55
Chisel plow	$26.79	$47.67
Moldboard plow	$43.57	$58.09

— BASF

Innovative No-Tillers

In 1986, Washington State University researcher Don Dillman pointed out that a farmer with a well-equipped shop was more likely to adopt no-till. He found these innovative farmers had the desire, resources and ability to play around with improved equipment modifications.

Dillman's studies indicated farmers who adopt no-till were more likely to be venturesome, willing to take risks, have more education, an ability to think through ideas abstractly, earned higher farm incomes and farmed larger acreages. He also found no-tillers tend to be younger than the general farm population, farm with family members, worry little about what the neighbors think and were willing to travel to seek out new cropping information.

Farm Shop Rigs

Faced with drills and planters that weren't getting an accurate seeding job, many no-tillers decided to spent hours in their farm shops creating their own rigs or making extensive modifications to existing equipment.

Here are a few examples of numerous drill and planter innovations and modifications created in farm shops during the 1970s and 1980s to make no-till seeding more accurate and effective. ❋

NO-TILL SOYBEANS WITH A COMBINE

October 1992

One way to double the number of hours you put on your combine each year might be to use it to no-till your soybeans. That's what Purdue University engineering students did by mounting a 15-foot Great Plains no-till drill on the front of a rubber track equipped Claas combine. Materials to hook the no-till drill to the combine wouldn't cost more than $3,000.

The students were able to get up to 45 pound per square inch of downpressure on the no-till drill with the combine, says Alan Van Nahmen, a senior marketing representative with Claas. With four hydraulic cylinders providing a 15,000-pound lift capacity, the rig could handle a 25-foot drill.

Van Nahmen says the front mounting provides extra visibility, which could reduce operator fatigue. Most importantly, there should be little additional wear to the combine's separator components while no-tilling beans.

Still another option would be to double-crop soybeans at the same time wheat or barley is harvested by

SPRINGTIME COMBINING. Soybeans were seeded with a Great Plains 15-foot no-till drill mounted out front on this rubber track equipped Claas combine.

mounting air seeder units on the rear of the combine. Or you could no-till wheat or a cover crop while harvesting beans. ❋

WILD IDEA BECOMES REALITY

For 15 years, Elliott Carter dreamed about building a self-propelled planter. After spending 1,000 hours in his farm shop, the Garnett, Kan., grower turned the idea into reality in 1973 for no-tilling 200 acres of corn, 100 acres of soybeans and 100 acres of grain sorghum.

Carter started the project by spending $300 on a used Massey Ferguson 92 combine. He then offset the wheels to match the desired 30-inch row spacing.

A used IH Cyclo unit is mounted under the old combine platform and a homemade seed hopper holds 11 bushels of seed. Two planter fertilizer tanks were modified to work as gravity-fed insecticide containers while old cultivator gauge wheels serve as press wheels.

Up front, he mounted a pair of narrow fluted offset tandem coulters 2 inches apart on one bracket with anti-friction bearings so they can follow a moderate curve on his terraced ground. After testing several coulter setups on 500 acres in 1972 with a traditional no-till planter, Carter found the 1-inch coulters left more loose dirt and sliced through stalks better than wider coulters.

"Just 2-2.5 inches deep is all I want the coulters to run," says Carter. "If you run them too deep, you leave an air space where the dirt isn't packed as well."

One 500-gallon fiberglass tank is filled with water so Carter can apply 30 gallons per acre of through Acra-Plant shoes with the seed. This allows him to plant at the best depth regardless of available moisture since he's providing the needed moisture. ✿

DOUBLE-DUTY RIG. One 500-gallon tank contains herbicides that can be applied with the planter. For later weed control, a folding hydraulic boom is used to spray a dozen 30-inch rows. The booms are easily replaced with row markers at planting time.

FARM SHOP SPECIAL. An old combine frame, a IH Cyclo planting unit, 12 narrow fluted coulters, Acra-Plant shoes, a pair of 500-gallon tanks and an assortment of parts were turned into a 6-row, 30-inch no-till planter by Kansas farmer Elliott Carter.

Frank Lessiter

NO-TILLING CORN WITH A PUSH-STYLE PLANTER

HOME-BUILT RIG. Al Buening created this 12-row 30-inch no-till corn planter from a retired Gleaner combine.

Handy in the farm shop, Al Buening created a 12-row no-till corn planter from an old Deutz-Allis N6 Gleaner combine. The Glenwood, Ind., no-tiller uses the rig to apply insecticides, herbicides and fertilizer while planting.

Buening utilized as many combine parts as possible and built a frame to hold the engine at the rear of the machine. A stainless steel tank is divided into a pair of 475-gallon compartments that hold liquid starter and herbicides. A bridge hitch built from 12-inch by 12-inch tubing with l/2-inch sidewalls holds the planter. The tubing also serves as a reservoir for 50 gallons of hydraulic fluid.

The 12-row 30-inch no-till corn planter is equipped with White planting units, Yetter trash wheels, Yetter row markers and Rawson coulters. It also includes planting, spray and hydraulic monitors and a Raven spray controller.

It took Buening two winters of farm shop work to build this unit, which was originally designed as a ridge-till planter.

"Everything is out front so you can see what's being done," he says. "I like being able to put on fertilizer with the planter and being able to vary my plant populations on the go." ✻

NEAT HYDRAULIC FLUID STORAGE. The tubing in the bridge hitch serves as a reservoir for 50 gallons of hydraulic fluid.

TERRIFIC VISIBILITY. The 12-row front-mounted bridge hitch for the no-till planter offers great operator visibility.

RIDGE TILL TRACTOR EARNS HIGH 'GRADES'

UP-FRONT PLANTING. This experimental tractor from Hiniker offered new ridging, planting and cultivating possibilities for ridge tillers.

While it looks much like a country road grader it's a revolutionary new tractor geared strictly for ridging, planting and cultivating.

The "one-of-a-kind" unit was designed by the Hiniker Co. of Mankato, Minn. The four-wheel drive 145 horsepower tractor allows a planter or cultivator to be mounted up front in a throwback to the old days when farmers used front-mounted cultivators on their tractors.

Farmers who tested the unit say it stays on the ridges since all four wheels provide power and a balanced load. The front-mounted system provides plenty of down pressure that is readily transferred and adjusted from the tractor cab to the cultivator or planter. About 40% of the tractor's weight is over the 60-inch front wheels while the other 60% is over the 120-inch rear wheels.

While the tractor was never marketed or not, it's a novel cost-saving machinery idea that was of interest to many ridge-tillers and no-tillers.

NO-TILL LOOSE ENDS...

VACATIONING NO-TILLERS HELPED ENGINEER NEW NO-TILL DRILL

One early December morning in the mid-1970s, I answered a phone call from the chief engineer at Haybuster Mfg., in Jamestown, N. Dak. He explained the company wanted to become a player in the no-till drill market.

A few days later, he flew to Milwaukee and we discussed some ideas. I suggested one of the best ways to get grower feedback would be for him to travel to Hawaii in early January for our upcoming no-till conference where he could visit with hundreds of no-tillers.

The engineer joined us in Hawaii, talked with no-tillers and went home with fresh ideas on how to engineer a no-till drill. Six months later, Haybuster introduced their first no-till drill.

NO-TILL INNOVATIONS FROM FARM SHOPS

In the early days of no-till, many growers weren't satisfied with the available equipment needed to make the new tillage concept successful. Gathered from a number of *No-Till Farmer* issues, these two pages feature a few 1970s and 1980s innovations and modifications that were created in farm shops to make no-till seeding more effective. ❊

NO-TILL-AGE®

How Tillage Impacts Residue

Tillage System	% Surface Cover
No-till	72%
Chisel plow	29%
Moldboard plow	5%

— University of Wisconsin

Frank Lessiter

No-Tillers Know 'One Equipment Design Doesn't Fit All Situations'

"There's money to be made in agriculture, and **no-tillers can make more money than anybody else**..." — *Moe Russell*

Nothing has played a bigger role in the acceptance of no-till since the late '60s than the continual improvement in equipment. Back in the early days, much of no-till's innovations were "built from scratch" or modified from conventional rigs in farm shops across the country.

But as no-till acceptance continued to gain ground in the '70s and '80s, engineers at a few manufactur-

ers started to develop planting and drilling ideas geared specifically to the needs and wants of no-tillers.

While full-line farm machinery manufacturers continued to design equipment to meet the needs of "all farmers" regardless of cropping conditions, location or tillage systems, it was the shortline manufacturers that led the way in capitalizing on new opportunities in the no-till market.

Over the past few decades, we've

seen many new ideas developed specifically for the no-till market. Even so, some manufacturers still operate with the mentality that when it comes to equipment design, "one size fits all." This approach is rooted in a mindset of continuing to sell high profit items such as high horsepower tractors that can pull extra-wide tillage tools, something that doesn't align with the lesser horsepower needs of the no-till market.

AMISH NO-TILLERS ENJOY HUGE LABOR SAVINGS

Bill Haddad

NO-TILLING WITH REAL HORSEPOWER. Instead of making a half dozen or more trips across their fields with conventional tillage, some Amish farmers are getting their crops in the ground with a single trip with a no-till planter or drill. Their reduced need for labor represents a tremendous savings.

Frank Lessiter

BIG-TIME NO-TILL PROMOTER. Since 1969, Bill Haddad, right, has participated in thousands of no-till meetings with farmers, retailers, extension and conservation staffers throughout Ohio and other states.

When farming with real horsepower, there's a limit to how many acres you can get over in a day. That's why the acceptance of no-till by Amish growers has had a tremendous impact on their labor needs.

Instead of walking a half dozen or so times across a field behind horses pulling tillage tools and a planter or drill to get their crops in the ground, doing it all in one trip definitely saves on labor needs.

Among the no-till veterans working with Amish farmers has been Bill Haddad of Danville, Ohio. Starting out in 1969 promoting paraquat with Chevron and later with other ag chemical firms, he's held thousands of no-till meetings with farmers, retailers, extension and soil conservation staffers over the past five decades.

Years ago, Haddad helped launch no-till with the Amish farming community around Millersburg, Ohio. Among the largest Amish communities in the U.S., these photos demonstrate how growers made no-till successful with real horsepower while dramatically trimming labor needs and saving energy and soil. ❋

Bill Haddad

SAVING ENERGY. No-till helps the Amish farm in a more timely, efficient and reduced cost manner. Note the bumper stick pasted on the back of the horse.

With bigger equipment and larger equipment investments, there's an ever-increasing emphasis on cost control in farming. As machinery costs continue to escalate, this is an area where no-tillers enjoy a major benefit in holding down costs.

How Many Acres Will 1,500 Gallons of Diesel Handle?

Tillage System	Acres
No-tillage	652
Disc-plant	254
Chisel plow	211
Moldboard plow	192

— November 1979 No-Till Farmer

NO-TILL-AGE®

April 1983

Labor Needs for Tillage Systems

Tillage System	Minutes Per Acre	No-Till Savings in Minutes Per Acre
Moldboard	73	37
Chisel plow	63	27
Disc	50	14
Rotary-till	52	16
Rotary strip-till	50	14
Till plant	44	8
No-till	36	—

— University of Nebraska

NO-TILL LOOSE ENDS...

April 1984

NO-TILL SAVES TIME, BUT DON'T WASTE IT WATCHING TV OR DRINKING COFFEE!

When it comes to being successful with no-till, George Kapusta is convinced there's a critical need for improved management. But the Southern Illinois University weed scientist says farmers can't afford to waste the valuable time they save by no-tilling.

"We must sell growers on making profitable use of their time," he says. "When not plowing and discing, they should be planting and not sitting in the coffee shop or in front of the TV set."

Kapusta suggests no-tillers follow a half dozen key management tips:

1 Avoid heavy, poorly drained soils when no-tilling.
2 Use a narrow row spacing, such as 15 inches for no-tilled soybeans.
3 Get good seed coverage with no-till.
4 Make sure no-till planting is done on a timely basis.
5 Identify specific weed problems.
6 Apply full rates of herbicides.

MORE EQUIPMENT NOT ALWAYS BETTER

If Darryl Smika has any advice for young farmers, it's to avoid investing in a big line of equipment.

"I see no future in it because in the coming years we'll be doing less and less tillage — in fact, I can see the time when we may not till at all," says the USDA agronomist at Akron, Colo. Even now, not tilling is a good strategy.

"In a reduced till, fallow-wheat rotation, yields have been averaging 7-8 bushels per acre greater than conventional stubble mulch farming," he adds.

Trim Tillage Trips

Smika says applying atrazine and a contact herbicide after harvest eliminates two tillage operations in the fall and two more trips the following spring.

"Overall, costs with reduced tillage are less than with conventional farming. Plus erosion control is no doubt better," he explains.

Another advantage is less need for large ca-

TOO MUCH IRON. Here's part of the minimum tillage machinery inventory used by a central Illinois farmer in the 1980s. By moving to no-till, many growers have dramatically sliced machinery investments.

pacity equipment. "If weeds are controlled chemically, you've got little need for doing any tillage," he says.

"Even after herbicides wear off during the fallowing period, there's less need for large equipment since herbicides dissipate at different rates on different fields.

"Plus if you're a large operator, and if your land is spread out, there's a good chance all your fields won't get rain at the same time — meaning you won't have to work all the fields at the same time."

Yields are Exciting

While Smika has been boosting wheat yields with a reduced till program, wheat yields in a complete no-till program are even more exciting.

"With a complete no-till fallow program starting in 1976, winter wheat yields at Akron have averaged 53 bushels per acre compared to just 40 with a tillage-fallow program," he concludes. ✳

"I can see the time when **we may not till at all...**"

'SEWING MACHINE' IDEA TURNED INTO NO-TILL 'SOWING MACHINE'

Even though they're only disturbing a small amount of soil with no-till coulters, some veteran no-tillers feel that's still more tillage than is needed.

Several years ago, no-till founding farther no-tiller Harry Young Jr., of Herndon, Ky., told the *No-Till Farmer* editors he could see the time when no-till planters would work much like a sewing machine. Instead of continually working a 1-3-inch width of soil in the row area with a no-till coulter, he expects to see the development of a "stitch-like" seeding mechanism. This concept would only work a very small area every few inches along the row where seeds would be planted.

Sewing or Sowing?

Well, the "sewing" idea for a no-till planter or drill is already here, thanks to an invention by Texans Arnold Topham and Harry Wilson. They've come up with a no-till planter with a unique planting mechanism that injects seed and fertilizer into the ground with a "stitching" action similar to a sewing machine.

"This is the first truly no-till planter," say the Santa Fe, Texas, investors. "Absolutely no tillage is necessary. The planter simply pierces the ground and deposits the seed and fertilizer like a sewing machine pierces a piece of cloth."

To provide positive depth control, a sled-type shoe on the no-till planter unit rides along the ground and can be easily adjusted for seeding depth.

Spacing between plants within the row can be

> "The spacing between plants can be varied from **2 inches to 2 feet...**"

varied from 2 inches to 2 feet with a simple gear change. Spaces between the rows can be as close as 4 inches to provide extremely narrow rows, which are ideal for no-tilling forages, small grains and maybe soybeans.

This "narrower-than-narrow" row spacing is possible since room is not needed for no-till coulters or planter openers.

Since the no-till planter injects seed and fertilizer into the ground instead of tilling the entire narrow row area, the unit requires much less horsepower to operate than a typical no-till rig.

Labor, fuel, and equipment costs can all be reduced. The no-till rig also saves on seed and fertilizer costs since the planter has the ability to accurately drop exactly the right amount of seed and place the correct amount of fertilizer next to the seed.

Seed is Injected

Each planting unit features alternating injectors "loaded" with seed and fertilizer via tubes from planter boxes. Ground-driven injectors drop to a pre-set depth to place seed or fertilizer in the row area.

Another big advantage is that the ground surface is not broken up, as is the case with conventional, minimum tillage, or even regular no-tillage, which results in considerable moisture savings.

Editor's Note: As far as I know the "sowing machine" concept never drew enough interest to reach the reached the manufacturing stage. ✳

SAVE 35% IN MACHINERY COSTS WITH NO-TILL

Need more ammunition to sell neighbors on the many benefits of no-tillage?

If so, you may be interested in these machinery investment figures worked up by Paul Schaffert of Indianola, Neb. Our "No-Till Farmer of the Year for 1977" penciled out both conventional tillage and no-tillage machinery costs for a grower in the western Great Plains.

While these costs are based on farming a sizeable acreage with the ecofallow system of utilizing chemicals instead of tillage for weed control with a wheat and fallow rotation, you could work out similar figures for no-till machinery needs in your area. ❄

Machinery Needed in Western Nebraska — 1977
Black fallow tillage (conventional tillage)...

Equipment	Cost
Combine, 6-row corn head and 20-foot grain head	$ 80,000
Two tractors, 130 hp equivalent	$ 70,400
Planter, 6-row	$ 6,800
Disc, 21-foot tandem	$ 9,250
Tri-flex, 24-foot with five blades	$ 8,200
Duckfoot, 20-foot with sweeps	$ 6,000
Plow or ripper, 6-bottom or 5-chisel	$ 5,100
Cultivator, 6-row	$ 4,600
Rotary hoe, 6-row	$ 3,100
Total machinery investment	$193,450
No-tillage (ecofallow system)...	
Combine, 6-row corn head and 20-ft. grain head	$ 80,000
One tractor, 100 hp	$ 30,000
Planter, no-till for 6 rows	$ 10,400
Sprayer, 20-foot spray boom	$ 3,500
Total machinery investment	$123,900

— *No-Tillage Farming textbook*

NO-TILLERS EVALUATE SEED COVERERS, SEED FIRMERS

REBOUNDERS VS. KEETONS. Specific farmer-to-farmer discussions during up to 80 No-Till Roundtables at the National No-Tillage Conference program help attendees learn how other growers make specific techniques work.

During the 2000 National No-Tillage Conference, one of the 80 No-Till Roundtables focused on the pros and cons of using seed firmers (Keeton) and seed coverers (Rebounders). Here are a few key points from this discussion:

1 The Keeton acts as a seed firmer and comes in contact with the seed.

2 The Rebounder does not contact the seed as it's designed to be a seed covering attachment.

3 The Rebounder will scrape up dry soil, which will end up surrounding the seed.

4 Firming seed in the bottom of the trench is critical for no-till success.

5 Concerns were voiced about the Keeton's "drum-stick" effect in moist soil where mud builds up on both sides of the unit.

6 While other no-tillers admit mud can build up, they haven't seen it lead to reduced yields.

7 If the Keeton unit builds up with mud along the bottom edge, it means you are no-tilling in extremely wet conditions. Wait until the soil dries.

8 Excessive buildup of moist soil with a Keeton attachment can be created by running the center coulter too deep in a three-coulter zone-till system. When you bring up too much mud with the center coulter, wet soil sticks to the Keeton seed firmer.

9 The Keeton units got high marks from some attendees who used them on no-till drills, especially older no-till drills, which don't have good trench closing abilities. This is important with "fluff and plant" drills and coulter cart drills.

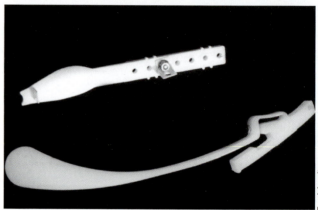

WHICH WORKS BEST? The Rebounder (top) and the Keeton seed firmer (bottom) both have a valuable role under no-till seeding conditions.

10 Several attendees had used Keeton units on John Deere 750 no-till drills with good results.

11 Rebounders work well with Case IH Cyclo no-till units and help avoid vertical seed bounce.

12 Without a Keeton seed firmer, air pockets develop under the seed, which can lead to poor germination.

13 If you want to no-till fast, use the Rebounders. If planting speed isn't a concern, go with the Keetons.

CONTROLLING EQUIPMENT COSTS CRITICAL TO NO-TILL SUCCESS

March 2001

Roughly one-third of the farmers in the U.S. are "so well-heeled" that they are financially secure for at least a decade, Moe Russell says. However, the Panora, Iowa, farm management consultant warns that roughly half of the remaining growers could be forced out of business by financial failure.

The remainder, he says, "will have an opportunity to make more money in the next 10 years than they did during the past 25. There's money to be made in agriculture, and no-tillers can make more money than anybody else."

Russell maintains the difference between farming's winners and losers is clear. The people who make a lot of money are doing things the also-rans are not — effectively managing marketing, equipment costs, inventory control, agronomic ideas and input costs. While all are important, Russell says too many farmers spend the majority of their time trying to manage input costs, although it has the least potential benefit.

Tracking Costs

Equipment represents a large portion of farm operation expenses, Russell notes. "Machinery ownership costs can amount to 25% of the balance sheet value of equipment," he says. No-tillers should know how much they invest in equipment as a per-acre cost.

To calculate that cost, divide the balance sheet value of all of your farm equipment (taking depreciation into account) by the number of acres farmed. Among the small, medium and large farms he analyzed in 2003, equipment investment ranged from $78 to more than $760 per acre, Russell says, evidence of the wide disparity between operations.

No-Till Advantage

No-till enjoys a profitability edge over other systems due to its reduced demand for equipment and labor.

A study at Iowa State University determined that eliminating tillage saves no-tillers $20 per acre each year. However, Russell calculates the savings at $10 per acre to allow for extra spraying with no-till.

If the savings sound less than dramatic, Russell

> "No-till gains an edge in profitability due to its reduced demand for equipment and labor..."

notes the $10 per acre figure, all by itself, raises the average no-till farm's return on equity (ROE) by 2.27%, a significant step toward the 12%-18% he recommends as the target for profitable farms. He says the average ROA for all Fortune 500 companies during 2002 was only 2.3%.

A widely accepted belief is that one way to improve operational efficiency is to spread your equipment use across as much land as possible, driving down per-acre costs. While Russell says that the idea reduces per-acre costs, it does not necessarily increase profitability. Timeliness and labor costs must also be considered.

Let Some Land Go

In truth, he says, farming some fields simply isn't cost effective. He recommends tracking yields from each of your fields for 3 years, then deciding whether to continue farming them. If poor-producing land is leased, negotiating lower lease payments might be an option, as would not renting the land.

Emphasizing the point to attendees at the 2003 National No-Tillage Conference, Russell recalled a conversation with a grower who intended to seek lower lease payments on a few unproductive fields.

"I asked him if he was afraid of losing the land to someone willing to pay the current price, and he said, 'If I lose it and my neighbor gets it, I win twice!'"

Another equipment consideration is keeping machinery in good operating condition. Russell recalls a planting variability study conducted by Iowa's Heartland Cooperative, which looked at 59 farms over 2 years.

"By eliminating skips and doubles, they increased corn yields by up to 17 bushels per acre," he says.

Managing equipment can lead to big differences in profitability.

"Successful businesses — including farms — make small incremental changes to improve themselves," he says. "Management is always tweaking the operation and improving it slightly." ✿

"Use a drag chain to **cover up your mistakes...**"

THREE TIPS FOR BETTER NO-TILLING

March
2001

Based on his own no-tilling experiences and working with numerous growers as a crop consultant, No-Till Innovator Ed Winkle shares a few equipment ideas you can put to immediate use.

What About Spider Wheels?

The Blanchester, Ohio, no-tiller likes spider wheels on a no-till planter because they leave loose soil over the seed. "Let the seed float a little in the slot with a spider wheel," he says. "If you leave that soil loose, you'll get good soil-to-seed contact that will push moisture into the seed and help germination.

"These wheels help provide the perfect soil density that will let the seed germinate and emerge at the most rapid speed possible without creating problems.

"We often no-till when it is cold and damp and I want the spider wheels to warm up the furrow by 4-6 degrees within a couple of hours."

Winkle pulls a 30-inch looped drag chain behind each spider wheel. The chain drags a small amount of soil over each row, evens out soil density over the seed and leads to more even emergence.

What About Gauge Wheels?

Winkle believes the Case IH gauge wheel is superior in no-till as it prevents hairpinning, which occurs when residue is pushed down into the planter slot and seed ends up on top of the residue instead of being placed in moist soil. "Since I moved the coulter 2 inches off the row to split the residue and get fertilizer down and moved residue off the row area with a row cleaner, I don't have hairpinning situations anymore," he says.

"This gauge wheel dishes out the slot, lifts the soil and explodes the soil. You may have to modify them a little to install on John Deere, White or Kinze no-till planters, but they definitely work."

Winkle says most other opener brands were designed for conventional tillage where seed is pressed into the V-slot.

"If your no-till planter that's equipped with these spider wheels is properly adjusted to drop seed at the right spacing, you can press all of the seeds equally and evenly in the V-slot," he adds. "If you don't, use a drag chain to cover up your mistakes."

Harrowing Experience

Like many no-tillers, Winkle has pulled a harrow behind a no-till drill. He finds it provides more of a cosmetic look to no-tilled fields rather than boosting yields.

"We've done better with a harrow mounted backward on a no-till drill and it has worked well for effective residue management," he says. "However, using a harrow behind a no-till planter hasn't boosted our yields." ❀

NO-TILL LOOSE ENDS...

NO-TILL SOLVES DENVER AIRPORT'S DUST WOES

When the Denver airport was built in the mid '90s, regulations called for controlling blowing dust from nearby farmland. Thanks to no-till, there was soon a big improvement in air quality.

Guy Swanson recalls that Box Elder Farms managers Paul and Connie Warner used two Yielder 2520 no-till drills pulled with a pair of high horsepower Steiger Tiger tractors to reduce concerns with blowing dust from fields located near the airport.

Swanson, the president of Exactrix Systems in Spokane, Wash., and a member of the family that developed the Yielder no-till drills, says land around the Denver International Airport continues to be no-tilled more than 30 years later. ❀

Drills, Planters that Left
the No-Till Scene

"A number of planters and drills **were introduced in the 1970s and 1980s by manufacturers who thought they had a head start** on overcoming seeding concerns specific to no-till..."

Here are a few tools from the early days of no-till that are no longer around.

These photos represent some of the planters and drills introduced in the 1970s and 1980s by manufacturers who thought they had a head start on overcoming the seeding concerns that were specific to various no-till conditions.

At the same time, most full-line farm equipment suppliers were still preaching the gospel that their "all-purpose" drills and planters were designed to work with all tillage systems under all kinds of residue and environmental conditions.

But the no-till market showed a number of drill and planter manufacturers that they were wrong. Many

no-tillers quickly recognized that specific planter and design designs, engineering, attachments and innovations were needed to effectively seed under their cropping conditions.

Dads and grandfathers may recall seeing some of the machines that follow on the next several pages that were used in North and South America and in England during the early days of no-till. ✺

"The no-till market **proved a number of drill and planter manufacturers** were wrong..."

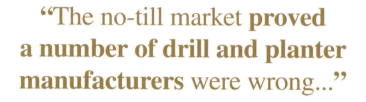

NO-TILL-AGE®

March 2016

How Crop Diversity Impacts Corn Yields

1900 to 2006 comparison with continuous corn

Crop Rotation	Corn Yield Increase
Continuous corn	—
Corn, soybeans	7%
1st year corn after 4 years alfalfa	15%
2nd year corn after 4 years alfalfa	10%
3rd year corn after 4 years alfalfa	8%
4th year corn after 4 years alfalfa	7%
Corn after oats/wheat/red clover hay	16%

— *Penn State University*

NO-TILL-AGE®

How Closing Wheels Compare

(Trials conducted with no-tilled corn from 1997 to 1999 on the Larry Thompson Farm, Todd County, Ky.)

Attachment	Average Corn Yield
Martin closing wheels	145
Conventional closing wheels	140

— Research conducted by Opticrop

NO-TILL LOOSE ENDS...

June 1999

THE NO-TILL QUITTER'S LIST

1 I can't wait an extra 3 days to no-till corn behind my conventional-tilling neighbors.
2 The fields are just starting to get fit for no-tilling when rains come in the spring.
3 The soil gets harder every year with no-till.
4 I still have erosion with no-till.
5 Broken no-till stalks will wash into lower areas of fields.
6 Getting consistent no-till stands is extremely difficult.
7 No-till corn always looks ugly until it's 2 feet tall.
8 No-till yields have dropped compared to yields for conventional-tilling neighbors.
9 I'm just too frustrated with no-till results.
10 It's easier to plow. ✳

Most of the 'Big Guys' Spent Their Time Bad-Mouthing No-Till

By the early 1970s, shortline farm equipment manufacturers had recognized no-till wasn't going to be the passing fad Case, John Deere, International Harvester, Massey Ferguson and other majors were hoping for.

When no-till arrived on the horizon in the 1960s, there was little in the way of specialized equipment to get the job done. As a result, early-day no-tillers had to be adept at modifying existing planters and drills in farm shops.

By the late 1960s, Allis-Chalmers had introduced the first true no-till planter, and American agriculture has never been the same. They were the only major line farm equipment manufacturer to show much interest in no-till for a number of years.

On the other hand, other major

farm equipment players — such as Case, International Harvester, John Deere, New Holland, Massey Ferguson, and Ford — had bad-mouthed no-till for years. They saw no-till's growth as a threat to the continued sale of tillage tools and the bigger horsepower tractors needed to pull them. ❀

INNOVATIVE NO-TILL ROW CLEANER DESIGN EVENTUALLY LEADS TO NEW BUSINESS OPPORTUNITY

Tough experiences with no-till on the family's home farm led to a number of innovative planter attachments that impacted the entire farm equipment industry.

By Howard Martin, Martin Industries, Elkton, Ky.

GAME CHANGERS. Martin closing wheels have helped thousands of no-tillers boost emergence, improve stands and produce higher yields.

Harry Young created quite a stir with his pioneering no-till efforts during the 1960s. I was fortunate to live just one county east of his farm and attended several of the field days held at his farm. Harry was a great teacher and host.

It wasn't long after a friend bought an Allis-Chalmers planter and I rented it to plant some soybeans. After that successful experience with no-tilled soybeans, I mounted a bar on the front of my John Deere planter and attached wavy coulters.

No-Till Worth a Try

Like all planters in that era, the depth-gauging wheel trailed behind the opener disc and the furrow wall was not controlled as it was being opened. To overcome these concerns, I added a depth tire with a large center rib (similar to what was used on the Allis-Chalmers planter) and was able to get some seed-to-soil contact even though there was very little

> "Whacking weeds **was not what I had planned to be doing at age 45...**"

soil left in the row area to cover the seed. That's because the no-till coulter and seed disc had thrown most of the soil out of the row area.

The first year I no-tilled corn, the planting season was dry and my stands were good. The following year was a wetter spring and my no-till corn stands were very poor. That corn should have been replanted, but by the time it was dry enough, it was time to plant soybeans.

It was obvious that my soil type was not suitable for no-till corn using the equipment available at the time. Unfortunately, the next several years were spent wearing out a variety of large horsepower tractors and tillage tools in our corn fields.

Tough Times in 1980s

The 1980s were not kind to many grain farming operations. Being highly leveraged with adjustable rate credit and marginally productive soils, it

SHOP TALK. Howard Martin started developing several no-till planter attachments in his farm shop during the 1980s.

During this time, the Deere planter factory manager called to tell me they were not going to build my row cleaner. He said it was a good idea, but not good enough to add to the Deere product line.

He wanted to know who I would suggest they license it to so they could get back some of their engineering expenses. My suggestion was to contact Yetter Manufacturing and they ended up licensing my row cleaner from Deere.

A New Start

I was glad when 1990 arrived. The past decade had been very stressful for our family and I was looking forward to a time of less pressure. Soon after I received a license from Deere to use their row cleaner patent, which was ironic since it had been my patent. I'd escaped the frying pan only to fall into the fire!

My family again came to my rescue and we were able to establish our Martin row cleaner name in the market and secure enough loyal customers to start a successful business. Both our banker and the Internal Revenue Service were soon happy.

My accountant warned me that earning enough income to be required to pay taxes was going to be very painful. Since we previously had little money, I told him it was OK, as the government would be letting me keep me at least 50 cents of each dollar of income.

only took a couple of droughts to turn our farm's balance sheet red. We ended up surrendering our best farm ground and most of the machinery, except for keeping enough tools to farm our remaining small acreage.

My family stood with me as they had done through a number of earlier tough times. Our sons went to work in town, and my wife, daughter and future daughter-in-law cooked and served BBQ at our farm machinery auction to earn needed cash.

Soon after the auction, my wife bought a used commercial lawn mower and soon she had more work than she could handle alone. Before long, I was recruited to operate the string trimmer. However, whacking weeds was not what I'd planned to be doing at age 45.

No Royalty Checks

At that time, it had been almost 5 years since I had sold my row cleaner patent to Deere. Due to confusion and indecision, they had not produced any of my row cleaners. My vision of large dollar royalty checks arriving in the mailbox was growing dim, especially with a daily dose of salty perspiration drenching my eyes from tackling lawn weeds.

> **"I received a license from Deere to use their row cleaner patent, which was ironic since it had been my patent..."**

We sold our first Martin row cleaners in January of 1991 and added closing wheels in 1994. A unit-mounted fertilizer opener and other attachments were later developed.

By 2000, we had relocated the business to our county seat town of Elkton and moved into a state of the art facility that was within sight of a newly opened four-lane highway.

I take a great deal of satisfaction from having played a part in the no-till movement and finding new ways to help farmers prevent soil erosion. I treasure all of the friendships made while dealing with no-till customers across the country.

BASAGRAN HELPED MANUFACTURER DEVELOP DRILLS FOR NARROW ROW, NO-TILL SOYBEANS

John Tye shares why this soybean herbicide was key to getting started with drilling no-till soybeans, and how the introduction of no-till equipment by the major manufacturers helped the practice become more accepted.

By John Tye, The Tye Company, Lockney, Texas

"My initial reaction was that no-till will never work..."

Frank Lessiter

DEMOS CRUCIAL. In the early days of no-till, field demonstrations were essential so growers could see drills perform under high residue conditions.

My initial reaction to no-till was "this will never work. You've got to plow the ground and kill the weeds, or the weeds will overtake it."

I wasn't the only one who was skeptical of no-till. Both farm equipment and chemical dealers were in the "this will never work" mindset. They questioned weed control and whether the herbicides would work.

While farmers were also in the same mindset, they were willing to listen because they were open to doing something different, if it meant cutting costs or increasing yields. It was, "I'm willing to listen, but it's going to take some convincing."

BASF introduced Basagran herbicide in 1970, which you could use to control weeds in no-till soybeans. The company had been looking for new ways to plant soybeans, because their research indicated growers could get better yields with narrower rows. But obviously if the rows were narrow, you couldn't cultivate.

With no-till and Basagran to control the weeds, there was no reason not to try narrow rows. BASF realized there was going to be a big opportunity if this idea worked.

We initially worked with BASF on narrow-row beans in conventional tillage. That worked well, but

NO-TILL INNOVATOR. John Tye, shown at left, with Lessiter Media president Mike Lessiter says his company led the way when it came to developing early day no-till drill innovations.

we could both see a real advantage to no-till narrow row — initially for double-cropped soybeans.

So we got together with BASF and configured some drills with crude, homemade coulters to no-till soybeans in 8-inch rows. It worked, and that's how we got started in no-till.

We then branched off into no-till pasture renovation at the request of Texas A&M University agronomists. They wanted to reseed grassland and pasture ground, but didn't want to rip up all the ground before seeding it, which was the current method.

So we put a couple no-till seeders together for the Texas A&M people to try. That's when we came up with the idea that you needed fairly small no-till drills, because the tractors were light on horsepower.

Manufacturers Catch Up

The earliest folks in manufacturing no-till equipment were Allis-Chalmers, who had no-till planters, and the Fleischer folks in Nebraska, who were promoting ridge-till and their Buffalo planters and cultivators. Both companies were ahead of their time, because we didn't yet have the ability to control weeds with over-the-top chemicals.

When Roy Applequist with Great Plains got out of the bearing business and into the implement business, no-till was among several options he looked at. For a long time, it was just Great Plains, Crustbuster and us in the no-till drill business.

No Deere Green, No Red Case IH Tools

In those days, John Deere and Case IH ignored no-till, as it didn't fit in with their program of marketing larger horsepower tractors, moldboard plows and wider and wider tillage tools.

John Deere later introduced a no-till pasture renovation machine that was developed by staffers at the University of Kentucky (See page 251). It had all these gear boxes, fast rotating coulter-like things that looked like saw blades and it was expensive. The maintenance and upkeep were horrible, but this was Deere's first run at no-till.

We did some work with John Deere after they decided they really needed to be in the no-till business. One of our first moves was to take a regular drill and hook it on the back of a carrier with a big bar and coulters. B.G. Schluetter in Illinois had patented the idea, so we paid him a royalty and made it marketable and manufacturable.

> "Once John Deere signed onto no-till, farmers **thought it must be OK...**"

That's when John Deere came to us and said, "We'd like you to private label that for us, too. We really need to be in the no-till business and we've got some new stuff coming up, but it's not going to be here in time and we need to do something." So we wound up making a few of these units for them, but never put an agreement together.

There appeared to be a big internal battle going on at Deere — the marketing folks wanted a product to be in the market immediately, but the engineering folks wanted to wait until they could develop their own. It appeared that the engineering side won.

A couple years later they brought out their first John Deere genuine designed and built no-till drill. And once John Deere signed onto no-till, farmers thought "no-till must be OK."

NARROWER-ROW SOYBEANS STARTED A NO-TILL EFFORT

*The Great Plains Manufacturing founder shares how the company got its start
with no-till drills, and the adjustments they've made to their equipment lineup over the years.*

By Roy Applequist, Great Plains Manufacturing Co. Salina, Kan.

INITIAL FIELD TEST. Great Plains Mfg's. first product, a 30-folding grain drill, is about
to get its first field test during the fall of 1976 — the year that Roy Applequist founded the business.

Great Plains Mfg.

"We ended up rebuilding each drill, which was one of the best things the company ever did..."

We first learned about no-till when Allis-Chalmers came out with a planter with no-till coulters. To my knowledge, that was the first commercially available no-till planter out there in the late 1960s.

As people were talking to us about no-till soybeans, we decided there was a market for a no-till drill. We built our first no-till drill and showed it to farmers at the Milan No-Till Field Day in Tennessee around 1980. I think CrustBuster and Tye were also there, and introducing no-till drills.

The no-till movement had started, but it didn't really grow rapidly because there was also a lot of interest at the same time among many growers in trying conventionally-tilled narrow-row soybeans.

At first we were selling more narrow-row soybean grain drills than no-till grain drills.

In the early days of no-till, farmers were really curious about soil penetration. Growers wanted manufacturers to prove their no-till equipment could penetrate ground as hard as a gravel road.

Our no-till drills had always penetrated really well because we had the needed weight plus a coulter running in front of the row units. We could plant just about anywhere, but it wasn't always the smart thing to do. A farmer shouldn't be planting if he doesn't have reasonable soil conditions.

Moving Gauge Wheels

Several years after we developed our no-till grain drills, we added folding no-till drills in 24- and

Mike Lessiter

GRASSROOTS GROWTH. Great Plains founder Roy Applequist holds the original grain drill model he constructed after asking 100 central Kansas farmers what they needed.

30-foot widths. With these units, the coulter was frame-mounted directly ahead of the opener.

This style of no-till drill started selling really well and in just a few months we had sold well over 200 units. These were built with the gauge wheels on the box ends and the transport wheels were mounted in the front.

It was a heavy machine that penetrated really well, but it was difficult to turn in the field. It created ruts when making 180 degree turns. This was not much of a problem in no-till, but this was not acceptable in conventionally-tilled fields.

Because of this problem, we ended up rebuilding every one of these drills by taking the gauge wheels off the ends and moving them in front of the drill. We set up locations in Illinois and Indiana where we had sold the largest volumes of these drills. Farmers would bring their drill to the location and our crew would switch the gauge wheels to the front.

This was an important and expensive "fix" for us, but it had a big impact on the future of Great Plains. It was one of the best things the company ever did as it demonstrated we'd stand behind our products no matter what.

Overseas Expansion

No-till really changed our company's focus on equipment. It became an important part of our business in the late 1980s and well into today

In the 1990s, we began selling our equipment overseas. The first major no-till market was in the Ukraine and Russia. At the time, Monsanto had an excellent package deal where farmers in those countries could purchase seed and chemicals along with a sprayer, no-till drill or no-till planter. We had a lot of the no-till drill business and Kinze had a lot of the planter business.

This whole no-till system package drastically increased the farmer's yield, so it was a really strong program and we sold hundreds of drills. Today, Great Plains still sells a broad range of models that continue to include no-till drills.

No-till is still a major part of our business, but in some cases we've evolved away from true no-till. We've seen a lot of situations where a grower can add a very light tillage operation that more effectively manages residue, creates additional soil tilth and leads to higher yields. In order to accomplish this, we developed the Turbo-Till line and then followed with our Turbo-Max tools.

Creating that mulch on top eliminates a lot of runoff in a no-till field. As a result, we believe a shallow tillage pass can be a good thing in no-till.

Editor's Note: *In 2016, Kubota acquired Applequist's company for $430 million, marking the Japanese tractor manufacturer's first major entry into other ag markets.*

April 1993

NO-TILL-AGE®

How Implement Sales Mix Changed

(Share of total tillage tool sales)

Implement	1967	1990
Moldboard plows	31%	2%
Discs	46%	24%

— Deere & Co.

TIME, KNOWLEDGE ABATE INITIAL SKEPTICISM TO NO-TILL

The former leader of DMI (acquired by Case IH in 1998) remembers farmers' early skepticism about no-till and conservation tillage, and how the tillage landscape has changed dramatically from the 1970s in central Illinois.

By Bill Schmidtgall, DMI Inc., Goodfield, Ill.

"Early on, you could see no-till start to evolve..."

People were very skeptical when no-till first came on the scene.

I remember a lot of negative comments, such as that if your soil was hard, you could no-till for 3 or 4 years and then your soil tilth was going to magically reappear.

Farmers were saying, "Yeah, right, I don't think I want to suffer for 4 years to see if that radical statement is right. I've only got 15 or 20 years of profitability on my watch."

Major Worries with No-Till Yields

But farmers were looking for ways to reduce costs without reducing the bottom line. While the chemical people were pushing no-till pretty hard, many farmers were still skeptical of what their yields were going to be. In reality, these concerns

TRIMMING CROPPING COSTS. DMI helped many farmers find new ways to cut costs with newly-designed tools that fit all segments of the growing conservation tillage market.

probably ended up helping drive better farm recordkeeping for many growers.

The longer no-till has gone on, the more it's been studied and the more people have learned how to effectively manage it with different soil types. You could see it soon start to evolve and take on some critical mass.

Strip-Till Had a Role

Where I'm from in central Illinois, I see conservation tillage everywhere today. Strip-till today seems to be a replacement for some of the massive no-till acres of the past. Growers that are strip-tilling like how the soil warms up and the more accurate fertilizer placement it provides. ❀

MAJOR WORRIES. Bill Schmidtgall (right) recalls many negative comments when no-till was getting started in the Corn Belt.

WHEN 4-ROW NO-TILL PLANTERS SOLD FOR $3,000

Even at that price, a dealer could make a decent profit unless the farmer talked him into including extra seed plates or drive sprockets.

By Ronnie Holt, Martin Industries, Elkton, Ky.

4-ROW POPULARITY. Ronnie Holt recalls 4-row no-till planters being the most popular units in 1979, as the 6-row no-till rigs weren't able to fit through extremely narrow farm gates.

On October 10, 1979, which also happened to be my 29th birthday, I went to work at Planters Hardware. This was the Allis-Chalmers dealership in Hopkinsville, Ky., which was located just 11 miles from the historic "First No-Till" roadside marker on the Harry Young farm at Herndon, Ky.

Harry was the first farmer in the area to try no-till and his son, John, and grandson, Alex, continue to operate the family farm today.

Four Rows Most Popular

In 1979, Allis-Chalmers introduced the all-new 333 No-Til planter. It could no-till corn in 30-inch rows and soybeans in 15-inch rows. The most popular configurations were a 4-row frame with four 30-inch corn units and seven 15-inch soybean units, or a 6-row frame with six 30-inch corn units and eleven 15-inch soybean units.

I was surprised at how easy these no-till planters were to sell. We sold more 4-row than 6-row

planters in 1979, as farmers couldn't get the 6-row planter through narrow gates and barn doors.

In 1979, I remember selling a four-row no-till planter with no-till coulters and a marker bar for around $3,000. I'd make a decent profit unless the customer talked me out of extra seed plates or drive sprockets.

Each row unit was individually driven by a sprocket on the closing wheel and another sprocket on a shaft under the seed hopper. The size of the sprocket determined the plant population.

To eliminate skips and double drops, the seed passed through a seed plate in the bottom of the hopper. The farmer would bring a coffee can of seed corn into the dealership, dump it out on the parts counter and we'd match a seed plate to the size of his seed.

Country Roads and Seed Corn

We also sold a 12-row Allis-Chalmers planter that year, a wing-fold model with 6 row units on the main frame and 3 additional units on each wing. The customer stopped by our dealership after the first day of planting and was bragging about how many acres he had planted that day.

He was in again a few days later and was mad as could be since the plastic seed hoppers had pop-top lids. The farmer and his brothers had just spent the last hour scooping seed corn off the highway. The pop-top lids had come off the planting units on the folded wings and he was losing seed corn as he drove down the road.

There was no auto-steer in those days, but you might catch some wise guy with a tarp strap tied from the steering wheel to the throttle lever. The closest thing to a yield monitor would be the landlord sitting in his pickup truck counting every time the combine unloaded grain. ✺

SCIENTISTS BATTLED OVER NO-TILL ACROSS THE U.S. AND CANADA

The former marketing executive with Allis-Chalmers recalls the opposition no-till proponents once faced in the U.S. and Canada, and the pivotal roles No-Till Farmer and Kentucky no-tiller Harry Young played in the no-till movement.

By Walt Buescher, Allis-Chalmers marketing executive, West Allis, Wis.

RESIDUE CONCERNS. Some growers felt they needed to knock down residue before no-tilling corn in the early days.

Having spent 39 years in the farm equipment industry, I was fully aware of the fact that new machines are not unanimously accepted everywhere immediately. Thank goodness they aren't, as the factory couldn't keep up if a deluge of farm equipment tool orders had come from everywhere at the same time.

Folks Pretended No-Till Wasn't There

No-till practically became standard equipment in Kentucky from the outset and Virginia followed close behind. The Deep South pretended no-till wasn't there, as did many other parts of the country. In fact, University of Illinois agronomist George McKibben at the Dixon Springs Agricultural Center in southern Illinois had a running battle

"Those old Iowa State guys have 30 years of obsolete and outdated research they don't want to throw away before they retire..."

going on regarding the pros and cons of no-till with the university folks at the Champaign/Urbana campus.

One issue of *Prairie Farmer* had an article entitled, "No-Till Fails Again." A professor from Illinois said his no-till plot came out 65 bushels poorer than his conventional till. The next time I saw George, I asked about this testing.

I said, "George, why would a PhD admit in the public farm press that he failed worse than any farmer in the state?" George replied, "Doggoned if I know, particularly since I had some no-till fields that beat his conventional plot."

Iowa also had a running no-till battle. At a field day at Ames, an elderly set of professors said they had 30 years of data to back up their recommen-

NO-TILL PIONEER. Walt Buescher, right, promoted no-till at hundreds of Allis-Chalmers meetings in the 1970s. He's shown accepting an award for his achievements from *No-Till Farmer* Associate Editor John Studer.

dations for conventional tillage. At the next stop on the field tour, a group of young professors told us, "Those old Iowa State guys have 30 years of obsolete research that they don't want to throw away before they retire. It's outdated. Listen to us. No-till and minimum-till is the future."

No-Till's Better But ...

At a Purdue field day, a professor told people on a wagon, "This is our no-till plot. It has fractured the soil structure, it has done everything that should not be done and it has out-yielded the previous year's plot 4 years in a row. No questions, please."

Afterward I asked the professor to fill in the blanks. He said, "Were you up on that wagon?" I told him I was. He replied, "If I had seen you, I would have kept my trap shut."

Pennsylvania also had a civil war going on in regard to no-till. One issue of *Pennsylvania Farmer* told of dismal failure with no-till. From the outset, I figured no-till was a natural for the Soil Conservation Service (now the NRCS), but the government folks paid little attention for 10 years.

At the University of Guelph in Ontario, Jack Tanner and George Jones planted no-till corn on a rocky hillside of Canadian fescue — so rocky you couldn't jab a pitchfork into the dirt without hitting a rock. They never harvested less than 100 bushels of corn and I figured this was done with input costs of only $35 per acre in those days of $1 per bushel corn.

Ontario was low on heat units and there are precious few heat units in the field when there's still snow in the fence rows. But Tanner no-tilled and did real well, thank you.

Worked With Hundreds of No-Tillers

We worked with hundreds of farmers, none more prominent than Harry Young of Herndon, Ky. When we made an Allis-Chalmers no-till movie with the University of Kentucky, Harry starred in the film. He's the man who married paraquat and no-till double-crop beans to the utter amazement of Chevron, who hadn't even heard the word "no-till" at the time.

After they saw what Harry had done with paraquat, they began to toot the horn and beat the no-till drums.

When *No-Till Farmer* first came off the press in 1972, it was a voice crying in the wilderness. But it filled a need for insightful information and data when most of the other farm publications were hoping no-till was just a fad and that the idea would go away.

Today, you can't go to a county soil conservation annual meeting without seeing a farmer walk away with a trophy for his no-till crops. ❀

NO-TILL-AGE®
How Machinery Costs Stack Up

May 1988

Tillage System	Ownership Costs	Operating Costs
Moldboard plow	100%	100%
Chisel plow	80%	80%
Disc	50%	65%
Till-plant	35%	45%
No-till	30%	45%

— *Iowa State University*

EQUIPMENT, RESEARCH CRUCIAL TO EXPANDING NO-TILLED ACRES

The agronomist and owner of Needham Ag Technologies recalls helping growers adopt no-till wheat, the equipment challenges faced in making no-till work and improvements that still need to be made to move the practice forward.

By Phil Needham, Needham Ag Technologies, Calhoun, Ky.

SPREAD EVENLY. If residue is not spread across the full combine width, it can be a challenge for a no-till planter to achieve a uniform stand and to make no-till successful.

"Almost every piece of equipment is not set up to plant in heavy no-till conditions..."

When I first came to the U.S. in 1989, there wasn't a huge amount of no-till.

The practice faced a lot of challenges because most equipment available on the market at that time, and to an extent still today, wasn't suitable for no-tilling wheat in heavy residue. This was especially true in moist conditions.

There was also a lack of research that clearly showed growers how no-till wheat yields could be equal to or better than conventional tillage.

In addition, there were a lot of farmers out there that didn't want to no-till. They weren't comfortable with no-till, weren't confident that it would work and didn't have the research to back up its benefits. There are still many older farmers out there that are traditional and think they know what works and what doesn't when selecting tillage practices.

Many growers also didn't want to switch for financial reasons. It takes a different set of equipment to be successful in no-till and some farmers don't want to make that kind of investment.

Field Walks Led to New Firming Wheel

Over the years, I've I spent most of my time working as an agronomist. One of the most important things I've had to do is train and educate growers on how to set up equipment for no-till.

Almost every piece of equipment that leaves the factory is not configured and set up to no-till in heavy residue conditions.

We had to help growers understand the needs and requirements of their no-till equipment. This was especially true when it came to row cleaners to part the residue and getting enough weight and downpressure on the planter to make sure the openers stayed in the ground.

I started Needham Ag Technologies because I saw opportunities to develop new planting and drilling attachments that weren't being adequately filled by the manufacturer. I'm out in fields every day for weeks on end. When you're taking thousands of stand counts per year, you start to learn what works and what doesn't. You learn what seeding equipment performs better than other brands and which planter and drill modifications do better than others.

We started out by developing a simple firming wheel. We saw the factory-produced firming wheel on a John Deere drill, for example, didn't do a good job of pressing seeds to the bottom of the seed slot. So we developed an improved wheel that matched those criteria.

I've also done quite a bit of work with Martin Industries. I still do consulting work for them and our company is one of their dealers.

Harvesting Challenges

Another equipment challenge in the 1990s and the early 2000 years was that few combines could evenly spread residue across the full harvesting width of the header. I contacted most manufacturers and tried to encourage them to design and develop an attachment for the back of the combine that would evenly spread residue. If you can't spread residue evenly at harvest, you probably aren't going to be successful with no-till.

NOT FOR EVERYONE. No-till will work for most growers in most areas, but not necessarily in every farming operation, says Phil Needham.

"There are many valuable nutrients that are found in residue..."

In those days, a Case IH or John Deere combine with a 30-foot head would only spread residue 20 or 25 feet wide. And if the residue was tough, it might spread it only 15 or 20 feet. So no-tillers ended up with concentrated residue bands behind the combine with little to no residue in other areas.

Avoid Residue Bands

That double amount of residue compared to no residue was a huge problem. We couldn't cut through the heavy residue at seeding time and the soil stayed cold and wet, while the area between the combine passes was easy to no-till into. Variable plant emergence was the result based on the volume of residue.

Sometime around 2007, I was at a combine research and development facility and met with their engineers and marketing guys. One of the guys commented that all the acres around their combine plant were tilled. He asked, "Why would we want choppers and spreaders to do a more even job, when the fields are being tilled?"

My response highlighted two things. No. 1, there are a whole lot of no-till acres across the country. And No. 2, there are a lot of nutrients in residue.

So even if a grower is still conventionally tilling and the residue isn't spread evenly, then he's increasing nutrient levels within the residue in those heavy bands behind the combine. It's still important to spread residue even in a conventional tillage system since the nutrient content of the residue can be considerable.

We've still got these challenges today when it comes to educating manufacturers and trying to get them to build equipment that works more efficiently under heavy no-till conditions. Few manufacturers build a planter or drill that's designed to no-till into heavy residue and moist soils straight out of the factory. ✣

Manufacturers have definitely improved the residue management abilities with their machines as more people get into no-till. While it's still not great, it's definitely improved. Yet bigger capacity combines are taking on even wider headers and once again the choppers are having a hard time keeping up.

John Deere's Role

I don't think any equipment improved no-till more than the John Deere single disc opener. Before this unit came out, there weren't any good no-till drills that could effectively no-till a crop like wheat or soybeans into heavy residue. Guys were using hoe drills or conventional disc drills, yet neither worked very well in no-till.

The John Deere single disc opener came out in the early 1980s after Roundup Ready soybeans were introduced. This combination was a huge game changer in expanding the U.S. no-till soybean acres.

No-Till Wheat Takes Off

Wheat was a big challenge for no-till in the early 1990s. Many growers didn't have a no-till drill, and those that did weren't confident they could use it to no-till wheat. They were concerned about the yield lag associated with no-till wheat, especially when drilling into corn stalks.

University of Kentucky agronomist Lloyd Murdock came up with the idea of doing some wheat field trials. The small grain grower groups in Kentucky helped fund a no-till vs. conventional tillage comparison. I helped with the trials, which the university oversaw.

What we did was take farmers' fields and split them into blocks to compare no-till wheat to conventionally tilled wheat. We did 20-25 replicated locations in the state over 5 years.

We concluded the yield difference between no-till

NO-TILL-AGE®

What No-Tillers Want in No-Till Planters

March 1998

Row Width	% Farmers
12 inches	3%
15	9%
20	3%
22	3%
24	3%
28	3%
30	70%
36	3%
48	3%
Number of Rows	% Farmers
4	3%
6	26%
8	36%
11	3%
12	26%
16	6%

— No-Till Farmer reader survey

and conventionally tilled wheat was almost nothing. But the costs were almost always less with no-till, mainly because of the reduction in tillage expenses.

That gave us the ammunition to go to growers and say, "Look, over 5 years at 20-plus locations, we've concluded that there is no yield lag for no-till wheat."

The majority of wheat in Kentucky is now no-tilled. That's huge, because there was almost no no-tilled wheat in Kentucky when we got started. Almost all of the soybeans that are double-cropped behind wheat are now no-tilled as well.

Improvements Still Needed

No-till will work for most growers in most areas, but it doesn't work everywhere. It's a tool the industry can use to maintain yields, reduce costs and more than anything else keep soil in the fields to stop erosion.

We need to expand and improve the number of acres being no-tilled. To boost that number, we've got to educate more growers and do more research — to encourage more growers to take no-till to the next level.

We also still need manufacturers to design, develop and market seeding equipment that's more suitable for no-till, as there are still numerous deficiencies. We still need manufacturers to build combines that do a better job of spreading residue.

Fields Must Drain

Drainage also has to be improved in certain higher rainfall areas. If you don't have a well-drained field, no-till probably won't provide great results.

I encourage growers to consider a cover crop. In many areas no-till and a cover crop will help keep the soil covered, improve soil biology and accelerate a grower's farming system to get better no-till results. But again, we need more research so we can tell farmers, "Look, if you plant this cover crop on this soil type ahead of that crop, these are the kind of results you can expect." ❀

NO-TILL-AGE®

What No-Tillers Want in No-Till Drills

April 1998

Row Width	% Farmers
6 inches	3%
6 2/3	9%
7	3%
7 1/2	3%
8	3%
10	3%
15	70%

Number of Rows	% Farmers
10	4%
12	8%
15	37%
20	27%
24	4%
30	12%
33	4%
45	4%

— *No-Till Farmer reader survey*

DESIRE TO MAINTAIN YIELDS LEADS TO STRIP-TILL DEVELOPMENT

When the founder of Progressive Farm Products, witnessed low no-till corn yields, the No-Till Innovator invented a row unit that would implement the best of both worlds — tilling a narrow strip for the corn and leaving the rest of the ground untouched.

By Rich Follmer, Progressive Farm Products, Hudson, Ill.

STRIP-TILL INNOVATOR. Richard Follmer designed the first strip-till rig for Midwestern growers dealing with cold soils.

"We fooled corn into thinking a field cultivator had gone through the field..."

I came up with the idea for strip-till when farmers who were no-tilling came to me with low corn yields due to poor stands. I thought the poor yields were due to the amount of residue covering the soil, which left the ground underneath cold and damp.

In 1988, I started working on a row unit that allowed a grower to till a narrow strip that would allow the soil to warm and dry out quicker before planting. The idea was to fool the corn plant into thinking a field cultivator had gone through the field.

Our first design was manufactured in 1992 and the results were tremendous. It brought corn yields back up to where they were comparable to conventional tillage.

Challenging Critics

I remember being a speaker in the final session at the 1995 National No-Tillage Conference. I ended up afterward answering questions for 3 1/2 hours out in the hallway. The reaction was unbelievable.

We did get some push back from no-till purists. They felt we were disturbing the soil and not protecting the residue. But farming in 30-inch rows, we only tilled a 10-inch strip and still had 20 inches of undisturbed soil. The main benefit was better yields.

Another time I was speaking at a meeting in Illinois and had some hecklers. I asked one guy how long he had been no-tilling and he said 15 years. Then I asked him how his yields were and he said they were doing fine.

"I knew we had hit a home run with strip-till when Jim Kinsella adopted it…"

I said, "Are they really? You're probably the only one in your neighborhood that'll say that."

He asked what I meant and I said, "Well, unless you made a test strip in each of the 15 years you've been no-tilling, and borrowed a field cultivator from a neighbor to make a couple rounds in the middle of a field to compare the yield, I'd say you're losing 30-40 bushels of corn per acre."

He got mad in front of the whole crowd and started heckling me. I explained how he could afford to buy a bigger horsepower tractor and strip-till bar for an extra 40 bushels per acre of corn.

A year later, I was at the Louisville Farm Machinery Show and he came up to me, shook my hand and told me he was the guy at the meeting in Illinois who had argued with me.

I'm thinking he's going to really nail me now, as he's come back to prove his point, I'll never forget it what he told me. "I owe you an apology. I did what you said to prove you wrong. I told everybody I was going to prove you a liar."

Instead, he had harvested an extra 38 bushels per acre in the rounds that were cultivated. And he asked me where he could buy a strip-till bar.

Yields Hurt No-Till Adoption

I think the anticipation of low yields were the main reason that no-till adoption was so slow.

I remember some no-till purists in our central Illinois neighborhood that didn't want to run an anhydrous bar and disturb the soil. They applied nitrogen on top of the ground and never worked it into the soil.

Their corn yields suffered because the nitro-

FILLED AN IMMEDIATE FARMING NEED. Progressive Farm Products evolved into becoming a major player in the fertilizer application and strip-till equipment business.

gen was volatilized in the sunlight when they didn't place it in the ground. A lot of no-tillers were suffering because of the way they were applying nitrogen. That gave no-till a bad name and slowed its growth.

You have to realize that when no-till started, farmers in the south were doing it to conserve moisture. I went and talked to John Bradley at Tennessee's Milan Experiment Station who told me they were no-tilling because it didn't rain much rather than saving trips over the field. But farmers up north adopted no-till because they could farm with less equipment, fuel and labor.

Strip-Till Takes Off

Strip-till helped farmers who were leery of reduced tillage. We had several growers in our area that didn't want to give up tillage, including me. A lot of my friends were going to no-till and talked about how a farmer could get rid of most of their equipment and just keep a planter and a small horsepower tractor. That was fine for them, but I didn't want to give up any yield.

When I decided I wanted to move into no-till, I used strip-till as the way to get into the practice. I've no-tilled a lot of soybeans, but have never no-tilled corn without using strip-till.

I knew we had hit a home run with strip-till when Jim Kinsella adopted it. He was a no-till purist, but adopted the strip-till system after he visited our factory in 1992. He saw what was going on and recognized strip-till is part of no-till and not a separate practice.

Editor's Note: *In 2010, Follmer sold his 29-year-old company to Denmark's Kongskilde Group.* ❁

DESIRE TO PRESERVE CLOVER PASTURES LAUNCHED NO-TILL SUCCESS

John Baker explains how the need to convert volcanic ash into productive pasture ground in New Zealand led scientists to investigate the feasibility of no-tilling grasses.

By John Baker, Cross Slot No-Tillage Systems, Feilding, New Zealand

RESIDUE MANAGEMENT. Featuring a horizontal seeding slot, the Cross Slot system places crop residue back over the seed slot while retaining 70%-90% of the residue on top of the ground.

Gavin Porter

"No-till must focus on capturing and sequestering as much atmospheric carbon as possible..."

Edward Faulkner was probably the first person in the world to publicly question plowing with his now prophetic statement made in his 1943 book, *Plowman's Folly:* "No one has ever advanced a scientific reason for plowing."

In reality, the first recorded scientific efforts to seriously investigate the feasibility of sowing seeds directly into unplowed soils started with a group of scientists and consultants in the 1950s in my home country of New Zealand.

I began studying no-till in 1967 after joining Massey University as a farm machinery specialist

2 years earlier. My research continued until 1995, when the core members of my team and I left the university to form Baker No-Tillage Ltd., with the purpose of turning the science behind no-till into usable hardware.

This led to the invention of our Cross Slot drill, as well as an early opener design known as the Baker Boot.

The aim of our team was, and still is, to make no-till as biologically fail-safe as possible and open the door to no-till becoming the mainstream method by which the world produces its food.

Clover Drives No-Till

Pastoral-style farming is the backbone of agriculture in New Zealand. Years ago, converting native bush and scrub land into productive pasture ground was a high priority in New Zealand.

One of the last frontiers in this conversion process was an extensive elevated plateau of volcanic origin found in the central area of the country's North Island. Due to the volcanic ash, the easily worked land was low in fertility. It was initially cleared by mechanically crushing and burning the scrub.

Mixes of grasses and clovers were then spread over the ashes by hand and later by aircraft. Nature was left to do the rest.

Because of the low fertility, the soil favored self-fertilizing clovers more than the nutrient-demanding grasses. As a result, the areas soon became clover dominant, which wasn't a bad thing as the clovers naturally improved the soil fertility. However, there was a need to find ways of reintroducing grasses without destroying the clover.

No-tilling — or "over drilling" and "sod-seeding," as it became known when applied to pasture renovation in New Zealand — was the obvious answer.

"No-till will become the most common method for sowing arable crops..."

To achieve this, Melvyn Cross, a senior lecturer in farm machinery at Massey University, determined in 1957 that there were a half dozen requirements for developing a successful no-till machine:

1 It should leave a plant-free track, about 1 1/2-2 inches wide with a uniform shallow depth through all types of vegetative cover and soil conditions.

2 It should open this furrow continuously from one end of the field to the other with all coulters. This implies the use of independently-mounted coulters, which will follow surface variations.

NO-TILL GLOBETROTTER. John Baker has been a student of no-till since researching the concept in 1967. His work eventually led to development of the Cross Slot system in New Zealand.

3 It should distribute seed and fertilizer uniformly along the furrow at required rates and deal with a wide range of seeds — including small, large, smooth and hairy seeds.

4 Loose or standing surface vegetation should not cause blockages, and if possible, the seed should be covered with soil.

5 Obstructions such as roots or stones should present no problems, but the machines must have a reasonably high working speed.

6 It would be desirable if the machine or its components could be used for other work as well such as overdrilling, so the initial cost may be spread over a number of jobs.

No-Till Here to Stay

The science and practice of no-tillage has long passed the "point of no return" since its inception in the 1950s. It is here to stay and is within sight of becoming the most common method on the planet for sowing arable crops.

But before that happens, I believe it must pass through yet another development phase. It must focus more on capturing and sequestering as much atmospheric carbon as possible into the soil. ✺

HEAVY RESIDUE IN NARROW ROWS LED TO EQUIPMENT INNOVATIONS

No-till led Marion Calmer to an easier, more profitable and environmentally friendly way of farming, along with the adoption of 15-inch corn rows.

By Marion Calmer, Calmer Farms, Alpha, Ill.

TRIM SEEDING RATES. Marion Calmer has harvested excellent soybean yields with populations of only 75,000 seeds per acre.

"No-till has made farming easier, more profitable and more environmentally friendly..."

I'll admit that when I planted my first no-till soybeans into corn stalks in 1985 that I thought, "This is no way to farm." But when I saw them come out of the ground, I wasn't as critical.

I tried those first no-tilled soybeans in a research plot to compare them to other tillage systems. When I harvested the strips, the no-tilled soybeans were my highest yield. I thought, "What? Maybe I ought to try to do a little more of this."

It was in the mid to late 1980s when we started doing strip trials, and the drought of 1988 was one of those moments in time where I was struggling

to find moisture in the tillage plots. But in the no-till plots, we just pulled in to plant and went right through the field. I thought, "This was too easy."

With the drought, the strip trials proved no-till was the best way for me to farm. It avoids the frustration of having to deal with dry soils during a drought year.

There was more of a learning curve for me in no-tilling corn into soybean stubble. I remember the frustration of having a weaker looking crop early in the growing season. Cold, wet soils led to that, and of course the neighbors noticed the poorer corn crop.

"We needed to do a better job of sizing the residue..."

Attending the National No-Tillage Conference kept me going. It's that once-a-year chance for no-tillers to recharge, remember we're on the right path and can make this system work. The feeling at the conference is always that since there are colleagues who are successful at it, then I should be able to adopt the system and be successful as well.

For me, no-till has made farming easier, more profitable and makes our farming operation feel more environmentally friendly.

Overcoming Residue Struggles

No-till also had a huge impact on my corn head business after I switched to no-tilling in 15-inch rows for my corn and soybeans in the mid-1990s.

I started growing a few more bushels per acre, but had a lot more residue in the 15-inch corn. It was a struggle to get all of the stalks to flow through a 15-inch-row no-till planter. Where we were no-tilling corn into bean stubble, there was still residue left from the two previous growing seasons.

ONFARM RESEARCH PAYS. Marion Calmer puts out hundreds of no-till plots each year and has shared results from his onfarm research over the years with thousands of National No-Tillage Conference attendees.

I was growing residue faster than I could get it to decompose. And the genetic modifications with the new *Bt* corn hybrids made the stalks even harder to decompose.

I was getting hit from both sides. The stalks were healthier and I was growing more of them.

They say necessity is the mother of invention, so we needed to figure out a way to do a better job of sizing the stalk. I learned at the National No-Tillage Conference that other guys were having the same problems.

I was personally motivated to develop a stalk roller that did a better job of residue management. Dealing with this problem was exciting, as I had an opportunity to do something for fellow farmers who were having the same concern.

Faster Residue Decomposition

Sizing the residue allowed it to decompose faster, which resulted in an increase in earthworm populations since the soil was healthier. The earthworms could more easily move small pieces of residue around and feed off of the pith of the stalk.

Developing this stalk roller was rewarding because so many farmers call in from around the Corn Belt and say, "We can't easily no-till into standing cornstalks, so we have to do one tillage pass." Now they say, "We don't have to do anything and just pull right in behind the combine if we're no-tilling wheat

Calmer Farms

MORE RESIDUE, MORE CONCERNS WITH NARROWER ROWS. When Marion Calmer shifted to no-tilling corn in 15-inch rows in an effort to boost yields, he found he had to deal with large amounts of residue from *Bt* hybrids that were still remaining from previous years.

or seeding cover crops."

Resisting Change

Change isn't easy for anyone, but we tend to get set in our ways as we get older and change becomes more of a challenge. Agriculture is America's oldest industry — the average age of the American farmer is probably in the mid 60s. Since it's really challenging to embrace change as we age, I think that's one reason no-till didn't catch on more quickly.

The other aspect is the simple economics of growing a profitable crop of corn and soybeans has become more of a challenge. As a result, some people resist and resist until they don't have any choice but to adopt no-till. It takes a long time to change the habits of the American farmer.

New technology has definitely helped the adoption of no-till. There's not only new cropping technology for farmers, but there's also new technology in the transfer of information.

With the Internet and social media, I can sit at our conference table and somebody from South Dakota can pick up their phone and show me some new information that just got posted about no-till. Back in 1994, you'd have to wait a whole year before you heard somebody talk about a new idea at the National No-Tillage Conference.

Helping No-Till Spread

I think there will be smaller increments of improvement being made to the no-till system in the future. The big jumps and gains for those of us that are 100% no-till have already taken place, and now we're simply fine-tuning the system from year to year.

I was part of one of the first ag teams to go to Cuba after President Obama opened the borders up. Riding on the tour bus in Cuba was like taking a step back in time to 30 years ago.

I was looking out the bus window at all this tillage equipment, the dry soil and dried-out crops, and I was thinking, "Good God, somebody's got to get down here soon and teach these people how to no-till."

Our next challenge will be expanding the percentage of no-till acres worldwide. Sometime in our lifetime, hopefully every crop acre in the world will be no-tilled. ❀

EARLY EQUIPMENT LOOSENED SOIL WITHOUT DISTURBING IT

The creator of Thurston Manufacturing recalls the earliest equipment the company designed for no-till and why the practice took time to spread.

By Layton Jensen, Thurston Manufacturing Co., Thurston, Neb.

LITTLE SOIL DISTURBANCE. Layton Jensen's first experience with no-till took place in the late 1970s when Thurston Manufacturing developed a sub-tiller implement that loosened the soil without actually disturbing the soil found on top of the ground.

No-till was slow to catch on. I think it was because the chemicals weren't keeping up with the technology in the beginning and mistakes were made that made the practice look bad.

When Roundup came into play in the mid 1970s, it really lit the no-till fire.

Overcoming Concerns

Our first experience with no-till was when we were developing our sub-tiller implement in the late '70s. We could see it had a good chance of fitting into a no-till system. Even though it loosened the soil, it didn't disturb the soil on top or leave much of a mark in the field.

In our own family farming operation, I experimented with building a 4-row twin-row planter on 36-inch rows. It was a pull-type unit equipped for carrying liquid fertilizer and herbicide tanks and an IH air drum. It included a hitch for towing an anhydrous trailer and eight Acra-Plant planter units following a sub-tiller shank and soil leveler. Essentially, we were strip-tilling and planting twin rows on each side of the strip in 1979. ✺

STRIP-TILL WINNERS. Thurston units have captured top honors four times in the strip-till category of *No-Till Farmer's* Product of the Year award program.

STARTED NO-TILL BY SEEDING INTERSTATE REST AREAS

Seeding prairie establishments along Interstate highway rest stops got Truax drills started in the no-till market.

By Paul Walvatne, Truax Co., New Hope. Minn.

Truax Co.

SEEDING SPECIALTY CROPS. Winning government contracts to seed prairie grasses along some of the nations' Interstate highways is how the Truax Company got started in the no-till drill business.

My first encounter with Jim Truax was in 1975 when he was tackling native seeding projects contracts in the Lady Bird Johnson wildflower program days. I remember seeing him use a Nesbit drill to seed a 5-acre prairie establishment area at one of the Interstate 94 highway rest stops in Minnesota.

Three years later the sand dropseed, which was a very drought-tolerant native warm-season bunchgrass that he had drilled in that highway rest area (at a rate of about 10 million seeds per pound),

> **"Seeds were so small that they ran out even when the channels were closed..."**

expressed itself nicely in a narrow path. Being such a small seed, it had filtered down and out of the mix even when the drill's seed channels were closed. We laughed many times about that one!

With his mechanical experience, Jim knew he could build a better no-till drill and has been at it for over 40 years. Lots of folks use Truax drills to no-till forages, pollinator species, desert sage brush, cover crops, intercrops and native plant seedings. ✳

DIGGING IN THE DIRT WITH SCREWDRIVER LEADS TO SEEDING INNOVATIONS

*As canola became popular in the early '90s in western Canada,
there was a demand for more accurate seeding equipment.*

By Norbert Beaujot, SeedMaster Corp., Emerald Park, Saskatchewan

SeedMaster

MORE ACCURATE SEED METERING. SeedMaster has focused on providing more accurate seed metering and fertilizing tools to make no-till much more effective with a wide variety of cash crops grown in western Canada and in other cooler climates of the world.

"Accurate seeding is critical in a fairly dry climate with a very short growing season..."

My ideas with finding ways to improve seeding started over 2 decades ago on the family farm when I dragged a screwdriver through the soil.

My ag engineering background and that experience led me to believe seeding shouldn't be complicated and that there had to be a better way of getting wheat, oats, barley, canola, sunflowers and other crops in the ground under western Canada's dry conditions.

Since that time, I've been trying to perfect seed metering, getting seed placed accurately in the ground, placing fertilizer where it will do the most good and improving other aspects of precision seeding.

Jeff Lazewski

NO-TILL WAS NOT A GIMMICK. Norbert Beaujot says major farm equipment manufacturers in the 1970s and 1980s looked at no-till as a fad that they really hoped would go away.

"An enormous amount of no-till technology out of Saskatchewan…"

Canola Brought Changes

Air seeders were popular and worked well with wheat in western Canada in the late '80s. But by the early '90s, canola was becoming an important crop in western Canada. However, the available seeding equipment for canola was terrible in regard to seed depth, fertilizer placement, packing, emergence and making the most efficient use of available surface moisture to get no-till canola off to a good start.

In 1992, we introduced the industry's first hydraulically-activated ground-following individual row opener and used it to seed 1,200 acres on our own farm. It left a terrible mess, but the crop came up evenly.

Most earlier-designed cultivator and disc style openers from other manufacturers were difficult to adjust for depth. As a result, we developed a long arm with the gauge wheel in the back, which acts as the packer wheel, that offers precise seed depth within each row. Thanks to hydraulics, we push the same amount of force against the ground with each packer wheel, which leads to more consistent depth.

Recognizing the importance of proper fertilizer placement, we also developed a two-knife system that works on the same long arm to accurately place seed and fertilizer in separate bands.

Over the past 10 years, we've focused on accurate seed metering, which has placed us in the forefront of the no-till movement here in western Canada.

"Gimmick" Becomes Commonplace

As western Canada growers live in a fairly dry climate and struggle with a short growing season, accurate seeding is critical. As a result, many Canadian growers years ago recognized the value of shifting to no-till.

This led to the development of an enormous amount of no-till seeding technology in Saskatchewan that is now used around the world that included SeedMaster, Conserva Pak, Flexi-Coil and Seed Hawk.

My brother and I started Seed Hawk in 1992 and about 10 years later he took over the firm, which was later sold to Vaderstad. Since I still had a great deal of passion for designing seeding equipment, I started SeedMaster in 2002.

Most of the bigger machinery companies in the '70s and '80s looked at no-till as a gimmick, something they hoped would go away so they could continue to sell high horsepower tractors, cultivators, discs and other heavy-duty tillage tools.

Instead, no-till has become the conventional way to farm in many parts of the world and is a big part of our shortline farm equipment business. ❋

NO-TILL LOOSE ENDS…

October
1976

DETERMINING MICE OR BIRD DAMAGE

Here's a guide to determine mice or bird damage: "Mice only eat the corn kernels," says Joe Newcomer, a University of Maryland agronomist. "A bird may leave a telltale sign of a funnel-shaped spot where he works to pull the seedling up." ❋

No-Till, the Feds and
Washington Politicians

"Farmers still cultivate too much — they've got to go to minimum tillage or zero tillage whenever it's feasible."

— President Jimmy Carter

Over the years, the topic of no-till has come up numerous times in the nation's capital among politicians, lobbyists, association staffers and government staffers.

It's often mentioned not so much as a way for farmers to trim costs and boost incomes, but more as a way to curb erosion or effectively bring idled acres back into production.

This chapter summarizes a few activities that occurred over the years in the nation's capital that relate to the advancement of no-till. They range all the way from the White House, to the halls of Congress, to U.S. Department of Agriculture offices and even presidential politics.

In January of 1980 during his last year in office, President Carter discussed ways of reducing energy costs in a White House interview with editors of the *Des Moines (Iowa) Register*. Reprinted in the February 1980 *No-Till Farmer*, President Carter said in the interview, "Farmers still cultivate too much — they've got to go to minimum tillage or zero tillage whenever it's feasible."

CONGRESS LEARNS ABOUT NO-TILLAGE FROM *NO-TILL FARMER*

June
1973

Chase Ltd.

NO-TILL TESTIMONY. Texas Representative Robert Poage, chairman of the U.S. House of Representatives Agriculture Committee, greets me following my 1973 testimony before both urban and rural Congressional committee members on the impact no-till would have on the future of American agriculture.

My work was cut out for me. I only had 10 minutes to fully explain the many advantages of no-tillage to a group of urban and rural Congressmen with limited knowledge about this relatively new reduced tillage concept.

My reason for being in Washington, D.C., on April 3 of 1973 was that I had asked to testify before the House Agriculture Committee on the important role that no-tillage was already starting to play in American agriculture.

I'd welcomed this opportunity to testify before the House Agricultural committee. With its many benefits for solving many of today's farm problems, no-tillage needed to be fully understood by our nation's lawmakers.

Need to Move Fast

To testify before a Congressional committee, you first had to get your name on the committee testimony roster. Then you showed up with 100 copies of your prepared statement. With limited time to tell the no-tillage story, I needed to have my thoughts well organized and down on paper.

Each speaker was allowed 10 minutes to share his or her ideas with the ag committee members. This means you can get over about five pages of single-spaced information, providing you talk fast.

The day I testified about no-tillage, committee members also heard about dairy support prices, honey production, cotton production, wildlife management and comments on the proposed farm bill.

Any Volunteers?

More farmers should take advantage of the opportunity to appear before both the Senate and House Agriculture Committees. Senators and Representatives I have talked with say they wish more practicing farmers (and I assume fewer farm organization officials or editors like myself) would appear before their committees. ✳

'WHAT DO YOU HAVE AGAINST ADVERTISING?'

In December of 1973, several members of the *No-Till Farmer* staff sat down with Secretary of Agriculture Earl Butz while he was in Chicago for a speech. After he leafed through the latest issue of *No-Till Farmer*, Secretary Butz jokingly asked, "What do you have against advertising?"

In those early publishing days, advertising support for no-till was practically non-existent. With only a few exceptions, most ag suppliers weren't yet sold on the concept of no-till. In 1972, no-till made up less than 2% of the total row cropped land that year.

A number of our readers voiced the same lack of advertising concerns. Impressed with what we were doing by providing valuable no-till information, these growers recognized that farm magazines needed advertising support to survive. Farmers were worried how long we could sustain publishing *No-Till Farmer*.

Right to be Concerned

The advertising support picture did not improve and the bigger publishing company (Reiman Publications) was forced to toss in the towel in 1974 on the model of providing a free magazine to 60,000 growers who were already no-tilling or interested in learning more about the practice.

The original circulation goal had been to secure the names of around 30,000 U.S. farmers who would be interested in learning more about no-till. When we asked farm equipment dealers to provide names and addresses of growers in their areas who they felt would be interested in learning more about no-till, we were overwhelmed when we ended up with over 60,000 names.

With big-dollar printing and postage bills, we moved in 1974 to a paid subscription publication model that continues to this day. It continued as a monthly 8-page

paid subscription newsletter after my wife and I purchased the publication in 1981.

Then in 1988, we shifted to a 17-times-a-year format where we published two issues each month from January through May and monthly issues during the remaining 7 months of the year. Interestingly enough, this frequency change came at the request of advertisers who wished to place 2-and 4-page equipment and pesticide advertising inserts in the publication during the key decision-making season.

By 1998, suppliers were starting to recognize the size of the growing no-till market (14.1 million acres in 1988) and the value of the no-till market.

When the advertising support dried up again in 2000, we returned to monthly distribution. We expanded the page count at the same time to 16 pages.

The first rekindling of interest in again producing a magazine among readers and suppliers took place in 2006 when we published the first *Conservation Tillage Guide* magazine for *No-Till Farmer* subscribers. That's when advertisers in our newly-acquired *Farm Equipment* (dealer universe) and *Farm Catalog* (high-income, high-acreage growers) were practically begging us to connect them with our unique and prequalified 100% paid no-till audience.

There was some risk to doing this on our part, but one advertiser encouraged us with a sight-unseen advertising schedule. "I trust that you guys would never do anything half-assed," said the marketing director at a long-time shortline farm equipment manufacturer.

Today's monthly *No-Till Farmer* paid subscription publishing format includes eight newsletters and four magazines that offer over 650 pages a year of in-depth information on all aspects of no-till. ✺

WHERE'S THE ADVERTISING? In 1974, Secretary of Agriculture Earl Butz studies *No-Till Farmer* with publisher Roy Reiman.

Frank Lessiter

NO-TILL GETS MORE ATTENTION AMONG SOIL CONSERVATIONISTS

Washington's interest in no-till increased dramatically in the 1980s and 1990s as several veteran no-tillers took over the role of Chief of the Soil Conservation Service (SCS), which is now known as the Natural Resources and Conservation Service (NRCS) in Washington.

The first was Peter Myers from Matthews, Mo. who served as SCS chief from 1982-1989 in Ronald Reagan's presidential administration.

A firm believer in the merits of no-till, Myers, who farmed in Missouri's southeast Delta region, would have been no-tilling if it wasn't for the furrow irrigation system used in his 1,100-acre operation.

"I would hope people realize the fallacy of rolling that fresh dirt in the fall and even in the spring," says Myers, an ardent promoter of less tillage.

"We reduced tillage as a fuel economy measure and the more I did, the better I liked it. We don't even use chisel plows much anymore and we no longer own a plow."

Myers believes no-till provides farmers with a low-cost self-help conservation tool.

"To me, the minimum tillage concept in times of a poor economy is something a farmer can do with the equipment he already has," he says. "He doesn't have to go out and buy a new planter and may only have to adapt his

NO-TILLER IN WASHINGTON. Missouri no-tiller Peter Myers served in President Ronald Reagan's administration as Chief of the Soil Conservation Service from 1982-1989. He credited reduced tillage with not only saving valuable soil, but also significantly boosting farm income with the family's 1,100-acre operation.

planter for no-till. But still, it's a much better way to go without having to put in an expensive terrace system."

Myers hopes the SCS will encourage further research by the land-grant universities on cost efficient new technologies such as no-till.

It's refreshing to hear the new SCS head talking about the tremendous accomplishments no-tillage can make in solving our erosion problems. (More on the next page).

NO-TILL BOOSTS INCOME WHILE SAVING SOIL

Peter Myers, the chief of the Soil Conservation Service (SCS), is a genuine farmer (cropping 1,100 acres at Matthews, Mo.) who believes in the many benefits of no-tillage. We think you'll find his answers to our No-Till Farmer questions of special interest.

Have you used no-tillage on your own farm?

Myers: I struggled with no-tillage and botched it 10-12 years ago. Like everybody in the early days, we learned from our mistakes. We now have better machinery and herbicides to make no-till work. I've since gone to conservation tillage on our farm to save time and money.

We furrow irrigate, which makes no-till difficult. If it weren't for that, we'd be doing considerable no-till.

Why is less tillage so important these days?

Myers: If we can convince farmers to leave residue on the soil instead of going to bare ground, then we're halfway home. The most important factor in reducing cropland erosion is to leave crop residue on the soil surface. Without the mulch that nature provides through crop residue, erosion is nearly impossible to control.

Even expensive structures, such as terraces and diversion ditches, may not always function properly when the soil is not protected with residue.

Why should farmers shift to no-tillage?

Myers: An important short-term benefit is the satisfaction that comes from a sense of accomplishment and good soil management. A farmer doesn't want to squander his soil — he wants to protect it while making an adequate income. No-till permits him to do this.

What will make farmers shift to less tillage?

Myers: Economic considerations will be the deciding factor because no-till often costs less than conventional tillage. In parts of Missouri, for example, conventional tillage results in annual soil losses of 40 tons per acre. Conservation tillage cuts loss down to 12 tons and costs the grower $3.50 less per acre than using conventional tillage.

How fast is minimum tillage and no-tillage catching on today?

Myers: Adoption rates have been relatively high. Reduced tillage is being accepted faster than any other practice in the history of farming.

Estimates for 2010 (based on earlier projections from our No-Till Farmer acreage surveys) put the number of acres in conservation tillage methods at around 95% just 27 years from now. At that time, more than 50% of the U.S. cropland will be in no-till. I believe these figures for the year 2010 are still attainable.

Will we see changes in government cost-sharing options?

Myers: Both the SCS and ASCS are assessing various cost-sharing options. We are looking at the elimination of cost sharing for terraces and watersheds if no-tillage could provide the same soil-saving results.

What's the future for the moldboard plow?

Myers: The new machinery, chemicals and techniques that have made conservation tillage possible will make the moldboard plow as obsolete as the scythe and butter churn.

I haven't owned a plow in a dozen years. In fact, we don't chisel plow any more since the heavy clay on our farm is loosening up.

The moldboard plow is a tool of the past since we don't need to turn the ground over like we used to do. Today's farmers will junk the plow or trade it in on no-till coulters.

There are a few isolated areas where farmers probably need a breaking plow, but the quicker we can get rid of them, the better off we will be.

We have to start doing less tillage or we'll lose all our land. As far as I'm concerned, you can take all the plows in the country and put them in the scrap iron pile. ❄

MEET BILL RICHARDS ... THE GODFATHER OF NO-TILL

NO-TILL'S PIONEER. Ohio no-tiller Bill Richards served as Chief of the Soil Conservation Service from 1990-1993. A huge Washington booster of the soil-saving and profit-boosting benefits of no-till, Richards and his sons made a commitment in 1979 to never again till any fields on their farm.

Somewhere along the line, Bill Richards earned the title of "Grandfather of No-Till."

He grew up in agriculture as the son of a successful farm equipment dealer before getting into farming. Having no-tilled since 1962 and with his sons running the family's 3,000-acre corn and soybean operation at Circleville, Ohio, Richards extensive work with no-till ended up taking him to Washington, D.C. He was appointed chief of the National Resources Conservation Service (NRCS), then known as the Soil Conservation Service (SCS), and served from 1990-1993.

Few Reasons to Till

In earlier days, his Ohio State University professors taught him that the only reason to till the ground was to control weeds. Since atrazine was introduced at about the time he started farming, he began working toward eliminating tillage as early as 1958. His first few reduced tillage ideas, such as mounting a two-row planter directly behind the moldboard plow and attaching a four-row planter behind a disc, were disasters.

Richards kept at it and worked with International Harvester product engineer Ralph Baumheckel to build a 6-row, 30-inch till-plant tool and eventually got to where he was no-tilling successfully. The planter had a shovel that threw trash to the middle, whiskers that sorted soil from residue, a covering disc that closed the trench and a packer wheel to firm everything down followed by the planter unit.

While other area farmers were going over their fields five to seven times. Richards was making

"No-till taught us to manage more acres with the same men and machines **and be more profitable...**"

only one pass and was able to no-till 1,000 acres of corn with only one helper by 1965.

"We had proved we could do it and we were off and running with no-till," he says. "Over the years I've come to understand that residue is what really makes no-till work. You need to have residue to cover the soil, prevent erosion and provide the organic matter needed to build the soil."

Selling No-Till to Farmers

President George H.W. Bush's administration brought Richards to Washington because they were in the middle of applying new residue requirements and needed help getting farmers to comply with 1985 Farm Bill conservation rules. While he had worked closely on numerous conservation plans with SCS staffers at the family's farm in Ohio, he was now sitting on the other side of the desk.

Richards spent 1990-1993 preaching the benefits of no-till to everyone from producers to USDA Secretaries of Agriculture. He explained how leaving residue on the surface was the key to no-till success and how there wasn't any scientific reason for farmers to till.

The geographic areas where Richards had the most trouble getting farmers to comply were places where the deepest, best topsoil is found. This included the rolling hills of Iowa and the Palouse region of Washington where farmers had topsoil so deep that they didn't see any yield difference if some of their soil blew or washed away.

Richard learned farmers in those areas were fighting mad about the new standards. However, he was able to go there as a farmer and say, "This will work. It's working for us now, we know how to do it and we can help you.

"Success was gradual, but we got the job done. If you look at erosion charts of the U.S. from 1985-2005, you can see a big dip in total erosion. If you lay a no-till acreage chart of the same time period along side, you can see what caused that dip."

U.S. Falling Behind

Though no-till has made great strides in the U.S., Richards says the conversion has been very gradual when compared with other countries, especially in South America. "No-till was born here, yet if you look at other countries you'll see we're falling behind," he says.

THE NO-TILL MONSTER. Even though no-till got its start in the U.S. Bill Richards says we've now fallen behind other countries in its adoption.

"When I go to other countries, they tease me that Yankees get government payments so they don't have to no-till. In their countries, they must no-till to be profitable.

"In some ways we have unleashed a monster in teaching the world to no-till. It's brought millions and millions of fragile acres into successful production that normally wouldn't be used. We have expanded our competitor's capacity to produce."

Since turning over most of the operation to his sons and becoming the senior farm advisor, Richards spends a lot of time spreading the word about no-till.

"I feel like continuing to share my no-till knowledge is my public service at this point," he says. "No-till has been my life and I'm pretty proud of how far it has come. I've always spoken out for agriculture and had fun doing it. I hope I inspire others."

TOTALLY RIDICULOUS ... IF YOU'RE RUNNING A FLUFFING HARROW, NRCS STAFFERS MAINTAIN YOU'RE NOT REALLY NO-TILLING

If you ran a fluffing harrow over some of your ground last spring just before planting, you may be surprised to learn the Natural Resources Conservation Service (NRCS) no longer considers those fields to be no-tilled.

The NRCS definition of no-till for many years required at least 30% residue left on the soil surface. But in 2006, NRCS quietly put through a new definition that rules that operating any type of full-width tillage regardless of depth no longer qualifies for use on no-tilled acres.

> "Banning the use of tools that help **overcome serious residue concerns is ridiculous...**"

That means using these tools will disqualify you for no-till government program payments. And if you're in a state where extra government program dollars are doled out for no-till, you won't be happy when some of your acres are reclassified as mulch tillage, even with more than 30% residue on the soil surface.

Lots of Double Talk

It appears that NRCS may be talking out of both sides of its mouth. It apparently is not relying on its new no-till definition when making in-the-field evaluations to determine tillage usage.

Attendees at the 2009 National No-Tillage Conference used fluffing harrows on 15% of no-tilled acres. So if you accept the USDA figure that 63 million acres were no-tilled in 2004, this means 9.5 million of those acres would no longer qualify as no-till.

While many veteran no-tillers use fluffing harrows, the use of these tools is even more important to growers transitioning to no-till. These growers often rely on low-disturbance tools to encourage earthworms and microflora to effectively break down crop residue. The units are also valuable in applying manure and handling corn stalks under high-yielding conditions.

On the other hand, trips made with vertical-tillage, rip-

Paul Mahoney

FLUFFY FIELDS. The National Resource Conservation Service banned the use of fluffing harrows under no-till conditions by the end of the 2000 decade. This was despite the fact that attendees at the 2009 National No-Tillage Conference, such as Minnesota no-tiller Paul Mahoney (shown here)were using these tools on 15% of no-tilled acres to more effectively manage residue and warm the soil for earlier planting.

pers or strip-till units still qualify as no-till since they don't represent full-width tillage passes. Besides rotary harrows, these USDA rules eliminate the use of Turbo-Till, AerWay, Gen-Till and other full-width, low-disturbance tools in meeting the more stringent no-till qualifications.

The revised no-till criteria from NRCS also states that a Soil Tillage Intensity Rating (STIR) value shall include all field operations performed during the cropping interval between harvest of the previous crop and harvest or termination of the current crop. The STIR value for no-till shall be no greater than 30 and should have a score of under 20 for carbon sequestration.

The STIR system assigns a value for each trip made across a field and a lower cumulative score is better. Just a combine, grain wagon and truck in a field earns a STIR value of 5 due to compaction concerns.

Guesses Don't Count

While STIR values for various tillage operations are based on Agricultural Research Service studies, apparently no work has been done on rating fluffing harrows, and guesses were used to set the values.

In addition, there doesn't appear to be any scientific evidence to indicate that running this equipment less aggressively, especially under high-residue conditions, damages the soil.

The revised definition will also have an impact in other areas. For instance, the Chicago Climate Exchange will likely not allow no-till carbon sequestration contracts when full-width, low-disturbance units are used. Look for the National Corn Growers Association to ban these tools in the no-till/strip-till category of their yield contest.

Yet there's still hope that these situations could be evaluated on a case-by-case basis.

What do I think about banning the use of these full-width, low-disturbance tools that have helped growers overcome serious residue concerns and no-till more effectively for many years? It's totally ridiculous. ✾

NO-TILL-AGE®
June 1991

What's Impact of Conservation Tillage?

Management Benefits, Concerns	% of Farmers
Reduces soil erosion	80%
Saves money	71%
Saves time	56%
Causes weed problems	56%
Results in higher costs	47%
Requires special equipment	50%
Requires more chemicals	37%
Requires different techniques	33%
Requires different timing	33%
Requires different equipment	33%
Leaves more trash	29%
Conserves moisture	25%
Reduces soil compaction	22%
Requires change	21%
Requires wind erosion	21%
Reduces yields	20%
Unsustainable for hilly fields and/or heavy clay soils	17%
Requires re-education	17%
Reduces water runoff	16%
Boosts insect populations	14%
Lets you plant earlier	10%
Early germination concerns	7%

— Interviews with 208 Corn Belt farmers by IMR Systems

118 YEARS VS. JUST ONE 1 YEAR

August 1972

When we think of the many benefits of no-tilling, we're reminded of the comments made by an Indiana Soil Conservation Service worker back in 1972.

At the level of acceptance at that time in his area of the Hoosier State, he indicated it would take 118 years to control erosion with strip-cropping, terraces and other conservation practices.

Or he says area farmers could switch to no-tillage and enjoy the same progress in controlling erosion in just one year's time. ❋

NO-TILL-AGE®

December 2015

Why Growers Use Conservation Practices*

Conservation Practice	% of Acres
Integrated pest management	4%
Nutrient management	6%
Buffer practices**	9%
Contour farming, strip cropping	9%
Terraces	15%
Grass waterways	23%
No-till	35%

*Includes corn, soybeans, wheat, barley, cotton and grain sorghum acres
**Includes filter strips, riparian buffers and field borders

— USDA's Economic Research Service

WASHINGTON FOLKS START TO SEE THE LIGHT WITH NO-TILL

May 1982

For 10 years, we've been giving no-tillage top priority as a way to help farmers make more profit. And while no-tillage has made a great deal of progress during these past 10 years, it's also been disappointing at how long it's taken many farmers, educators, suppliers and government workers to see the real advantages of this tillage practice.

That's why we found a recent USDA news release of special interest. During the past couple of years, there's been a great deal of pressure coming out of Washington pushing for increased emphasis on no-tillage and minimum tillage. And as far as we're concerned, that's good.

In a recent report by Norman Berg, Soil Conservation Service chief, reduced tillage was given top priority in his "most wanted" list of nine research projects.

First on his list was the need for accelerated educational efforts in regard to conservation tillage.

Second was his desire to expand research to de-termine the potential and limitations of various soils, plants and machines for conservation tillage. There's no doubt — such research is badly needed.

Several other "most wanted" research projects also dealt to some extent with the move to less tillage.

10 Years Too Late

While we salute these research priorities with reduced tillage, it's too bad that they hadn't been in place 10 years ago. If they had, both farmers and the whole country would be in better shape today — thanks to the many benefits of moving to less and less tillage on more and more cropland.

Much of the money currently being poured into water and soil conservation projects around the country could have been saved if we'd taken these steps to move much earlier to more use of reduced tillage.

There's no doubt — minimum tillage, and especially no-tillage,— can save this country millions of dollars in the future. ❋

NO-TILLAGE TOPS LIST OF MAJOR AG DEVELOPMENTS

December 2012

When it comes to ranking the most important developments in American agriculture over the past 75 years, a panel of conservationists recently placed no-till right at the top of the list.

This discussion took place a few weeks back as the Conservation Technology Information Center (CTIC) in West Lafayette, Ind., celebrated its 30th anniversary. Back in 1982, over 40% of the nation's cropland was eroding at an alarming rate and most growers were still relying on moldboard plows and heavy discs for tillage.

The group's goal was to provide a central clearing house where farmers, suppliers, government agencies and organizations could find the latest conservation information. The group has lived up to its goal, as CTIC serves as a model in terms of what can be accomplished with conservation on farms around the world. (The organization's official name was originally Conservation Tillage Technology Center, but members soon decided the word "tillage" limited funding support from a number of groups and government agencies who weren't yet on the no-till bandwagon.)

Right from the start, the group saw many advantages of getting more farmers to no-till. In 1982 when CTIC become a reality, 11.5 million acres were no-tilled, an increase from the 3.3 million acres no-tilled in 1972 when *No-Till Farmer* was started. A USDA survey places the no-till acreage at 88 million acres in 2009.

During the CTIC anniversary meeting, no-till was named as the most significant conservation development since World War II. Other nominations also show the importance of taking a no-till systems approach.

No-Till Keeps Growing

Veteran no-tiller Bill Richards from Circleville, Ohio, placed no-till at the top of his list of developments in American agriculture. Richards and his sons have been no-tilling for over 45 years and the former chief of the Natural Resources Conservation Service (NRCS) says no-till lets them boost yields while dramatically trimming costs.

"Planting is getting much easier with no-till and the practice has allowed the country's farmers to provide more affordable food and fuel to the nation," says Richards. "A key no-till benefit is that rain falls on residue rather then on bare soil, which dramatically reduces soil erosion."

Mechanization Was the Key

Bruce Knight has been no-tilling for 10 years on the family's South Dakota farm. "We saw the real benefits of no-till during this year's drought," says the conservation consultant and former chief of the NRCS in Washington, D.C.

Knight says the major change in his family's farming operation occurred when his Dad sold the work horses.

"With mechanization, farmers didn't have to use 20% of their crops to feed the horses," he says. "The result was that we could provide more food at a reasonable price to feed a growing nation."

Other areas named by the panelists as being critical to the development of no-till include improved crop genetics and traits, GPS technology, weed control, farm equipment safety and water management.

With tightening crop supplies, Knight says more no-till acres are needed to provide the world with reasonably priced food, fiber and fuel.

"We must improve ag production to be sustainable which is economics, productivity and being socially responsible," he says.

That's something thousands of no-tillers have been doing for decades. ❁

NO-TILL LOOSE ENDS...

April 1984

TAKE THE NO-TILL PLEDGE

Out in Iowa, the Jones County Untillage Committee came up with an "Untillage Pledge Card" for area farmers to sign. It reads:

"I will till no soil UNTILL its time. If I don't have a darn good reason for fall tillage, I will wait UNTILL spring. If it's planting time and I still don't have a good reason, I will try NO-TILL."

The first year, 140 farmers pledged not to fall till 10,680 acres of bean stubble. That represented 20% of the soybean farmers and 31% of the bean acres grown in the county. ❁

FIRES SCORCH NO-TILL PROFITS

*Since no-till residues sometimes go up in smoke,
this has become one of the most popular No-Till Farmer articles.*

As a no-tiller, you know crop residue is crucial to the success of the system. It protects the soil against wind and water erosion, provides food for earthworms and microorganisms, and holds valuable nutrients.

Yet, it's hard to assign a dollar value to crop residue that can go up in smoke during field fires. Hopefully, you'll never have to, but there are some steps that can be taken to soften the blow.

Field fires are not an uncommon occurrence as the no-till acreage continues to expand, says Purdue University agronomist Pete Hill. Unattended burning backyard barrels, roadside ditch fires and machinery fires are common causes. It doesn't take much for a no-till field to go up in smoke, especially a corn field.

"Residue is a tremendous fuel," Hill says, "especially in dry periods or when there's low relative humidity."

In the fall of 1995, which was very dry throughout the Midwest, 15 field fires were reported in Hill's local newspaper over a 2-day period.

"That crop residue is worth a lot of money," he says. "Think about how it conserves soil and improves the soil's physical properties. It really does have a dollar value, but we're at a stage where we just don't know enough about it, so you see $30-$150 an acre being asked by farmers who file an insurance claim."

125 Bushels on the Ground

Hill cites the example of a fire in White County, Ind., that burned 800 acres and traveled 7 miles. "The dynamics of this fire were incredible," he says. "Corn husks burned back and could not support the weight of the ears, which dropped straight to the ground.

"To say the farmer faced volunteer weed control the next year was an understatement. There were over a million corn kernels per acre on the ground."

The sandy soil was so dry that the fire burned the corn cobs from end to end and left small piles of kernels in the field. In situations where the fire left the soil completely bare, Hill says it's best to leave the field alone rather than discing or plowing the ground.

"Burying the kernels would probably promote volunteer corn," he says. "In this instance, there were 125 bushels of corn on the ground and nothing else. Winds were gusting up to 40 mph and all the ash was blown off site. The farmer was left with the soil surface prone to wind erosion with absolutely no protection."

What was the true impact of this field fire?

"The residue cover is gone," Hill explains. "You've lost your soil surface protection, organic matter and all the carbon in the organic matter has gone up in the atmosphere. You've likely lost all the nitrogen in the residue. You may have lost all the potassium and phosphorus in the soil if you've lost the ash.

"However, if the ash is still left on the surface, phosphorus and potassium will actually be more available for plant growth. You've also lost the food source for earthworms and the microorganisms.

"In almost every situation, there is always poor weed control. Whether herbicides were applied before or after the burn, it's almost always havoc for weed control."

Managing Scorched Fields

If a field burns up in the fall, Hill recommends getting it covered as soon as possible.

"Plant a cover crop that establishes quickly to get that organic matter replaced," he suggests. "Or spread a straw-type manure on the surface to get that organic matter back in the ground and get the soil surface protected."

With heavy rainfall and a bare soil surface, Hill says you'll likely have thick and hard crusting problems that can seriously impact your stands and yield potential.

If you have a field fire, Hill recommends explaining to your insurance agency where you lost money.

"Talk about soil quality, reduced yields and the need for additional weed control," he says. "I don't have a known economic value. You can replace lost nutrients. That's fair to ask of your insurance people."

There are short- and long-term effects on yields, says Hill, who recommends doing yield checks for 3 years after the fire to monitor what's happening in the system.

In the short term, Hill says corn yields may increase. However, the long-term impact won't be good since organic matter and the nutrient cycle are disturbed.

Most insurance companies will pay $30-$150 per acre for no-till field fire claims, but Hill says most payments end up in the $50-$80 range.

"Filing an insurance claim is a wonderful thing," he says. "I applaud the no-tiller that thinks of the true value of his or her crop residue." ❁

No-Till Myths, Lies and Predictions
that Didn't Come True

"Even after decades of success, **no-till myths and untruths are still bandied about.** You've probably heard these and other unfounded claims **still being spoken by tillage enthusiasts…**"

In the early days of no-till, myths and outright lies were constantly spread about the reduced tillage practice. Plenty of laughs and jokes were thrown at no-tillers by farmers who felt the only way to farm was to bury the residue, cultivate for weed control and maintaining that the ideal seedbed was a smooth, clean field.

Here's a sampling of some of the outrageous comments our editors have written about over the years in regard to no-tilling. ✳

UPROOTING OLD TILLAGE MYTHS

March 1990

Despite many benefits, some farmers believe no-till farming not only requires more herbicides but also reduces yields.

Many farmers feel they are already reducing soil erosion as much as is needed by using conservation tillage. As a result, they see no reason to try no-till.

Moldboard Plow Myth

Over 50% of farmers surveyed in eastern Nebraska by University of Nebraska ag engineers indicated they were using conservation tillage. These farmers have basically parked the moldboard plow and have reduced tillage over the past 15 years in order to save fuel and labor.

However, Nebraska ag engineer Elbert Dickey says intensive tillage systems are still being used, which destroy too much of the soil-protecting residue cover.

He says reducing the number of trips across a field or changing tillage tools leads many farmers to believe they have adopted conservation tillage. However, this is part of the problem since conservation tillage is defined by the amount of residue cover that is left on the soil surface — not by your tillage tool choice.

Dickey says a conservation tillage system leaves at least 30% of the soil surface covered with residue after planting. By using this definition, only 5% of fields surveyed in eastern Nebraska had 30% or more protective residue cover.

"Even though these farmers are not moldboard plowing, very few are

NO-TILL-AGE®

No-Till's 40 Biggest Lies

December 1986

Over four *No-Till Farmer* issues in 1994 and 1995, we listed what were considered to be the biggest untruths still being spread about no-till, along with a few more gathered from no-till enthusiasts.

1. I'll have to plow every 4 years to loosen the soil and control weeds.
2. Grandpa says no-till will never work.
3. I've always plowed, but will be no-tilling the entire farm next year.
4. All those herbicides will poison the soil.
5. You'll never get a planter through that stuff.
6. Two of the most beautiful things in the world are peacocks and well-tilled fields.
7. No-till is for lazy farmers.
8. With no-till, we've saved so much soil that the EPA and the NRCS will now get off our backs.
9. My banker will lend money for a plow or boat, but never for a no-till planter.
10. They can't make five trips with their ATV sprayer cheaper than I can plow, disc, field cultivate, spray and cultivate.
11. No-till requires more phosphorus and potassium because of reduced rooting.
12. You can't no-till successfully into wheat stubble.
13. No-till will reduce the number of wildlife on your farm.
14. Our landlords will never let us no-till.
15. No-till will work only if it rains.
16. You can't cultivate to control weeds with no-till.
17. Yields are always lower with no-till.
18. No-till corn won't work in clover sod because it's too wet.

19 No-till corn won't work in clover sod because it's too dry.
20 Nature demands that the soil be stirred every fall and spring.
21 No-till always yields less than plowed ground.
22 No-till requires corn hybrids with good early vigor, which have less yield potential.
23 You can't make no-till work with cold soils.
24 No-till is just for small acreage farmers.
25 Nothing will grow in our fields unless we plow and disc six times.
26 I'll never no-till a single acre
27 You can't use a rotary hoe with no-till.
28 No-till is a passing fad.
29 No-till takes too much time.
30 No-till fields always have more weeds.
31 Carefully manicured fields always yield best.
32 If we no-till, we'll lose the farm.
33 You have to cultivate to stir the soil.
34 No-till is simply a conspiracy on the part of the federal government to sell more ag chemicals.
35 No-till causes weed and insect resistance.
36 There is as much dignity in tilling a field as in writing a poem.
37 No-till farming is very easy.
38 No-till caused all of my farming problems.
39 No-till allows more pesticides to wash into streams.
40 No-till is a sure way to go broke.

using conservation tillage as defined by residue cover," he says.

Chemical Myths

When you ask farmers about herbicide use, you'll probably hear them say that conservation tillage, especially no-till, requires more herbicides.

Yes, a no-tiller would have a field full of weeds if no herbicides were applied until planting time. These weeds would gobble up valuable soil moisture.

But by applying herbicides a few weeks ahead of planting, many farmers could switch to no-till and use the same herbicide program that was used with conventional tillage.

Yield Myths

Dickey says studies have shown yields from all tillage systems generally are about the same. However, in rainfall-limited areas or low rainfall years, yields are generally higher with a conservation tillage system.

More moisture is available because the surface residue acts as a mulch to reduce evaporation. In addition, making one or more trips with tillage tools can definitely dry out the soil.

Sure, both no-till and conservation tillage require more management ability. But no-tillers have proven a number of commonly accepted beliefs about tillage are definitely myths.

COMBATING THE MYTHS OF NO-TILL

When it comes to change, myths can overshadow reality and hold back growers from adopting new technology. Even after more than four decades of successful no-till, it's still common for folks to talk about, and be influenced by, unfounded claims rather than successes.

Alan Mindemann finds most myths are based on what folks don't know or fully understand about no-tilling. The veteran no-tiller from Apache, Okla., says what many non-believers think is fact generally turns out to be fiction.

University of Nebraska ag engineer Paul Jasa says the myths he hears about no-till run the gamut of topics, including weeds, compaction, heavy residue, lower yields, insects, diseases and soil quality. He's spent years of research disputing them.

David Lobb says a major myth is that wind and soil erosion are major causes of degraded topsoil. Instead, the University of Manitoba soil scientist maintains the culprit is tillage erosion. Over time, extensive tillage pushes more soil down the slope than up the slope. This results in extensive soil losses at the top of the hills and excessive soil accumulation at the bottom of the slopes.

Lobb says research shows more than 75% of soil loss at the top of slopes is caused by tillage erosion. With 15%-25% of fields classified as hilltops, moderate to severe erosion occurs after decades of tillage.

Another myth is that by switching to no-till, you will stop all soil erosion and allow the topsoil to immediately restore its productivity.

Even with large quantities of crop residue on the surface, no-till may need several decades or centuries to restore severely eroded topsoil. ✷

WHAT FOLKS 'DON'T GET' ABOUT NO-TILL

Below are a dozen of the most common myths we've heard about no-till.

1 Weeds will take over in no-till and the herbicide costs will kill me.
2 With continuous corn, I won't be able to handle the heavy residue and my planter will plug. I'll need to burn residue or invest in a harrow.
3 Since it's always cold and wet under the excessive residue left by no-tilling, tillage is needed to warm up and dry out the soil.
4 Insects and diseases will be a critical issue. Tillage kills the insects, destroys their habitat, decomposes residue and reduces disease.
5 Cover crops are a waste of forage, nutrients, moisture and dollars.
6 I'll have to invest in a ripper due to compaction concerns with no-till. The combine and grain carts will compact the soil.
7 Soil must be tilled to raise crops. Roots can't penetrate untilled soil and yields will drop with no-till.
8 Residue needs to be incorporated to improve soil organic matter.

9 I'll have to till the soil so it can soak up water and eliminate costly runoff.
10 Since no-tilled ground turns hard, tillage will be needed to loosen the soil and smooth out harvest-time gullies and ruts.
11 Switching to no-till means buying a new no-till drill and sprayer. Since I'll still need to use my existing tillage equipment, why take on the additional equipment expense?
12 I'll need to add non-mobile soil amendments and use tillage to incorporate surface-applied fertilizer and lime.

What Do You Think?

Overcoming the myths of no-till means being able to adopt a different management system. It's essential to have the proper mental attitude that will allow you to continue to expand your no-till knowledge every day. It's the only way to set yourself apart from farmers who practice excessive tillage and believe the unfounded myths about no-tilling. ✷

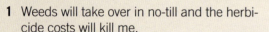

UNPACKING THE MYTHS OF NO-TILLAGE

Unfortunately, folks who don't believe in no-tillage come up with many reasons as to why it won't work. Myths are what Purdue University agronomists Jerry Mannering and Don Griffith call them. The idea that any one tillage system works well under practically every condition is one of the major myths concerning reduced tillage, they say.

"Although farmers are concerned about soil erosion, they are primarily interested in ways to increase profits," say the two agronomists.

Research and farmer experience over the past 20 years have identified major factors which influence reduced tillage success. They include:

◆ **Soil drainage.** As natural soil drainage improves, the need for tillage decreases.

"While conservation tillage systems, including no-till, are as good as or better than moldboard plow tillage on well-drained soils, no-till yields are reduced when planting into heavy crop residues on poorly-drained soils," they add. "Lower soil temperature and excess wetness early in the growing season are hurt by heavy residues."

◆ **Previous crop.** The positive influence of having a crop rotation is even more important when no-tilling poorly drained soils. This may be due to the improved soil physical properties and less residue after soybeans, which means the soil will be drier and warmer when planting.

◆ **Length of growing season.** The tillage systems that leave most of the soil surface covered with residue have generally been successful in the southern half of the Corn Belt and in states further south. Farther north, reduced yields have often occurred with residue systems compared to clean-tilled systems. Reduced soil temperature, caused by the residue cover, has a greater impact on plant growth and yield in more northern areas.

◆ **Pest problems.** Weed, insect, disease and rodent control problems are more likely with no-till. It's because pesticides can't be incorporated, cultivation is often not possible, weed seeds are concentrated near or on the soil surface and residues act as a host or cover for insects, diseases and rodents.

But the agronomists say control is possible by changing practices and increasing pesticide rates.

◆ **Proper nitrogen fertilization.** Surface application of nitrogen is common with no-till. "Where there is no means of incorporating nitrogen, it is subject to loss through volatilization and other means," say Mannering and Griffith. "This can be avoided by banding the nitrogen below the soil surface, either as anhydrous ammonia or as liquid." ✲

No-Tilling in 1972 vs. What's Predicted for 2010 and 2025
(Percentage of U.S. Row Crop Acres)

Tillage System	Actual 1972 *	Estimated 2010 **	Estimated 2025 ***
No-till	2%	54%	54%
Minimum tillage	13%	41%	35%
Conventional tillage	85%	5%	11%

* 1972 No-Till Farmer survey of NRCS state agronomists in all 50 states.
** 1975 estimates from USDA technology assessment report.
*** 2025 estimates from 2011 panel of no-till industry leaders.

FED'S PREDICTION THAT 54% OF ROW-CROP GROUND BY 2010 WOULD BE NO-TILLED WAS TOO OPTIMISTIC

In 1975, USDA staffers predicted we'd see more than half of all U.S. cropland being no-tilled by 2010. The report estimated that by 2010, some 41% of the ground would be minimum tilled and only 5% conventionally tilled.

As you can tell, those predictions didn't take place, and Glover Triplett argues the 1975 projection was overly optimistic.

"The adoption of hybrid corn — from its introduction until 95% of the corn was grown this way — took 30 years," says the retired Ohio State University and Mississippi State University agronomist. "Today, no one questions the increased value of corn's improved germ plasm and this only involved changing the source of the seed.

"Adopting no-till is much more complicated."

No-Till Resistance

Triplett says big differences in soil characteristics and crop response held back no-till acceptance.

Another limiting factor may have been resistance to change. If intensive tillage was good enough for a father or grandfather, growers didn't see any need to switch to less tillage.

Ernest Flit, an area agronomist with Mississippi State University, believes a lack of appreciation for the need to conserve soil held back no-till adoption. Even today, there's still pressure from lending officers and absentee landowners to use excessive tillage.

Dick Wittman, a veteran no-tiller from Culdesac, Idaho, maintains social and cultural barriers had more to do with the lack of no-till acceptance than economic or environmental factors.

Still another factor is the increased cost of no-till planters and drills, and not knowing what to do with the sunken cost of conventional-tillage tools.

Others feel some early-day no-till experiences may have held back acceptance. As an example, once growers realized they could no-till without coulters, no-till seemed to take off, says Ed Winkle. But early on, most manufacturers simply hung coulters on existing planters and called them no-till rigs, says the Martinsville, Ohio, crop consultant.

Suppliers Weren't Sold

Joe Nester says equipment companies kept selling the philosophy that intensive tillage is good. As the Bryan, Ohio, crop consultant points out,

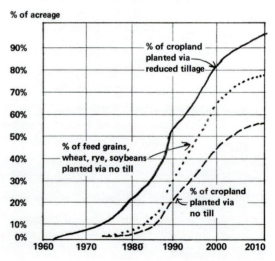

Reduced tillage on the move

% of acreage

- % of cropland planted via reduced tillage
- % of feed grains, wheat, rye, soybeans planted via no till
- % of cropland planted via no till

"From the introduction of hybrid corn until **95% of the crop was grown this way took 30 years,** and no-till is much more complicated…"

"Equipment companies kept selling the philosophy **that intensive tillage is good…**"

machinery manufacturers didn't fully understand how the physical and biological properties of the soil improve with no-till.

But Norm Widman maintains there's been more progress since 1975 than we may think. The national agronomist for the NRCS says minimum tillage is much more defined today.

"Based on the 1975 definition of minimum tillage, we've probably far exceeded the 41% goal," Widman says.

Another obstacle has been the increasing amount of residue due to higher yields, leading some no-tillers to size and lightly incorporate residue to allow earlier planting and more uniform crop emergence.

"No-till works, but it takes management and adaptive techniques to make it work in different climates, soil and cropping systems," Widman says.

Despite the fact that no-till adoption has been slower than the very optimistic prediction from 35 years ago, no-till acceptance has definitely caught on. And it will certainly pay an essential role in making American agriculture much more efficient in the years ahead. ✺

NO-TILL LOOSE ENDS...

April 2005

REMEMBERING THE NO-TILL BASICS

No-Till Farmer regularly delivers in-depth information on a single subject or takes a broad look at the practices on one farm. But it's always helpful to remember the basic principles of no-tilling. With that in mind, here are a few tips from Natural Resources Conservation Service agronomists.

- ◆ No-till needs earthworms. Use cover crops, legume rotations, high-quality lime or purchase earthworms if populations are low.
- ◆ Adjust the combine to spread crop residue evenly across the full width of your grain or corn head.
- ◆ If you no-till into sod, kill the grass in the fall.
- ◆ Weed and other pest pressures won't necessarily get worse, but will change when no-tilling. Make scouting the core of your pest management program.
- ◆ Apply herbicides in the fall in fields with heavy winter annual and perennial weed populations.
- ◆ Use a seed treatment if no insecticides will be used.
- ◆ Plant according to soil conditions, not the calendar.
- ◆ The planter is the heart of the no-tilling system. Proper maintenance and adjustments are critical.
- ◆ Equip your planter with row cleaners and wavy coulters rather than bubble coulters.
- ◆ Plant at 5 miles per hour or slower.
- ◆ Seek advice from successful no-tillers, not from growers who have simply tried no-till.

Frank Lessiter

How No-Till Changed Lives in the U.S.

"I asked subscribers to share their thoughts on how no-till has changed the way their families live and farm. Here are their stories, in their own words..."

The U.S. growers and other no-till leaders featured here believed enough in the benefits of no-till to take the time to personally relate how this reduced tillage system has impacted their lives.

It wasn't like they simply took a few minutes to answer a survey or poll question. Instead, they did some serious thinking and wrote me a personal message to explain how no-till has affected their operations and their lives.

SOUTH DAKOTA

IMPOSSIBLE TO PLANT WITHOUT NO-TILL

Ron Hesla,
Clay County, S.D..

I live in the northern plains prairie pothole region of South Dakota where my soils are predominantly heavy clay loams. I no-tilled for the first time late in the spring of 1993.

The 1992 harvest in our area is still referred to as "the slough of '92," due to record late-season fall rains followed by an early snow cover that remained on the ground all winter. Few fields were harvested without having to pull combines out of mud, water and snow. Harvest continued well into January of 1993.

The "slough of '92" soon became the "sea of '93," with untimely rains making it impossible to till corn stalks prior to planting soybeans.

As the late spring window for planting beans was about to close, I remember walking through a field of corn stalks on a hot, windy day to see if there was any hope for a new crop before the expected rains came. Although the ground was wet under the trash, the stalks and husks were dry and brittle.

With no-till, it was exciting to discover the double disc openers cut right through the dried husks and nicely placed seeds 4 inches off to the side of the standing stalks.

A "heckler" casually mentioned to me, "I don't know what you call what you're doing in your fields, but it isn't farming."

I still remember the comments from neighbors that year. One said, "I drove by your field of corn stalks and I see you didn't get your farm planted either." I replied, "Drive by later this summer to see if it looks any different."

Later that summer the same neighbor told me, "I was shocked to see beans standing above last year's harvested corn stalks."

Years later, a farmer renting ground across the fence from one of my fields stopped to visit. He said, "I've been watching you no-till and have been walking into my field, then across the fence row into your field and both seem to be doing equally well. I'd like to try what you are doing, but my landlords won't hear of it."

In 1993, I harvested an average crop of beans where there would have not been any crop if the fields hadn't been no-tilled.

Over the years, I've noticed trash breakdown was occurring faster and pushing my electric wire steel fence post "soil tilth checker" into the ground was getting easier. Earthworms were seen 90% of the times when I would check the seeding depth.

Some of the benefits I discovered included an earlier planting opportunity, fuel savings, equipment savings, ease of harvest and no dirt in my soybeans at harvest from tilled corn root balls or loose soil clumps.

There was smoother field travel. Our crop spraying service operators used to comment that driving back and forth through my no-till fields was like going up and down a super highway.

Why No-Till Matters: Without no-till, we would have gone a year in the early 1990s without growing any crops. ✿

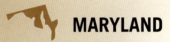

MARYLAND

TWEAKING COVER CROPS. Trey Hill finds seeding cover crops profitable on 10,000 acres.

FISHING FOR WORMS IN NO-TILLED FIELDS

Trey Hill,
Harborview Farms, Rock Hall, Md.

One day I took my 6-year-old son Clay out to get worms for fishing. When digging in a field, I let him know how much I appreciate the soils on our farm. I'm hoping that later in life these experiences will give him an appreciation for nature and beauty regardless of whatever career he follows.

As we dig for earthworms, we look at the plant root, holes, channels and the overall health of the soil. I've always viewed soil as a living organism that is a great microcosm for life. As we dig, we look at the older plants nourishing the new ones, holes made by the previous crop roots and the worms working each day to foster the process. We see where new roots are growing in their place and the radishes and the barley growing on top of the ground.

That particular day, my 6-year-old son looked up at me and said, "Dad, there certainly are a lot of

NITROGEN CYCLING. Maryland researchers are evaluating cover crop nitrogen values.

roots and worms here. It must be a pretty healthy soil."

While I've always thought of the soil as something very beautiful, I've found it difficult to articulate to people who only see what is on top of the ground. Yet my son at 6 years of age got it, demonstrating that the right explanation is possible when presented without prejudice.

Most people can appreciate the prisms of dew as the sun rises, the dancing of grain in the wind at sunset and small grain crops emerging into rows in the early spring frost. But it's difficult to show the cyclical philosophy, the intertwined complex beauty of the soil and the fact that there are both physical and philosophical layers as you look across a field.

Digging deep into the soil is a great example of what's going on in the world where people are not digging deep into other issues and only looking at what lies at the surface. The entire system is intertwined from the rain and sun above to far below the soil where the plant roots, worms and organisms live.

Why No-Till Matters: Seeing lots of roots and worms means you have a pretty healthy no-till soil. It demonstrates how the past has supported and nourished the new, while relying on each other to thrive. ❂

MICHIGAN

HAVE NO FEAR, NO-TILL IS HERE
Larry Bonnell,
Great Lakes Cover Crops, Pittsford, Mich.

After 16 years of no-tilling and using cover crops, I've seen my soils with an average of 3.5%-4% organic matter start producing 180 bushels of corn and 55-75 bushels of soybeans per acre.

Why No-Till Matters: When dry weather sets in, I have no fears about my crops. ❂

MONTANA

NO-TILL BOOSTS WHEAT VALUE. Willie Theisen says its due to eliminating summerfallow.

FOUR TIMES THE PRODUCTION WITH NO-TILL
Willie Theisen,
Scobey, Mont.

Here in eastern Montana, no-till has allowed farmers to double wheat yields over the past decade due to its ability to more effectively utilize limited moisture. At the same time,

no-till has eliminated the need to summerfallow ground every second year in an effort to conserve water.

No-till has allowed our growers to produce higher yields and increase their income by farming every field every year. Based on my math, that's a 400% increase in wheat production due to reduced costs, conserving moisture and avoiding the need for costly summerfallow.

As an example of how no-till conserves limited moisture, our son produced a 46-bushel-per-acre spring wheat crop, even though the ground received no rain from seeding until 3 weeks before harvest. Area growers using more extensive tillage settled for much lower yields.

Why No-Till Matters: Not only has no-till dramatically boosted our yields, but it's eliminated the need for summerfallow. ❂

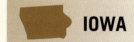
IOWA

BETTER SOIL HEALTH. Steve Berger gets it done with no-till, manure and cover crops.

IF NO-TILL WAS EASY, EVERYONE WOULD BE DOING IT
Steve Berger,
Wellman, Iowa

When we started no-tilling in the 1970s, we thought no-till was the

TAKES A TOTAL PACKAGE. Steve Berger found that relying on no-till alone wasn't enough.

ADDING COVER CROPS. Steve Berger says adding cover crops was a great decision.

gold standard for crop production. But we soon learned no-till by itself wasn't enough, as we were still losing organic matter in our corn and soybean rotation.

As part of the solution, we added cover crops and consider it be one of the best management decisions we've ever made.

Success with no-till, cover crops and efforts to improve soil health take time. It took thousands of years to create soil and you can't expect your soil profile to change in just a few years.

Why No-Till Matters: Along with cover crops and manure, no-till helps us improve our soils. We won't know if we've been successful until a farmer 150 years from now says, "Darn, they did a good job." ❋

ILLINOIS

15-INCH ROWS. Ultra-narrow rows boosted Marion Calmer's no-till corn yields by 5%.

MEETING THE PSYCHOLOGICAL CHALLENGES OF NO-TILLING

Marion Calmer,
Calmer Farms, Alpha, Ill.

In the early days of no-till, you needed a support team, as no-tilling for farmers was much like big city folks needing to see a therapist. You needed somebody to help you fully understand no-till, to listen to you whine about the results and to challenge you to solve the problems.

The National No-Tillage Conference has played a big part in my success. I've only missed one of these annual January events since Frank Lessiter started it in 1993. It's my one chance a year to get my batteries recharged for the coming crop year.

The speakers, topics and hallway networking help you find new ways to make no-till more effective and offer the confidence that you are on the right path. With over 1,000 no-tillers in attendance who

continue to innovate to overcome the cropping challenges, you realize you can be successful as well.

Why No-Till Matters: No-till has made farming easier, made it more profitable and more environmentally friendly. ❋

OHIO

THREE CASH CROPS IN 2 YEARS. That's what no-tillage did for western Kentucky growers.

HARVEST 1,800 ACRES FROM 1,200 ACRES

Howard Doster,
former Purdue University ag economist, Waynesville, Ohio

I remember meeting Kentucky no-till pioneer Harry Young, Jr., around 1971 and learning how he was successfully double-cropping no-till soybeans after wheat.

After I ran his farm budgets through the Purdue machinery analysis program, I realized he had enough machinery to be timely with a 600-acre central Indiana corn and soybean rotation. Yet by no-tilling, he was growing 1,800 acres of corn, double-crop soybeans and wheat on only 1,200 acres of south central Kentucky land.

I coordinated the Purdue University Top Farmer Crop Workshop for 30 years. We helped over 7,000 Corn Belt farmers, including many no-tillers, interpret the shadow price signals in over 25,000 economic analysis plans as they tested alternative crop rotations, machinery size, tillage systems and farm size. There was a "tillage" presentation, often dealing with no-till, at every workshop.

No-till was not an accepted practice when we started these workshops. Even though no-till looked good in Purdue test plots, and in the economic budgets based on this data, many farmers knew birds, mice and ground squirrels sometimes ate the corn seedlings out of the rows, weeds were often terrible and that no-tillers had fewer days suitable for fieldwork.

Purdue agronomist Don Griffith was a real no-till missionary, as he spoke to thousands of farmers throughout the Corn Belt and Argentina, sharing how he and his colleagues made it work. Another Purdue agronomist, Jerry Mannering, showed farmers throughout the Midwest and Romania how to no-till.

Why No-Till Matters: There's no other way Harry Young could harvest 1,800 acres of corn, double-crop soybeans and wheat from only 1,200 acres of south central Kentucky ground. ❀

LONG-TERM ANSWERS. These on-going no-till plots at Wooster, Ohio, were started in 1962.

No-tillage provides an opportunity to rapidly expand production while protecting the soil against erosion. Desirable soil conditions formed under permanent vegetation, increased organic matter and stable macrospores are retained with no-till.

There's also a long-term trend toward increased ag productivity. Larger equipment, fewer operations for soil preparation, less cultivation and increased use of herbicides have contributed to this trend.

Larger acreages are no-tilled with fewer workers. No-till has accelerated this trend by further reducing the operations used for crop establishment and will continue to do so. No-till also improves harvest efficiency by providing better traffic support on the ground during wet harvest seasons.

Why No-Till Matters: It gives you an opportunity to rapidly expand the acres you farm while dramatically trimming soil and water erosion. ❀

One of the core focuses of our farm centers around stewardship and sustainability. We hope to hand over the care and protection of the land to the next generation in as good or better condition than it is today.

Our goal is to continue to develop and refine new technology practices and to implement them in a way that ensures long-term stability and productivity. We try to demonstrate that we are good stewards of the land that we use to grow grains to feed the world.

OLD TOOLS, NEW USES. With ingenuity and winter shop work, Andrew Wyss turned an outdated cultivator and grass seeder into a rig (shown off by grandson Andrew Joseph Wyss III) to seed his Conservation Reserve Program acres and waterways for effective erosion control.

Over time, we've implemented no-till practices such as cover crops, filter strips and grass waterways to reduce erosion and add nutrients to the soil.

Farm shop work and modifying equipment is a task I thoroughly enjoy. During a recent long winter, I modified an outdated cultivator and grass seeder that had been in the barn for years. We use this rig to seed our Conservation Reserve Program acres and waterways to control erosion.

Why No-Till Matters: No-till is a simple yet effective tool that lets us improve our land now and for future generations. ❀

OHIO

NO-TILL HELPS GROWERS EXPAND

Warren Dick,
Ohio State University agronomist, Wooster, Ohio

INDIANA

LEAVING THE SOIL IN BETTER SHAPE

Andrew Wyss
Wyss Farms Enterprises, Fort Wayne, Ind.

KANSAS

BRIGHT FUTURE. With the seventh generation, no-till is dramatically improving soil health.

LOOKING BACK. The Sutherlands wished they'd used cover crops when they first no-tilled.

SHOULD HAVE SEEDED COVER CROPS MANY YEARS EARLIER

David Sutherland,
Lot 40 Farms, Le Roy, Kan.

Our family's 2,400-acre no-till operation is located in the rolling hills and Neosho river bottom ground of east-central Kansas. Our grandchildren are the seventh generation to live on this farm.

My great, great grandma Mary Sutherland and children came in a covered wagon to Kansas and homesteaded in Woodson and Cof-

fee counties. She later left her oldest child, Henry, on the family farm and moved on to homestead in Colorado, again in Oregon and finally made her last land run in Oklahoma. As time went on, the original 160 acres here in Kansas became 3,000 acres, including grassland.

MORE FAMILY TIME. No-till sharply reduced the labor needs with the family's 2,400 acres.

In the late 1960s, Dad tried no-till. Success was variable, as there weren't many chemicals to help with weed control and equipment was not efficient for planting in heavy residue and sod-like conditions. We struggled with no-till for a number of years until new solutions came along.

A few years after the National No-Tillage Conference (NNTC) got underway, we started attending and quickly became friends with farmers across the country. Their advice and experiences have been invaluable to our no-till success. Combined with the knowledge we learned from the NNTC sessions, we moved to 100% no-till in 1999.

Dad added cover crops in the late 1980s, but there wasn't much information available then. Since then, we've seeded winter peas, oilseed radish, buckwheat, hairy vetch, sunn hemp, annual ryegrass, cereal rye, tillage radishes

and crimson clover alone and in tankmixes.

The biggest mistake we made when we converted to no-till was not planting cover crops right from the start. If we had, our soils would be in much better shape today.

The switch to no-till has provided more free time to spend with our family, less equipment wear, fuel savings, soil savings and has had a favorable environmental impact on our farm.

Why No-Till Matters: The friends we've made over the years by attending the NNTC and all the learning that takes place there will be with us forever. Thank you to all for providing this unique no-till learning opportunity. ✻

NEW YORK

WINNING COMBINATION. No-till and cover crops together have dramatically improved soil quality and erosion control on the John Kemmeren family's dairy operation.

TOTALLY SOLD ON NO-TILLING AT AGE 12

John Kemmeren,
Bainbridge, N.Y.

I was introduced to no-till in the early 1970s at an Allis-Chalmers dealer's open house. Even though I was only 12 years old, I was hooked on no-tilling. Together, with my father, we started no-tilling in 1975.

Our farm has produced amazing crops throughout the years with tremendous soil health and erosion control. Cover crops have added tremendous value to our farm and we use them anywhere we can. We have no-tilled many crops and cover crops successfully and are always willing to try new things.

I'm also proud to have shared my knowledge and helped many others start no-till on their own farms.

Why No-Till Matters: My wife and I could not farm any other way. ✹

ILLINOIS

NO-TILL GREW QUICKLY. In the early 1970s, corn was often no-tilled into sodded ground.

NO-TILLERS ALWAYS EAGER TO SHARE EXPERIENCES

Darrell Smith,
retired Farm Journal field editor, Champaign, Ill.

I was fortunate that my ag journalism career spanned the formative years of no-till in the Midwest. I wrote my first no-till story about Ed

Hawley of Elizabeth, Ill., for the first issue of *No-Till Farmer,* edited by Frank Lessiter — who, incidentally, became my journalistic role model.

A few years later, my first significant article for *Farm Journal* was a no-till story about Warren Harms in Carroll County, Ill.

No-till's conservation benefits were exciting, and the technology was fascinating. But as with everything, it's the people that you remember most.

Looking back, I came to feel that Herman Warsaw, the first 300-bushel corn yield contest winner, was sort of a spiritual father to no-till, although he wouldn't have realized it. Herman's goal was not to gain personal recognition, but to prove that high yields could be produced with soil-saving conservation tillage.

Not far from Herman's McLean County farm in Illinois, Jim Kinsella shared Herman's desire to conserve soil and help other farmers. Jim took Herman's conservation tillage concept the rest of the way, proving that no-till could work in a corn and soybean rotation.

Jim's sense of stewardship, his desire to learn, his willingness to innovate and to share his knowledge and, frankly, his courage in the face of opposition could fill a book (and they should).

Among other things, Jim taught me to respect no-tilled soils by halting his tractor, leaping out of the cab and performing a magnificent war dance to convey a long-distance message: "Don't drive that truck on my no-till field!"

Many other no-till (and strip-till) farmers helped the concept grow. Everyone I met shared Warsaw's and Kinsella's sense of stewardship and desire to help other farmers.

Another thing they shared was a passion to understand their soil.

Although those early no-till farmers didn't have a word for it, what they were wondering about and studying was soil health — a term that's on everyone's tongue today.

So, although I was merely an observer, no-till impacted me, as well as farmers. By providing wonderful people eager to share their stories, no-till helped me make a living as a farm writer.

Seeing how no-till farmers produced crops influenced me in another way, too. Around 1990, I became a no-till gardener, and I haven't tilled my dirt since — and nobody messes with my night crawlers!

Why No-Till Matters: Early day no-till farmers didn't have a word for it, but what they were studying was soil health — a term that's on everyone's tongue today. ✹

IOWA

NO-TILL BOOSTS WATER, NUTRIENT EFFICIENCY

Jerry Hatfield,
National Laboratory for Agriculture and the Environment, Ames, Iowa

As a USDA researcher, my emphasis has been on how we can improve crop production efficiency through improved water and nutrient use. One of the unrealized aspects of no-till is that changes within the soil will dramatically increase both water and nutrient use efficiency.

The foundation for improved crop production efficiency is relat-

SOIL CHANGES. Jerry Hatfield has spent his career researching no-till soil concerns.

ed to how no-till began to change soil health. Through these observations, I began to understand the foundation to soil health was reduced tillage, which provided four essential components for an effective soil biological system: food, water, shelter and air.

These insights prompted the formation of what I call the soil aggradations climb. This is based on enhancing the soil, beginning with soil biology, organic matter and nutrient cycling.

This leads to improved soil structure and soil water availability and results in improved crop yields, profit and efficiency.

I credit these insights to producers who have been willing to share their experiences and ask pointed questions on what they can do to improve. These challenges have served as an incentive to begin to understand the processes on how no-till changes occur in the soil.

No-till reduces the amount of water lost from the soil surface to the air and makes that water available to the plant. No-till residue absorbs the erosion energy created by raindrops and allows more water to move into the soil.

Among the current problems we have in managing moisture is find-

ing ways to move more water into the soil. This has led to developing new ways to demonstrate how no-till boosts the amount of rainfall available to the plant by increasing the amount that penetrates the soil. This leads to more plant growth, greater production per unit of moisture and improved water use efficiency.

I love agriculture and try to do everything I can to help producers understand the fundamentals of how no-till improves all aspects of production. Thanks to all of the producers who have helped educate me over the years. Hopefully, I have given something back.

Why No-Till Matters: No-till has altered my thinking about how soil processes function and what it takes to reach higher levels in crop production. ✳

KENTUCKY

STARTED OUT WITH DOUBLE-CROP SOYBEANS

Joel Armistead,
Adairville, Ky.

We are mostly a no-till operation with some vertical tillage and low disturbance ripping as necessary while raising winter wheat, double-crop soybeans and corn.

My father, Raymond, purchased an Allis-Chalmers no-till planter in 1969. He added John Deere #71 planter units and Dickey-John row monitors. The planter was set up to spray herbicides on the back and apply dry fertilizer in the row. It let us start double-cropping soybeans, and later no-till corn and wheat.

No-tilling lets us handle over

MORE ACRES, LESS LABOR. Joel Armistead no-tills 1,600 acres with only two full-time folks.

1,600 acres today with only two full-time people. It has saved untold tons of soil and increased soil health for us. It would be terrible to lose certain chemicals and GMO traits as this would make it difficult for us to continue no-tilling in the future.

Why No-Till Matters: It has increased soil health, saved untold tons of soil and allowed us to make a good living with two people. ✳

OREGON

EDUCATING GROWERS. Donald Wirth's mission is to educate growers on cover crop values.

COVER CROPS GOT US STARTED WITH NO-TILL

Donald Wirth,
Saddle Butte Ag, Tangent, Ore.

We started working with the Oregon Ryegrass Commission in the late

DEEPER ROOTS, DEEPER PROFITS. Don Wirth checks the rooting depth of annual ryegrass.

1990s with on-farm plots in Indiana and Tennessee and in university trials where annual ryegrass was evaluated as a potential cover crop.

With the success from these plots of seeding cover crops in no-till, we formed Saddle Butte Ag in 2002 to market cover crop seed. The primary role for our half dozen employees is educating farmers about the many benefits of cover crops.

Why No-Till Matters: Along with cover crops, no-till has allowed growers to dramatically improve the health of their soils. ✻

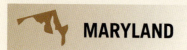 **MARYLAND**

SOLD ON NO-TILL, SO NEVER LOOKED BACK

Ray Weil,
University of Maryland,
College Park, Md.

I didn't experience no-till farming until I returned to graduate school in the mid-1970s at Virginia Tech. While my farming experience had been with full tillage, several other graduate students were researching no-till with cover crops and introduced me to the practice.

Once I moved to the University of Maryland in 1979, the farming community in Maryland was rapidly adopting no-till. A senior faculty member, Allen Bandel, was finding ways to overcome nitrogen immobilization and volatilization problems, which led to dribbling UAN solutions on no-till corn.

PLANTING GREEN. Soybeans were no-tilled into wheat in this 1991 relay cropping trial.

EARLY-DAY COVER CROP RESEARCH. In early 1980s, University of Maryland agronomist Ray Weil started working with cover crop trials.

I started researching no-till practices and its impact on soil quality in the 1980s and have not stopped since.

Why No-Till Matters: There's no better way to improve soil quality and crop profitability. ✻

 OHIO

25 YEARS OF SUCCESSFUL NO-TILLING

Edward Pohlman,
Maria Stein, Ohio

We started no-tilling in 1992 and the next year I attended the very first National No-Tillage Conference in Indianapolis. It took us 3 years to see the no-till differences and savings. We've never looked back!

Thanks to no-till, soil health immediately improved. Our corn yields increased from 125 to 180 bushels per acre and they have stayed very consistent over the years. Plus, the fuel, fertilizer and labor savings are great.

Why No-Till Matters: It leads to better soil health and reduced costs. ✻

 PENNSYLVANIA

NO-TILL'S BEEN VERY GOOD TO OUR FAMILY

Martin Greenleaf,
Kirkwood, Pa.

During the first 5 years my wife and I farmed, we had two full-time hired men and four tractors with over 100 horsepower. When one employee quit to farm on his own in the early 1970s, I traded five higher horsepower tractors for a new 80 horsepower tractor and no-till planter.

The following year, I let the other hired man go, upped my acreage from 500 to 1,000 acres and my wife and I did everything ourselves.

There were no more salaries, worker's compensation, health insurance or other employee costs. No-till made that possible.

We went from a farm that was barely making us a living to a much happier and more profitable cropping situation. Our neighbors were slow to switch to no-till, but that was fine and it left us better off economically. They eventually saw the light and jumped on the no-till wagon. Thanks to no-till, I had more spare time for things family members wanted to do.

Our yields have followed a long steady uptrend and have been satisfactory under hot, cold, wet and dry conditions. In 40 years, we only received crop insurance payments once, and that was after a 2-year drought that still netted us no-till corn yields of 96 bushels per acre. In the last 10 years, our overall average corn yield has dropped under 200 bushels only during three extremely dry years.

I honestly don't know if I could have stuck with farming without no-till. It's been over 40 years without using a moldboard plow, chisel plow, subsoiler or vertical tillage. We've made enough money to put a son through an Ivy League college, 5 years of medical school, a fellowship at Harvard and watched him become a successful career as a spinal surgeon. No-till has been very good to us!

Why No-Till Matters: I honestly don't know if we could have stuck with farming without no-till. ❄

NO-TILL-AGE®

December 2014

Big Fuel Savings with No-Till

(Trials conducted with no-tilled corn from 1997 to 1999)

Diesel fuel consumption...

Crop and Acres	Conventional Tillage	Mulch Tillage	No-Tillage
Corn (700 acres)	3,779 gal.	3,170 gal.	2,191 gal.
Soybeans (700 acres)	3,667	3,058	2,191 gal.
Winter wheat (400 acres)	2,127	1,244	908
Total fuel usage (gallons)	9,573	7,472	4,730
Annual fuel savings over conventional tillage (gallons)	———	2,101	4,843

Diesel fuel cost ($3.75 per gallon) ...

Crop and Acres	Conventional Tillage	Mulch Tillage	No-Tillage
Corn (700 acres)	$14,174	$11,891	$8,216
Soybeans (700 acres)	$13,754	$11,471	$6,116
Winter wheat (400 acres)	$7,979	$4,665	$3,405
Total fuel costs	$35,907	$28,027	$17,737
Annual fuel savings over conventional tillage	———	$7,880	$18,170

— National Resources Conservation Service

WISCONSIN

NO-TILL MAKING POSITIVE CONTRIBUTIONS TO CLIMATE CHANGE

Bill Rohloff,
Cambridge, Wis.

To sum up my appreciation for no-till, it has helped me achieve my goal of leaving our precious soil in much better condition than when I started farming. We only get to go around once in life and we only have the planet to sustain human life, so we need to find a way to make a difference.

Agriculture has been around since the beginning of time and has evolved into an industry to provide food and fiber for human civilization throughout the world. No-till represents one of the greater ag achievements of all time and greatly contributes to the sustain-

NO-TILL SINCE 1992. With no-till, Bill Rohloff is having a major impact on climate change.

MUCH BETTER ENVIRONMENT. No-till is helping Bill Rohloff make a difference in agriculture.

ability of the land we cultivate.

I started no-till in 1992 because of these benefits and have not looked back. By using no-till, we're making a positive contribution toward climate change as well.

A round of applause and sincere appreciation to *No-Till Farmer* and all who shared what they found, as I could not have accomplished what I have without the timely articles, information and knowledge I've acquired from reading this publication.

No-till will be here for another 25 years as will be *No-Till Farmer* and the National No-Tillage Conference. Congratulations to your great organization and continued success.

Why No-Till Matters: Thanks to no-till, we're all making a positive contribution toward dealing with climate change. ❋

WISCONSIN

NO MORE DUST, NO MORE CRUSTS, NO MORE CLODS, NO MORE ROCKS

Tom Burlingham,
Palmyra, Wis.

My father scolded and yelled at me every day for 2 months straight. That was in the spring of 1982 after Vern Davis drove his tractor and Buffalo planter 25 miles and no-tilled 15 acres of corn directly into a hay field on our farm.

It took until July for Dad to settle down — as the corn started looking great. Eventually, we got the whole farm off the tillage merry-go-round.

However, doubt would sometimes creep in, and it took years of

side-by-side trials to see the many benefits of no-till. We couldn't get a yield response from working the ground, so tillage earned its way right off our farm. No more dust, no more crust, no more clods to bust, no more rocks to pick. Everything clicks with no-till.

Why No-Till Matters: It let us get the entire farm off the costly intensive tillage merry-go-round. ❋

ILLINOIS

THERE'S PATRIOTISM IN NO-TILLING

Rich Reynolds, *Sheldon, Ill.*

My brother, Ron, and I took over the family farm in 1980 from a very skeptical father who felt intensive tillage was the only way to farm. But since erosion was a big concern, we looked for ways to improve our operation.

By attending several BASF meetings, we learned the new way of farming — no till. We learned a great deal from the successes of long-time Illinois no-tiller Jim Kinsella and figured out how to make his system work on our farm.

We used videotapes, pamphlets and articles to show landlords the many benefits of no-till. Today, none of them want us to go back to conventional tillage. Along with the help of cover crops, they recognize the long-term benefits of no-till and the sustainability it will bring to the farming community.

Why No-Till Matters: No-till and cover crops are a way to be patriotic and do what is best for God and our country. ❋

TOP YIELDS ON POOR GROUND. Thanks to no-till, 200-bushel corn yields are now common.

NO-TILL SETS UP FAMILY'S FARM FOR LONG-TERM SUCCESS

Mark Turner,
Livermore, Ky.

After a couple of years of no-till, it seems like some soils go through a rough spot, sort of a metamorphosis. Unfortunately, this is when some folks too quickly give up on no-till.

We no-tilled successfully for 3 years before being hit with a wet spring. That year, it seemed like all the nutrients were locked in the soil and our crops looked like we hadn't been fertilized at all. For a couple of years, our no-tilled corn looked like death warmed over and turned out less than ideal yields.

But then our soils balanced themselves and soon we again had great looking and high-yielding crops. We now no-till crops that are just as good as farmers who farm conventionally. Best of all, we've seen incredible yield turn-arounds on some acres.

As an example, we have a

gumbo soil that acts more like a sandy soil. My grandfather, dad and I all tried to farm that field without success, so it was seeded to pasture. If we moldboard plowed that field in the spring, it would end up being be too wet to plant in the morning and too dry to plant in the afternoon. We eventually started row cropping it for 1 year and then returned the field to pasture for several years.

Since we've switched to no-till and added cover crops, there's

COVER CROPS ARE KEY. Combining no-till with cover crops sets up farm for future success.

been an astounding transformation in that field. I thought it was a fluke the first year, but it's continued to yield phenomenally year after year, continually turning out 200 bushel per acre corn yields.

We're also excited to bring tobacco back into our no-till program. We strip-tilled this crop successfully for a number of years after trying no-till 10 years ago with poor results. With the latest plant-setting rigs, no-till tobacco has become a reality. It seems today like you can no-till practically any crop.

Harvesting 200 bushel corn off formerly poor ground is amazing and a tribute to the benefits and profitability of no-till and cover

NO-TILL TOBACCO. New equipment ideas led family to bring back this high-value crop.

crops. No-till will allow my son and grandson to become part of this farming operation in the future.

Why No-Till Matters: No-till and cover crops are doing a good job of setting up our farm for success 20 years down the line. ✳

SEEING IS BELIEVING

Garrett Duyck,
Soil Conservationist, Natural Resources and Conservation Service, Portland, Ore.

If you need proof that no-till farming works, look no further than the rolling hills of north-central Oregon. For decades, this region was dominated by winter wheat growers who relied on extensive tillage to control weeds during the summerfallow year. From the early 1900s through the 1980s, excessive tillage was the traditional way of farming in the area.

But over time, soil erosion became a serious threat. Our Columbia Plateau terrain is characterized by rolling hills of wind-blown silt called loess. On our steep slopes, these soils are very susceptible to water and wind erosion, leading to considerable acres of "highly erodible land."

Old-timers living in the region recall stories of roads, houses and other barns being buried in mud after major rains. Evidence of severe soil erosion can still be seen today in large gullies that were once only roadside ditches.

Beginning in 1935, the Soil Conservation Service helped farmers install structures to reduce soil erosion and prevent sediment from leaving crop fields. These practices included diversions, sediment basins, grassed waterways and terraces.

This work continued as Congress passed the highly erodible

NO-TILL-AGE®

How No-Till Influences Net Farm Income

November 2004

Type Of Income	Conventional Tillage	Mulch tillage	No-Tillage
Cash income	$118,000	$196,000	$211,000
Variable expenses	$68,000	$112,000	$112,000
Fixed expenses	$22,000	$31,000	$40,000
Net cash income	$28,000	$53,000	$59,000
Acres operated	460	644	913

— Economic Research Service, U.S. Department of Agriculture

2016 FIELD. With no-till, the terraces that were built 54 years earlier are barely visible.

1962 FIELD. The installation of costly terraces was the major way to control soil erosion.

> **"No-till and cover crops leave fields looking like a mess,** so the idea has to grow on you."
>
> —*Trey Hill, Rock Hall, Md.*

land compliance rules in the 1985 Farm Bill that spurred wide-scale adoption of soil conservation practices.

In the 1990s, NRCS and the Wasco County Soil and Water Conservation District (SWCD) encouraged farmers to adopt a new conservation tillage system called direct seeding (no-till). The goal was to minimize soil disturbance and preserve crop residues over the entire fallow period from harvest to

planting — dramatically reducing soil erosion.

Direct seeding was made possible with new planting equipment and herbicides to control plant growth during the fallow period, now referred to as chemical fallow.

More than 95% of the dryland grain cropland in Wasco County is direct seeded today, thanks in part to the NRCS and the Wasco County SWCD. Terraces installed in the

mid-1900s are barely visible today and soil erosion has been curbed significantly.

It's an incredible story of the evolution of no-till cropping systems and the success both farmers and conservation agents have achieved in this region.

Why No-Till Matters: Area farmers say if they had to return to conventional tillage, they'd rather stop farming because it's so taxing on themselves and the land. ❧

The World Gets on Track
With No-Till

"Switching to no-till was essential for the survival of South American growers since there are no farm subsidies and they must be competitive in world markets..." — *Rolf Derpsch*

Rolf Derpsch maintains the long-term conversion to no-till could have more impact than any other innovation on Third World food production.

The South American conservation tillage consultant from Paraguay spells out why growers around the world are on the right track by adopting no-till.

What had been a U.S. born innovation was quickly embraced by the world, with South America now ahead of North America in total no-tilled acres.

Tillage Vs. Residue

"Unfortunately, we have concentrated too much and too long on not tilling the soil instead of concentrat-

ing on crop residues as the main tool for no-till management," Derpsch says. Taking climate and socio-economic factors into account, he sees a tremendous potential in bringing no-tillage technology to Third World countries.

"Either you no-till or you sell your farm to the neighbor for economic or

BRITISH SAW NO-TILL BENEFITS. An intensive campaign to increase Great Britain's no-till wheat and pasture renovation acres led to a 25% increase in 1973. Growing from only 100,000 acres of no-till in 1967, the goal was to reach 500,000 acres by 1978 while enjoying a savings of $25 per acre.

ecological reasons," he says. "That's because studies in South America have shown significant economic advantages with no-till."

Derpsch points out that U.S. and South American growers disagree about a number of the remedies to solve compaction and water infiltration problems.

"South American farmers have learned that having a permanent soil cover, the use of strongly-rooted cover crops and diverse rotations allow the soil biology to develop and are key to counteracting soil compaction," says the educator who has spoken at several National No-Tillage Conferences.

South Americans emphasize the importance of soil residue to increase water infiltration. Many U.S. farmers, meanwhile, still think they need to loosen the soil with tillage to increase water infiltration.

"By continually switching tillage systems from year to year, the soil is constantly in the transition phase and farmers never experience the full benefits of the no-till system," he says.

South American growers were quick to recognize the value of leaving no-till residue on the soil surface.

"No-till with low residue levels won't give us full benefits," says Derpsch. "We must produce at least 22,000 pounds per acre of crop residue each year to have a good working system. Cover crops are essential in achieving a permanent cover."

South America has long been the global leader in cover crops, as growers quickly recognized the value of keeping the soil covered every month of the year. Brazilian researchers have shown as much as a 60% increase in no-till soybean yields when black oats were seeded as a cover crop.

No-Till Acres Continue to Grow

The number of no-tilled acreage around the world has grown tremendously. Data from the Food and Agriculture Organization of the United Nations indicates there were only 6.9 million acres of no-till throughout the world in 1973, with 70% in the U.S.

By 2013, the worldwide no-till acres had increased to 387.9 million acres, with only 27% of that ground being in the U.S. ❋

TREMENDOUS GAINS IN WORLDWIDE NO-TILLED ACRES

Here's how no-till's acreage grew in 40 years.

Year	No-Tilled Acres
1973	6,918,950
1999	111,197,421
2003	177,915,874
2008	263,180,155
2013	387,955,448

— Food and Agriculture Organization of the United Nations

NO-TILL ACRES AROUND THE WORLD

Finland 494,210
France 494,210
Russia 11,197,742
Ukraine 1,729,737
Kazakhstan 4,942,107
Canada 33,122,276
United States 88,001,639
Spain 1,957,074
Venezuela 741,316
Colombia 313,823
Brazil 78,606,692
India 3,706,580
Bolivia 1,744,563
Paraguay 7,413,161
Uruguay 2,648,969
South Africa 909,347
Australia 43,725,297
Argentina 48,726,710
New Zealand 400,310

2013 Data

Country	No-Tilled Acres	Country	No-Tilled Acres	Country	No-Tilled Acres
United States	88,001,639	France	494,210	South Korea	56,834
Brazil	78,606,692	Zambia	494,210	Switzerland	42,007
Argentina	48,726,710	Germany	494,210	Iraq	37,065
Canada	33,122,276	Chile	444,789	Sudan	24,710
Australia	43,725,297	New Zealand	400,310	Tunisia	19,768
China	16,481,928	Mozambique	375,600	Madagascar	14,826
Russia	11,197,742	United Kingdom	370,658	Hungary	12,355
Paraguay	7,413,161	Colombia	313,823	Morocco	9,884
Kazakhstan	4,942,107	Malawi	160,618	Uzbekistan	6,054
India	3,706,580	Turkey	111,197	Lesotho	4,942
Uruguay	2,648,969	Mexico	101,313	Azerbaijan	3,212
Spain	1,957,074	Moldova	98,842	Lebanon	2,965
Bolivia	1,7,44,563	Slovakia	86,486	Kyrgyzstan	1,729
Ukraine	1,729,737	Kenya	81,791	Netherlands	1,235
Italy	939,000	Portugal	79,073	Namibia	840
South Africa	909,347	Ghana	74,131	Belgium	667
Zimbabwe	820,389	Syria	74,131	Ireland	494
Venezuela	741,316	Tanzania	61,776		
Finland	494,210	Greece	59,305		

**2013 worldwide no-till acres
387,955,448**

— *Food and Agriculture Organization of the United Nations*

A CLOSER LOOK AT
NO-TILL AROUND THE WORLD

Here are a few highlights about the adoption of no-till in a number of countries as gathered from the archives of *No-Till Farmer*.

BRAZIL

No-till was originally perceived in Brazil as a soil conservation measure. It evolved into a sustainable production system with positive economic, environmental and social consequences to where it was used on over 78 million acres in 2013.

"No-tillage puts conventional tillage knowledge upside down," Rolf Derpsch says. "Anybody that wants to have success with no-till has to forget almost everything they learned about conventional tillage systems."

When several University of Kentucky agronomists and others headed to South America in the early 1970 to show farmers how to no-till, little did they realize that demonstrating the new technology would eventually turn Brazil into a major player in the world food market — and a serious competitor to U.S. grain producers. It's not likely they foresaw the eventual impact this would have on grain prices paid to American farmers.

In 2005, uncleared brush land in Brazil was selling for $30-$40 an acre and could be cleared for $125-$250 an acre. At the same time, highly productive ground typically sold for $700 per acre.

With irrigation, Brazil's growers are finding it's possible to no-till seven to nine crops within 3 years on the same ground.

Herbert Bartz, a no-tiller from Brazil, says both small and large acreage farmers have utilized farmer-driven research to make the no-till systems approach work. As a result, no-till is widely accepted as the farming system of the future among Brazil's university educators and researchers. Schools are now training technicians and pushing no-till programs at several levels while forming valuable partnerships with industry, farmers and research people. 🌼

AFRICA

Derpsch says a 1998 study on the potential use of no-till in Africa indicates no-till is the system of choice in regions where sufficient biomass can provide year round ground cover.

However, several ecological factors are working against no-till growth. These include low precipitation, low biomass production, short growing seasons, sandy soils with a tendency for compaction and the risk of waterlogged soils.

There's also a strong demand for using crop residues as livestock feed, uncertain future land use rights, poorly developed infrastructures for marketing, credit and education, market preferences for one crop such as corn and high farm management needs. 🌼

RUSSIA

The explosion at the nuclear power plant in Chernobyl in the 1980s left over 11 million acres of Russian soils so badly contaminated from radiation that none of this cropland has ever been farmed.

By the mid 1990s, tests showed the soils were still so radioactive that they could not be used to grow crops for human consumption or as livestock feed. Grass held the soils in place, but grazing was not allowed.

At the time, University of Manitoba weed scientist Elmer Stobbe felt no-tilling industrial grade rapeseed could bring these Russian acres back into production. In fact, Stobbe worked with Russian ag officials in an attempt to set up a trial to no-till several thousand acres of the radioactive soils. Unfortunately, there were serious concerns over erosion washing radioactivity from the ground into streams.

Around the same time, two North Dakota no-till leaders also met with Russian ag officials about cropping the land near the Chernobyl nuclear accident. Bob Nowatzki, former head of the North Dakota State University experiment station at Langdon, N.D., and veteran no-tiller Luther Bernston from Adams, N.D., felt canola and flax could be grown on the contaminated land.

Bernston felt rapeseed had the ability to extract heavy metals from the soil and leave them in the residue. On the other hand, growing cereal grains could allow plutonium to get into the food chain, which would still leave land around Chernobyl unsuitable for growing food crops. As a result, Nowatzki and Bernston urged officials to consider growing flax for oil usage.

Due to safety concerns and the political climate in Russia, none of these 11 million acres have yet been returned to production. ✻

CHILE

Carlos Crovetto, a long-time no-tiller from Conception, Chile, and speaker at several National No-Tillage Conferences, says his driving motivation for developing a successful no-till system was the need to triple food production within the next 50 years as the worldwide population doubles.

Crovetto found it took extensive efforts to make no-till work when farming slopes as steep as 30%. Yet boosting corn yields on these no-tilled slopes demonstrated what could be done with a total no-till systems approach, providing the farmer has the proper mindset and technology. ✻

UKRAINE

In 1992, Roger Denhart no-tilled 300 acres near Champaign, Ill., and also grew 600 acres of irrigated soybeans, sweet corn and field corn on the Karl Marx and C. Lenin collective farms near Odessa in the Ukraine. By the following spring, the goal was to no-till 50,000 acres of soybeans in the southern area of the country.

While Denhart and his financial partners in the Freedom Farms International project hoped to get a $17 million loan from the World Bank, Ukraine farm workers would farm the ground.

The goal was to split the 50,000 acres into 2,500 blocks (20 acres each) and use John Deere 750 drills to no-till soybeans. They expected to eventually double-crop soybeans behind wheat.

Denhart felt no-till was a perfect fit for this land that only receives 14 inches of rainfall annually. Center pivot irrigation systems would pull water out of deep irrigation canals, which were already in place.

Several years later, he explained to National No-Tillage Conference attendees how no-till was a difficult sell with the Ukraine's collective farming philosophy. While collective farm managers saw the labor saving benefits of no-tilling, they weren't willing to give up intensive tillage, which enabled them to keep 100 farm workers fully employed during each cropping year.

Denhart's group ended up growing 33,000 acres of soybeans, corn and wheat in the Ukraine. It also built a flour plant, a hog operation that turned out 20,000 hogs a year and a pork processing facility before the operation was sold in 2004.

A number of times over the years, Nila Martyniuk attended the National No-Tillage Conference. An American citizen born in New Jersey and the daughter of Ukraine immigrants, she worked as a representative of the Ukraine-based Agri Soyuz farming operation.

Started in 1996, this Ukraine farming operation relied heavily on no-till to farm more than 64,000 acres. When the Soviet Union broke up and communal farming practically vanished, individuals or families in the Ukraine were given the right to farm up to 25 acres of government owned land. As a result, Agro Soyuz rents ground from 1,200 landlords.

The operation also includes meat and egg production from 2,800 ostriches, milk from a 1,800 head dairy operation and a farrow-to-finish operation that turns out 190,000 market hogs per year.

Even with a devastating war taking place 130 miles down the road in the Ukraine, the ag operation is carrying on. Despite the unstable Ukrainian economy, the country's currency losing its value and hamstrung exports, the farming operation is surviving.

No-till was introduced to the operation in 2002 after Martyniuk and several company agronomists visited Dwayne Beck at the Dakota Lakes Research Farm in Pierre, S.D. When they saw what Beck was doing with no-till, they immediately called home and said, "no more tillage."

The Ukraine climate is similar to the High Plains with limited spring and fall moisture in a growing season that runs from April through October. The farm is located in the largest black soil region in the world with 2 feet of typical black soil that runs as deep as 30 feet in some areas.

Over the years, they've moved to no-till to trim costs and boost yields. The next step is adopting cover crops to gain still more benefits with no-till.

During a visit years ago to the

No-Till Farmer offices in Wisconsin, Nila and her Ukraine crew purchased and took home copies of every issue, book, report, white paper and National No-Tillage Conference proceedings we'd produced since 1972.

Their intent was to learn all they could about no-till under all types of conditions from the thousands of no-till knowledge pages we sent home with them. We provided them with valuable no-till information that could take a single person practically a lifetime to read. ❋

PARAGUAY

Even though Paraguay is a small country, 60% of its agricultural land is no-tilled. Widespread acceptance of no-till in Paraguay is due to farmer-to-farmer education, the heavy emphasis on educating small acreage farmers on how to no-till and finding new ways to obtain credit to purchase no-till equipment and supplies. ❋

UNITED KINGDOM

No-till was more popular in England in the '70s and '80s — when growers could burn straw — than is the case today. Once the British government banned straw burning, Jim Bullock says it put a halt to no-tilling with the no-till drills available at the time.

At the 2004 National No-Tillage Conference, the Worcestershire, England, grower shared his reduced tillage experiences on a farm that has been in the family for nearly 500 years. After successfully no-tilling in earlier years, he and his brother moved back to more intensive tillage due to the government's ban on residue burning.

A visit with Kansas and Oklahoma no-tillers at the conference convinced him modern-day no-tilling would work in the family's 850 acre operation. The brothers switched back to no-till due to its many economic, soil erosion, logistic and climate change benefits.

"The Kansas and Oklahoma growers were double-cropping soybeans into wheat stubble with the planter running right behind the combine in the same field," says Bullock. "I thought that if no-till can work under these conditions and with those large amounts of residue, then it could work for us."

By going back to no-till, the family trimmed machinery costs by 30%, cut fuel consumption in half and no longer needed to hire extra labor at harvest.

"We appreciate how our soils are changing," he says. "The soil is no longer an anchor to hold our crops in place while we pour on fertilizers and pesticides to get the crop to grow.

"We've now understanding what's happening in the soil and trying to work with nature — not against it." ❋

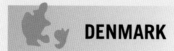

DENMARK

Lars Munkholm says the major no-till challenge for Danish farmers is the country's humid environment. Another drawback is the fact that most Danish fields have sandy loam soils when a mixture of sand, silt and clay particles would probably be better choices for no-tilling.

"Denmark is located in a cool and humid climate where soil compaction is a major problem," says the researcher at Aarhus University in Aarhus, Denmark. "Intensive soil loosening is typically needed to aerate the soil and stimulate drying of the surface soil."

In an 11-year study on two Danish farms, Munkholm and his colleagues found no-till created little soil disturbance. Leaving residue on the surface and seeding cover crops increased production and also led to the adoption of more diversified crop rotations that included wheat, barley, oats, rapeseed and peas.

The cover crops kept roots growing in the ground all year, which helped break up soil clumps and left room in the soil for much needed air and water. No-till also created a beneficial environment for soil microbes, fungi, ants and earthworms.

"Very few Danish soils have greater than 15% clay in the topsoil," he says. "The clay content varied a bit at both farm sites, which significantly affected a range of soil properties. Even though they were productive, these soils were probably too sandy as compared to the ideal situation for no-till."

Despite concerns, interest in no-till continues to slowly increase among Danish growers. ❋

ARGENTINA

The first farmers in Argentina started no-tilling in the late 1970s after exchanging ideas with Carlos Crovetto, a renowned no-till expert from Chile, and agronomist Shirley Phillips and soil scientist Grant Thomas from the University of Kentucky.

At the beginning, growth was slow due to a lack of experience, machines and limitations on the availability of herbicides.

A milestone in Argentina took place with the founding in 1989 of AAPRESID, the Argentinean association of no-till farmers based in Rosario. By 2013, 49 million acres — about 79% of all cropland in Argentina — was under no-tillage.

"A main factor in the rapid growth of no-till in Argentina was that machinery manufacturers quickly responded to the increasing demand for no-till seeders,"

Derpsch says. "At least 15 no-till seeder manufacturers in Argentina export their equipment."

Roberto Pieretti, a no-tiller from Argentina, says the "right" no-till system must be sustainable, competitive and profitable. With Argentina's highly erodible soils losing as much as 10 tons of soil for each ton of grain produced, farmers decided it was time to find solutions to their serious erosion concerns. No-till was the answer.

He says providing valuable residue cover is a key factor in saving the soil with no-till as it improves water management and protects the soil from the direct erosion

impact from billions of raindrops. More effective water use increased no-till profitability and boosted long-term sustainability.

By adopting a total systems approach, Pieretti sees no-till decreasing atmospheric carbon dioxide levels, enhancing water management, decreasing soil erosion, boosting soil organic matter content and expanding the biodiversity of crop production.

As a result, farmers in Argentina have increased their net income by lowering production costs. This has enabled them to protect the environment and be more competitive among farmers on a global scale. 🏵

MOUSE CONTROL NOT EASY IN NO-TILL CROPS

December 1986

While many no-tillers enjoy seeing and hunting wildlife, some animals occasionally cause problems.

This is often the case with mice and voles in no-till crops, says William Minser, University of Tennessee educator. These rodents live in runways under grass sod and in tunnels located a few inches under the soil.

They're usually not a problem in conventionally-tilled fields as tillage destroys their cover. But with the cover remaining intact in no-till fields, these rodents often eat planted seeds before they germinate.

Minser says mice and voles are often a problem when crops are no-tilled into old pastures or where a field hasn't been no-tilled for several years. When the rodent's habitat is undisturbed for a number of years, the population has increases.

If a grass field is being considered for no-tilling, Minser suggests looking for mice. Do this by parting the grass by hand in several locations around the field.

The presence of mice is indicated by 1-1 ½ inch wide under the grass. Close mowing and/or grazing for a couple of months before no-tilling should reduce mice numbers.

With no-till fields where mice are already a problem, plowing will eliminate mouse and vole cover and the animals themselves. Plow again if the problem should reoccur in the next few years.

Zinc phosphide-treated baits applied with seed when no-tilling should eliminate the problem. Unfortunately, tests have not been conducted proving whether or not this bait will contaminate the harvested crop. 🏵

AFRICANS TRY NO-TILL ON A VERY SMALL SCALE

While U.S. growers continue to think in terms of going to bigger and wider no-till equipment, farmers in some under-developed areas of the world have an entirely different idea of what mechanization is all about.

As you can see from these photos of equipment choices that were evaluated at the International Institute of Tropical Agriculture in Ibadan, Nigeria, African farmers in the 1970s were thinking about mechanization in a completely different sense.

Compared to using conventional tillage, there were considerable benefits to using these no-till units.

For instance, using the jab planter system required only 35 hours per acre of labor. By comparison, conventional tillage required 236 hours per acre of labor. In addition, yields were twice as high with the jab planter system. ✺

TRIPLE DISC PLANTER. This no-till planter features fertilizer banding unit with a bicycle chain drive. A disc coulter slices a slit in the soil and a "V disc" further opens the seed trench while a rear wheel compacts the seed. The rubber tires are for road travel.

NO-TILL STRIP-TILL PLANTER. Featuring a small gas-powered engine running on a single axle, fluted coulters mounted on a single axle slice through the mulch and till a 2-inch wide strip. A second worker follows with a hand-propelled planter to seed the crop. The rig was particularly suited for seeding in heavy residue.

NO-TILL JAB PLANTER. With this unit, an African farmer walks backwards, seeding one hill while compacting the previously seeded hill. With a 10-inch space between the seeded hills, a worker can seed 20 to 30 hills per minute.

Ray Jewardene

WIVES KEY TO NO-TILL GROWTH IN ASIA

Years ago at a Monsanto-sponsored conference on finding ways to expand the no-till acreage around the world, it was pointed out that women were the key to expanding the no-till acreage in Asia.

With thousands of Asian families farming small, terraced rice paddies, the wives were not only caring for the children, preparing meals and doing the daily housekeeping chores, but were also providing most of the rice paddy labor.

Meanwhile, the cultural norm included husbands heading into town each morning to relax, have coffee or a few alcoholic drinks and converse with other men in the community. The fieldwork was left to the women.

So, when it came to the labor saving benefits of no-till with Roundup, the village women were the ones most interested and receptive to adopting the newer crop production concept. ✺

'SYSTEMS' APPROACH BOOSTED FOREIGN NO-TILL ACCEPTANCE

For many years, U.S. farmers led the way in adopting no-till. But that's changed drastically as growers in South America and Central America have cashed in on the "total systems" approach to no-till.

Since these farmers pioneered the "total systems" approach to no-till, there's a great deal we can learn from their experiences. Again and again, farmers and educators in these areas stress the importance of a total system approach for successful no-till.

Intensive Tillage No 'Cure-All'

Jim Kinsella, a veteran no-tiller and strip-tiller from Lexington, Ill., maintains the U.S. is lagging in no-till development because our research is not systems oriented. He says tillage is often thought of as a "cure-all' to solve an array of intensive tillage woes.

Kinsella cites drainage as an example of being among the biggest problems holding back no-till success since moisture management is so critical to gain the full benefits of no-tilling.

Another concern is the fact that the essential interactions between nitrogen and tillage are not fully understood. As a result, there's a general tendency among farmers regardless of all tillage practices to over-apply nitrogen to corn fields.

John Hebblethwaite, former executive director of the Conservation Technology Information Center in West Lafayette, Ind., says the widespread no-till growth in Argentina and Brazil was due to significant improvements in soil quality. No-till was widely accepted because of increased profits due to more intensive cropping and the desire for sustaining a long-term farming system.

Hebblethwaite has also seen a significant growth of no-till in the Far East. By adopting no-till with rice and other crops, Asian farmers moved from growing only one crop to three crops per year on the same ground. This is due to cultural crop differences with no-till, significant labor savings and not having to wait on the availability of equipment to till the ground for rice production.

Bruno Alesii, conservation tillage manager with Monsanto, maintains acceptance and success of any no-till system depends on the farmer's attitude, plant stand, plant population, weed control, fertilizer source, nutrient placement, the community's no-till expertise and support, and on-farm field demonstrations.

Saskatchewan no-till advocate Russ Evans agrees that a systems approach is essential. He says the major barriers to no-till growth in western Canada are due to growers not fully understanding the increased management skills needed and the overall risk of change.

Attitude Really Matters

Keith Thompson, a no-tiller from Osage, Kan., believes the main limitation holding back no-till is due to "mindset."

He's convinced of the need for a "total systems" commitment. This creates serious concerns, however, with conflicting agendas among farmers, researchers, educators, government officials and environmentalists. It's particularly true with the contradicting views regarding the use of intensive tillage within universities and government agencies.

With the family's 2,600-acre cropping operation, Thompson says it took 3-5 years to bring about significant changes in soil properties after switching to no-till. To make an overall no-till system approach work, Thompson says farmers must diversify their crops and switch to intensively-managed rotations to maintain soil health through effective use of residues for conserving moisture.

The Thompson family has made effective use of cover crops in some years to not only fully utilize excess water but also to conserve limited amounts of water in other years. Improved soil health with residue management has led to successfully double-cropping and relay intercropping where soybeans are no-tilled into growing wheat several weeks prior to small grain harvest. ✻

Bob Linnell

Canadian No-Tillers Added Crops, Said
So Long to Summerfallow

"Are you trying to tell me everything
Dad taught me about farming was wrong?"

North of th U.S. border, you'll find dramatic no-tilling results in several western Canada provinces. Thanks to no-till, thousands of Canadian growers have eliminated summerfallow and diversified by adding oilseed and pulse crops to the traditional wheat-on-wheat rotation.

While no-till has caught on in the eastern provinces of Canada where corn and soybeans are the major crops, the biggest change came to the drier western areas where soil moisture is in short supply.

No-till was used on 46% of Canada's 86 million acres of cropland in 2006, a dramatic increase from only 16% in 1996. No-till acres in the three western Canadian provinces grew from 600,000 acres in 1991 to over 9 million acres by 2006.

The 2006 Canadian Census of Agriculture reported 60% of cultivated land in Saskatchewan was no-tilled, followed by 48% in Alberta and 21% in Manitoba.

With no-till offering more efficient use of water, more diversified rotations became common and led to a dramatic shift away from continuous wheat. Over the past four decades in Saskatchewan, the wheat acreage dropped by 47% while field pea acres increased from only 3,000 to nearly 3 million acres and the canola acreage jumped by an astounding 14-fold.

No Single Reason

As outlined in *Landscapes Transformed: The History of Conservation Tillage and Direct Seeding* that

NEW WEED CONTROL OPTIONS. The Canadian introduction of Roundup herbicide in 1976 and bigger spraying equipment led to a dramatic increase in the no-tilled acres in the western Prairies.

Flexi-Coil

describes no-till adoption in the three Prairie provinces, the move to no-till did much more than just change tillage and seeding systems.

The book points out that no single development was responsible for the dramatic growth of Canadian no-till acres in the late 1980s and early 1990s. Profitability alone doesn't explain the philosophical shift among farmers to accept no-till as a legitimate tillage system.

Murray Fulton at the University of Saskatchewan maintained no-till not only had a major impact on the crops being grown, but also led to development of a new ag machinery industry. It also changed farm labor needs and enabled western Canada

to continue to serve as an important international grain exporter.

Long-time Agriculture & Agri-Food Canada researchers Guy Lafond and George Clayton say several surprise factors led to no-till's rapid growth. No-till proved to be more successful than anticipated in cooler regions on a wider variety of soils

Other surprises in the late 1980s and early 1990s included the rapid development of air seeders designed specifically for western Canada, a 50% decline in Roundup prices and the introduction of new herbicides.

There were also fewer problems with diseases than anticipated. Still another surprise was how quickly Canada's acceptance of no-till was recognized worldwide as an important source of carbon sequestration.

No-Till Trend Started Early

Even though the widespread use of no-till in western Canada did not take place until the late 1980s and early 1990s, the idea behind less tillage goes back to the prairie prov-

inces' dust bowl days of the 1930s. Even in those days, farmers recognized extensive tillage was wasting valuable topsoil and was leading to poor usage of limited amounts of moisture.

In the 1970s and early 1980s, much of the western Canadian ground looked like the prairies of the 1930s, as wind and water erosion resulted in losses of millions of dollars. Research into low-disturbance direct seeding in the 1960s at western Canada research stations and universities laid the foundation for later acceptance of no-till.

Thanks to high grain prices in the mid-1970s, growers found the stamina and dollars to shift to no-till. No longer focused solely on farm survival, they started to think about long-term sustainability. Instead of trying to conserve limited moisture with summerfallow, growers placed the most emphasis on healthy soils and organic matter losses due to excessive tillage and its impact on nutrient and water availability.

On the chemical front, the Canadian introduction of Roundup in 1976 led to a dramatic increase in no-tilled acres.

Interestingly enough, no-till drills that worked well in the states didn't perform as well under western Canada's drier conditions. This led to western Canada shortline equipment companies developing no-till seeding equipment ideally suited for prairie conditions. The introduction of chaff spreaders and straw choppers to spread residue across the full width of the combine header helped overcome residue plugging issues that had led to serious seeding, disease and nutrient concerns.

Once farmers could evenly spread residue, Canadian manufacturers designed no-till drills and air seeders

"No-till drills that worked well in the states **didn't perform as well under western Canada's drier conditions..."**

with the needed residue clearance. This led to shallower and more accurate seed placement and the ability to fully utilize available soil moisture.

By leaving stubble as tall as 24 inches, no-tillers captured additional moisture from winter snows. This allowed them to move to annual cropping with cereal, oilseed and pulse crops while eliminating the need to conserve limited moisture with summerfallow.

Package Approach Essential

In the early days, the emphasis was on no-till equipment. Later on, growers figured out that successful no-till meant paying close attention to crop diversity, fertility management, weed control, equipment, economic viability and greenhouse gas mitigation.

Expanded rotations were key. For instance, Reduced Tillage Linkage agronomist Don Wentz cited a rotation used by a southern Alberta producer with over 20 years of no-till success. Farmers starting to no-till in the 1980s had the flexibility of being able to consider no-tilling the following crops in any given year:

Cereals: Spring-seeded crops include hard red spring wheat, durum wheat, utility feed wheat and occasionally barley. Winter wheat and triticale could be sold as a feed grain.

Oilseeds: Canola, flax and mustard. For growers who don't want to no-till canola, flax and mustard are good oilseed options.

Pulse Crops: Field peas, lentils and chickpeas.

Forages: Annual choices include millet, annual rye grass or barley grown for feed or silage. Perennial forages were grown in problem areas of fields, although they weren't part of the no-till rotation.

With these four cropping categories, Wentz says this producer

Frank Lessiter

DIVERSIFIED NO-TILL CROPPING OPTIONS. Tours of government, university and industry research plots allowed western Canada growers to see the benefits of using no-till to trap limited moisture while doing away with the need for summerfallowing much of their ground to conserve moisture.

created a flexible no-till rotation that suited the management of his farming operation and matched changing market opportunities.

New wheat classes and winter cereals gained ground, as did oilseeds such as canola, mustard and flax in no-till rotations. Warm-season crops like German millet, proso millets, fall-seeded rye, canary seed and other forages also gained popularity and stretched out the seeding season.

No-tillers also developed rotations to help rectify the resistant weed situation. As an example, no-tilled cereals gained acres as Alberta growers realized wild oats could be reduced with winter wheat.

Thanks to the moisture conserving efficiencies offered with no-till, growers diversified by adding oilseeds such as canola, rapeseed, mustard, soybeans, flax, sunflowers, hemp and safflower to their rotations.

Canola led the way with 12.4 million acres in 2006, a 33% increase in just 15 years.

Pulse Crop Bonuses

Canadian production of pulse crops (edible seed harvested from the pods of dry beans, dry peas, lentils and chickpeas) has dramatically increased since the 1980s, ranking the country among the leading worldwide producers and exporters.

Popular as a protein substitute, pulses contain twice the amount of protein compared to rice or cereal crops. Their low glycolic index is helpful in managing blood sugar levels.

Pulses were seeded on nearly 5 million acres in Canada in 2011, more than 11 times the acreage 30 years earlier. The proportion of Canadian farms no-tilling pulse crops increased from only 15% in 1991 to an astonishing 73% by 2011.

"Pulses were seeded on nearly 5 million acres in Canada in 2011, more than 11 times the acreage 30 years earlier..."

Compared to leaving ground in summerfallow, adding pulse crops to rotations improved water usage, boosted nitrogen availability and increased both yields and profits.

At Saskatchewan's Swift Current research center, a 3-year cropping sequence that included pulses along with wheat or barley increased total grain production by 35%, improved protein yield by 60% and boosted nitrogen efficiency by 33% when compared with a cereal grain and summerfallow rotation.

Data from a 25-year Saskatchewan study indicated wheat in a lentil and wheat rotation produced as much grain as a continuous wheat system with 29% less nitrogen. Compared with continuous wheat, nitrogen use efficiency in the lentil and wheat rotation was 80% greater in dry years, 97% greater in normal years and 36% greater in wet years.

Early on, John Bennett saw the acceptance of pulse crops as an important benefit of no-till. Despite the high price of Roundup and less than ideal seeding equipment, the Biggar, Saskatchewan, grower stayed committed to no-till because of improved productivity and reduced erosion benefits. Moving away from growing continuous wheat, his solution was to no-till field peas into wheat residue.

Other no-tillers saw the same diversification benefits, as shown by significant acreage increases with no-wheat, canola, flax and field peas.

Summerfallow's Demise

For decades, summerfallow was the traditional way to conserve moisture for the following year's crop in western Canada. Resting the land for a year also allowed the release of nitrogen into the soil in forms that could be readily used. However, the dust storms of the 1930s had demonstrated that a lack of residue cover and reduced soil organic matter led to dust clouds carrying valuable topsoil hundreds of miles from fields where it originated.

Brian McCone, an Agriculture & Agri-Food Canada researcher at Swift Current, Saskatchewan, wrote a chapter in *Landscapes Transformed* on the early no-till research that took place in western Canada.

With the development of new herbicides after World War II, major efforts for eliminating tillage trips to reduce excessive soil erosion were aimed at summerfallow. After paraquat was registered in Canada in 1962, chemical weed control on summerfallow ground became practical and became even more popular by the mid 1970s when Roundup was labeled. While expensive, Roundup was typically tankmixed with 2,4-D or dicamba to further improve broadleaf weed control in both annual cropping and summerfallow programs.

The sharpest decline in the summerfallow acreage took place in the early '90s when no-till took hold and growers recognized the advantages of diversifying into oilseeds and pulses.

Lucien Lepage of Montmartre, Saskatchewan, was an early adopter of no-till to curb wind erosion on the family's highly erodible light sandy soils. He found continuous cropping with cereal, oilseed and pulse crops offered the same yield benefits without needing to rest the land for a year.

While operating the family farm and working for the shortline equipment manufacturer Conserva Pak, Lepage made hundreds of no-till presentations across the western Canadian provinces. Summerfallow was a frequent point of contention among growers at these sessions, as the concept that tilled ground needed a yearly rest was well established.

By the mid '90s, summerfallow was no longer a necessity for many growers. The summerfallow acres dropped by 25% from 1996 to 2006 as no-till and annual cropping became more popular.

As the Alberta Reduced Tillage Linkages project was winding down its activities in 2009 after a 15-year run, a closing report indicated report yields were generally higher with no-till. Wheat yields were up 4%, barley yields up 6%, flax yields up 8%, pea yields up 5% and lentil yields up 13%. Thanks to no-till, net returns were 30% higher for wheat and 25% higher for field peas.

The report indicated labor costs were considerably lower with no-till requiring an average of only 3.5 trips across a field. By comparison, 5.8 trips were needed for minimum tillage and 7.5 trips with conventional tillage. With no-tilled wheat, growers saved 2 gallons of fuel per acre compared with using conventional tillage to seed wheat into summerfallow ground.

The 2009 report also pointed out that no-till had dramatically reduced the risk of soil erosion to where only 14% of Alberta's farmland was susceptible to water erosion and 30% of the ground to wind erosion. ❀

July
2002

PEA DISEASE WORRIES. Crop adviser Sandra Tallieu and Harvey Brink check peas to determine when to spray fungicide for more effective disease control.

DECISION TIME. When dealing with no-till disease problems, humidity, crop canopy and plant and soil wetness are critical considerations when reaching fungicide application decisions.

FIVE FEWER TRIPS SAVES DRIVING OVER 11,500 ACRES!

When Harvey Brink explained how no-till had helped trim labor and equipment costs, he maintained it saved driving across 11,500 acres per year.

Using an air seeder, the Bentley, Alberta, grower cropped 1,300 acres of his own ground and custom seeded another 1,000 acres in a typical year. Relying on no-till to make 5 fewer trips across every acre, this amounts to avoiding travel across the equivalent of 11,500 fewer acres each year.

For Brink, no-till meant reducing input costs, saving labor, making fewer costly tillage trips, increasing soil moisture and improving soil fertility. Like many no-tillers in western Canada, Brink went from raising a crop every second year with summerfallow and conventional tillage to no-tilling a profitable crop every year. ❀

NEW NO-TILL OPPORTUNITIES. Western no-tillers soon saw a growing demand for canola as a means to move away from growing continuous wheat.

CANOLA TO THE RESCUE FOR CONTINUOUS WHEAT GROWERS

During the 1950s and 1960s, bright yellow fields of rapeseed and mustard were a colorful sight alongside rural roads in western Canada. For hundreds of years, oil crushed from rapeseed pods was used as a fuel and oil for lubricating steam engines and marine parts. After World War II, Canadian growers looked at rapeseed as a food product, although its acidic flavor and greenish color put a damper on its popularity among consumers.

New Cropping Opportunities

Around the same time that no-till was catching on in the early 1970s, research work on rapeseed varieties being carried out by Keith Downey and Baldur Stefansson kept the colorful crop from fading away. These University of Manitoba researchers introduced a new variety of rapeseed that eliminated the anti-nutritional components (erucic acid and glulcosinolates) from the traditional rapeseed to assure its safety for human and animal use.

Named "canola," the crop is now among the most important vegetable oils in the world. The "can" at the beginning of the name stands for Canada while the "ola" at the end defines the oil's low acid properties.

Other scientists soon found new ways of processing canola that offered a light cooking oil that was healthy, flavorful and ideal for baking and frying many foods. In fact, the crop's benefits were strong enough that the U.S. Food and Drug Administration in 2006 recognized canola oil's high amount of unsaturated fat as a qualified health claim.

By the 1970s, numerous oilseed and pulse crops were being seeded in western Canada, of which canola was the most important. The continued adoption of no-till and the agronomic and economic benefits of canola dramatically changed crop rotation needs and led to the demise of summerfallow. Suddenly, no-till crop rotations on prairie farms included more than just growing wheat after wheat.

Today, U.S. food and fuel processors purchase over $700 million of Canadian-grown canola oil, seed and meal each year. In Canada, canola is grown on 16 million acres while the U.S. acreage is only 1.7 million acres.

No-Tilled with Air Seeders

More than 43,000 Canadian farmers grow canola, with most located in Alberta, Saskatchewan and Manitoba where the majority of the crop is no-tilled with air seeders.

Canola and no-till go hand-in-hand on the western Canadian prairies. As canola production continues to increase, so does the percentage of the crop grown with no-till. ✻

MAKING THE CASE FOR WINTER CANOLA

Guy Swanson believes winter canola could replace a considerable amount of the winter wheat acreage in the western U.S. The head of Exactrix Systems in Spokane, Wash., says replacing 2 million acres of stateside winter wheat with no-tilled winter canola could reduce soil erosion by at least 10 times when compared with winter wheat. With the loss of 1,500 pounds of topsoil for every bushel of exported U.S. wheat, Swanson maintains winter wheat is responsible for 90% of the soil erosion that takes place in Washington, Oregon and Idaho.

There's considerable demand for canola, as U.S. food and fuel processors buy over $700 million of Canadian-grown canola oil, seed and meal each year.

"When properly implemented, no-tilling Roundup Ready winter canola drops soil erosion levels well below the expectations of any other cropping system," says Swanson. "Winter canola also offers better returns than winter wheat and growers can earn bonuses for higher protein and oil content."

No-tilled or strip-tilled in 30-inch rows, winter canola not only offers a good profit margin, but can protect the soil over the winter much like a cover crop.

A properly fertilized winter canola crop will gross

WINTER CANOLA PROFITABLE. A properly fertilized winter canola crop will gross around $290 with a cash outlay of $90-$120 per acre.

around $290 with a cash outlay of $90-$120 per acre. Combining normally takes place 2 weeks ahead of wheat, which spreads out the harvesting season.

"Winter canola helps build soil quality and its early establishment protects soils that tend to erode under winter wheat production," says Swanson. "Adding winter canola to a no-till rotation is a paying proposition. No-tilled wheat and corn yields jump as much as 10% when winter canola helps overcome serious nematode concerns." ❈

NO-TILL STUBBLE PROTECTION. No-tilling or strip-tilling winter canola in some of the driest areas of the Great Plains offers needed winter crop and soil protection from high winter winds, cold, temperatures, snow and ice.

EFFECTIVE NO-TILL DEPTH CONTROL. Jim McCutcheon experimented with using an easily adjustable shank opener to replace the double disc seed and fertilizer openers. A cutting coulter could also be mounted ahead of the shank opener.

"When you borrow nutrients from the soil, you're obligated **to put the same amount back in the ground...**"

THE GRANDFATHER OF NO-TILL IN CANADA

Jim McCutcheon began no-tilling near Homewood, Manitoba, in 1974. He was among the first growers in western Canada to no-till his entire farm. I interviewed McCutcheon at several no-till meetings and my son, Mike, and I visited his 410-acre operation in 1981.

Even after 10 years of successful no-till, I remember him explaining why there still wasn't a drill on the market that ideally fit his no-tilling needs. As I wrote in a 1982 issue of *No-Till Farmer,* he indicated every

concern a farmer has with no-tillage is usually due to these issues:

- ◆ No-till drills and planters.
- ◆ Weed control.
- ◆ Straw and chaff management.
- ◆ Fertility.

McCutcheon got the idea in 1973 to try seeding directly into residue as a means of managing wild oats from University of Manitoba researcher Elmer Stobbe. He no-tilled three fields in 1974, no-tilled half his ground in 1975 and was 100% no-till by 1976.

When Mike and I visited his operation and presented him with our "No-Till Farmer of the Year" award in 1981, he was no-tilling winter wheat, spring wheat, rapeseed and flax.

Over a 2-year period, he only put 450 hours of work on his 80-horsepower tractor, including time needed to pull a PTO-powered combine. Besides conserving soil, he credited no-till with slicing fuel costs in half and dramatically reducing his machinery investment.

McCutcheon made no-till work with simple, low-cost modifications to existing equipment. He never had fancy equipment or a big acreage, but he certainly made no-till work.

That first year of no-tilling, he rented an ordinary double-disc press drill, rented an International Harvester model 100 double disc press drill the second year and bought a Kirschmann double disc press drill in 1975.

By 1976, he'd modified a 16-foot wide International Harvester model 620 double disc press drill by adding a 11-inch diameter cutting coulter ahead of each row, pulled a small drag chain behind each of the 32 rows

and added a fertilizer opener to run between each pair of rows. He later added 1/2-inch wide by 17-inch diameter coulters between the press wheels to work fertilizer into the soil.

McCutcheon continued to try new ideas such as single disc openers and ways of closing the seed furrow. He later abandoned banding nitrogen with the no-till drill, settling on a one-pass sidedressing application.

Even with all these drill modifications, he always said it's easy to get carried away with the frills. He told me the main thing is to place seed into moisture, and for him, that meant no-tilling into as little as 1 inch of moisture.

A founder and past president of the Manitoba-North Dakota Zero Tillage Farmers Association, McCutcheon passed away in 2015.

To McCutcheon, managing the soil was similar to borrowing money from a bank — you borrow nutrients in the soil to grow food and then you are obligated to put the same amount back in the ground. ✻

Frank Lessiter

SIMPLE BUT EFFECTIVE NO-TILL DRILL MODIFICATIONS. Several styles of single disc openers and double disc fertilizer openers were sometimes used on Jim McCutcheon's International Harvester 620 press drill when it came to no-tilling winter wheat, spring wheat, rapeseed and flax.

FAR NORTH FALL SEEDING PAYS WITH NO-TILL. Craig Shaw, left, explains to Reduced Tillage Linkages agronomist Rick Tallieu how fall seeding of several crops has expanded his planting and harvesting windows while leading to higher profits.

PAIRED ROW SEEDER. With a 44-foot wide Harmon air seeder equipped with triple fertilizer and seed boots, Craig Shaw applies anhydrous ammonia down the middle tube and a combination of seed and fertilizer in the outside tubes.

ROTATION CHOICES CRITICAL IN NORTHERN AREAS

October 2002

Despite cold winters and a short growing season, direct seeding in the fall was paying dividends for Craig Shaw when we visited his northern Alberta operation.

In fact, Shaw was farming so far north that many people believed fall-seeded crops couldn't survive with the area's extremely cold winter temperatures. Thanks to direct seeding, the Lacombe, Alberta, grower is making fall seeding work as he's boosted yields, trimmed costs, conserved moisture, put more dollars in the bank and spread out the seeding, spraying and harvesting workloads.

Think No-Till, Think Snow

While capturing extra moisture is valuable with direct seeding, Shaw says it's not as critical as relying on snow to protect winter wheat from freeze damage.

By stretching the time between harvest and fall seeding, direct seeding lets Shaw be timely with field work and provides more efficient use of labor and equipment in his 2,700-acre operation. With winter wheat, he's found seeding on the right date at the

BARRIERS TO NO-TILL

As described by western Canadian researchers, the challenges to no-till have not varied much over the years.

- ◆ Cost of the best seeding equipment.
- ◆ Cost of inputs, particularly herbicides.
- ◆ Soil and climatic factors.

- ◆ Equipment performance.
- ◆ Crop residue management.
- ◆ Fertilizer management.

— *Landscapes Transformed: The History of Conservation Tillage in Canada*

right depth with the right varieties are critical.

Shaw first tried fall seeding no-till wheat and canola in 1996. Among the surprises with winter wheat were fewer weeds, even with reduced herbicide costs.

"It's important for U.S. no-tillers to understand the difference in the way these crops are grown," says Shaw. "Roundup Ready canola direct seeded into barley stubble works well in rotations with fall-seeded wheat. We swath the canola around Aug. 1 and harvest around Aug. 15, leaving plenty of time to no-till winter wheat.

"Fall seeding leaves fewer acres to seed in the spring. That means I have a better chance of seeding at the most optimum time."

Flexible Rotations

His ideal rotation includes a broadleaf crop such as canola or peas followed by wheat and feed barley. Peas and canola have proven extremely useful in reducing residue levels.

Triticale is seeded in the fall after winter wheat while wheat, barley and peas are direct seeded in the spring. The farm has also tried direct seeding hulless oats and faba beans.

With a downturn in grain prices, Shaw is looking at ways to add forages to the rotation. As a value-added crop, he's growing 100 acres of timothy hay for export to Japan.

Even with a variety of no-tilled crops, Shaw isn't done experimenting. He believes rye has considerable potential for fall seeding and is convinced a blend of triticale and wheat could become a valuable livestock feed. Still another option is harvesting triticale as silage

in mid-summer and grazing the regrowth in the fall.

"As margins continue to shrink, we need to continue to make management decisions to improve yields and quality while reducing both fixed and variable costs," says Shaw. "Our experience has only been positive, and we'll increase our fall-seeded acreage."

Proper Packing Essential

Shaw direct seeds in 9 1/2-inch rows with a 44-foot-wide 4480 Harmon air seeder. He applies anhydrous ammonia down the middle tube and places seed and dry fertilizer with the two outside boots.

Shaw finds packer wheel selection isn't as important with fall seeding as in the spring. "While packing is critical with canola, snow can do much of the packing in a normal year," he says. "We seed as shallow as possible. We need the stubble to catch and hold as much snow as possible to provide needed moisture during the next growing season."

Shaw relies on semi-dwarf barley varieties to reduce lodging and to overcome excessive residue concerns. Shorter wheat varieties allow him to apply more fertilizer to push up yields.

While early seeding is critical, Shaw says the logistics of seeding 3,000 acres in a week's time just won't work. "Every planting day that we lose means less likelihood of getting the net returns we need to survive," he says.

As is the case with most no-tillers, it's the total package that makes the difference for Shaw. Fall seeding, experimenting with new crops, adding acres, saving labor and spending fewer dollars on equipment with direct seeding are letting Shaw farm more efficiently than ever before. ❀

NO-TILL TO THE RESCUE WITH FESCUE

When construction was taking place on a new golf course in the Wisconsin town of Erin, there was a concern about how to seed the fairways. While the thin layer of topsoil was ideal, the ground beneath it consisted of gravel and rock left behind by the glaciers.

Designing a course that truly fit the land, the golf course architects feared discing would bring rocks to the surface where they wanted to take full advantage of the glacial dunes and ridges avoid moving dirt.

The solution was to spray the fairway areas with Roundup to control the existing vegetation. Fescue grass was then slit-seeded directly into the dead plant matter.

"I've been involved with 1,400 golf course projects and built probably 300 courses and that's the first time I heard of doing that," golf course construction expert Bill Kubly told the *Milwaukee Journal Sentinel*. "That's why all the little bumps and rumps are all there in the fairways. It truly is natural."

All thanks to the no-till seeding technique that was used on this course that served as the site of the prestigious U.S. Open in June of 2017. ❀

Laura Rance

September 1999

WIDER ROWS PAY. Bob McNabb shifted from 7- to 12-inch rows when no-tilling small grains, pulse crops and oilseeds. The Minnedosa, Manitoba, grower uses the narrowest openers available on his 44-foot Edwards air drill.

September 2002

NO-TILL-AGE®

How Direct Seeding Openers Compare

(4-Year average yields in western Canada)

Opener	Chickpeas	Peas	Barley	Durum
Angle disc	22 bu.	36 bu.	31 bu.	21 bu.
Knife	20 bu.	31 bu.	28 bu.	20 bu.
Spoon	19 bu.	31 bu.	24 bu.	18 bu.
Sweep	17 bu.	32 bu.	25 bu.	18 bu.

— *Saskatchewan Soil Conservation Association, Aneroid, Saskatchewan*

Laura Rance

LESS SOIL DISTURBANCE. Bob McNabb credits the wider rows with reducing soil and residue disturbance while increasing microbial activity.

FOLLOWING A 21-YEAR NO-TILL ROTATION

Many Corn Belt no-tillers follow a 2-year rotation of corn and soybeans. So they may be surprised to learn Canadian growers Garry and Glen Meier follow a 21-year rotation in their direct-seeding program.

The two brothers from Ridgedale, Saskatchewan, direct-seed 4,100 acres besides custom farming 4,000 acres. Yields have reached as high as 120% of the area's average yields with a summerfallow program since they began direct seeding.

Forget About Crop Prices

"Our farming philosophy is to farm the soil rather than farming the markets," says Garry. "The sequence on our farm is basically unaffected by market prospects for a particular crop: 25% of our annual-seeded acreage will be canola, 25% pulse crops, 25% flax and 25% cereals.

Timing is very important. "Seeding into untilled ground allows earlier seeding in wet years and quicker emergence in dry years," he says. "Research has shown earlier-seeded crops generally yield better than later-seeded crops."

He says the mid-row banding capability of their seeder has contributed to healthier stands and offers a "no risk" method of applying all fertilizer at seeding time.

A pre-harvest application of Roundup has improved harvest timeliness by allowing the brothers to control days to crop maturity and to direct-cut crops which were normally swathed in the past in this northeastern corner of Saskatchewan. As a result, residue management problems are reduced and grain grades are generally higher.

"A well-planned rotation solves many weed, disease and insect problems," Garry says. "Direct seeding has given our crops enough of a jump on the weed competition, and when combined with a timely Roundup burn-off treatment, it has reduced the cost of or eliminated the need for an in-crop herbicide."

Trimming Costs

"Machinery is where we look for ways to reduce costs," says Garry. "We don't have a fond attachment for older machines and appreciate the comfort and efficiencies of newer equipment." ✺

NO-TILL-AGE®
May 1997

21-Year No-Till Rotation Works!

Year	Crop
1	Argentine canola
2	Hard red spring wheat
3	Peas or lentils
4	Flax
5	Hard red spring wheat
6	Polish canola under-seeded to alfalfa for seed or hay
7	Alfalfa seed or hay
8	Alfalfa seed or hay
9	Alfalfa seed or hay
10	Winter or spring cereal
11	Argentine canola
12	Hard red spring wheat
13	Peas or lentils
14	Flax
15	6-row barley
16	Argentine canola
17	Hard red spring wheat
18	Peas or lentils
19	Flax
20	Hard red spring wheat
21	Polish canola under-seeded to alfalfa for seed or hay

— Garry and Glen Meier, Ridgedale, Saskatchewan

TRIMMING COSTS. Bob Linnell says direct seeders in Saskatchewan are realizing up to 50% less fuel costs along with harvesting 10% higher yields. In addition, each direct seeded acre is removing the equivalent carbon dioxide that is released from burning 40 gallons of gasoline.

Bob Linnell

NO-TILL-AGE®

March 2010

Yearly Machinery Costs for 2,000 Acres in Western Canada

System	Machinery Cost per Year*
No-till	$48,130
Minimum tillage	$55,420
Conventional tillage	$63,720

Costs for tillage, seeding and spraying.

— *Alberta Agriculture Farm Survey*

No-Till? *Direct Seeding?* Zero-Till

WHAT'S IT CALLED IN CANADA?

It all depends on where growers are located in the vast neighboring country to our north where agriculture stretches from the Atlantic to the Pacific oceans.

In eastern Canada where growers enjoy higher quality soils and more moisture, farmers call it no-till. Following corn and soybean cropping programs much like their Northeast and Corn Belt neighbors to the south, growers in the Ontario, Quebec and the Maritime providences refer to the system as no-till.

However, it's different in the western prairies where direct seeding or zero-till is the more common terminology.

But regardless of the preferred term or location, it's the same reduced tillage practice throughout North America.

> "You have to be convinced **that tillage is as bad for the health of your soils as smoking is bad for your personal health.**"
>
> — *Art Cowan, Hartney, Manitoba*

NO-TILL-AGE®

September 1998

Why We No-Till 1,500 Acres
(Wheat, barley, oats, flax, canola and peas)

1	Improved economics
2	Less water erosion
3	Better control of available moisture
4	Saves cropping time
5	Limited manpower
6	Reduced machinery wear and tear
7	Improved soil structure
8	Improved perennial weed control
9	Easier field marking and increased dust control
10	Let's us find better things to do
11	Keep the neighbors wondering
12	Save soil for future generations

— *Gary Trofimenkoff, Veregin, Saskatchewan*

EXPANDED COVER CROP ACREAGE. Kevin Elmy uses a Bourgault 8800 air seeder to seed a wide variety of cover crop tank-mixes in his Salt-coats, Saskatchewan, operation. Along with packer attachments, he uses either a 1.75-inch hoe opener or a 2-inch spike, depending on the kind and amount of residue he's seeding into.

Laura Rance

How No-Till Changed Lives Around the World

"My grandfather told me that if I ever thought I needed to buy a moldboard plow, he would **send me to Brazil so I could fully understand why there's no reason to plow**..." — *Sarah Singla*

I asked our subscribers in foreign countries to share their thoughts on how no-till has changed the way they farm and live. Here's a sample of a few of their farming observations and updates.

MIX IT UP. Diversifying with crops, such as fava beans, helps break the disease cycles.

CONTROLLED TRAFFIC. These wheel tracks are followed with every no-till field operation.

NO-TILL MAKES BEST USE OF LIMITED MOISTURE

Allen Postlethwaite,
St. Arnud, Victoria, Australia

After our two sons came back to the farm in 1983, we sold the sheep, went 100% no-till and eliminated summerfallow. With our 6,000 acres, the farm looks and functions much differently than in 1982.

We've no-tilled wheat, barley, fava beans, chickpeas, lentils, canola, flax and sorghum. To no-till and grow a crop each year under our dry conditions, crop diversity is a must. Combining broadleaf, legume and grass crops in our

rotations helps us avoid serious issues with herbicide resistance.

Controlled traffic has helped keep our soils free of compaction, which allows more of our limited amounts of water to infiltrate into the ground. Protecting our soils from compaction has resulted in very mellow soils and better yields over the years.

Why No-Till Matters: Combining controlled traffic with no-till gave us a much better way to better manage with limited moisture. ❊

WHERE WIVES AND CHILDREN DO THE NO-TILLING

George Kaweesi,
Rural Enterprise Development Services, Uganda

Rural Enterprise Development Services is a Ugandan organization that helps small-acreage farmers become more efficient and more productive. We train, advise and guide small acreage farmers to adopt better farm practices and improve productivity through alternative technology options to bring about climate change adaptation.

With no-till, farmers have seen improved emergence of crops, less crop stress, more soil water retention and improved soil structure. There's been a substantial reduction in labor requirements, a major relief for the women and children who handle most of the farm work.

Within the first season of adopting no-till, yields increased by up to 100% with more efficient utilization of both on-farm and purchased inputs. Farm family nutrition has also improved

dramatically through the integration of high protein legumes. Farm revenues have increased with the development of off-farm business opportunities and increased marketing of crops not needed for family consumption.

Why No-Till Matters: It makes life easier and more profitable for the women and children who do much of the farming work. ❊

CROPPING LIMITATIONS. In 1972, Franke Dijkstra (right) started looking at no-till.

NO-TILL MORE THAN DOUBLED YIELDS

Franke Dijkstra,
Carambei, Brazil

We switched to no-till in 1976 and 3 years later got some much needed technical assistance from the late University of Kentucky agronomist Shirley Phillips, Agricultural Research Service scientist Wayne Reeves and Kentucky soil scientist John Grove, who made numerous trips to Brazil.

Before switching to no-till, we struggled with our sandy soils, sloping terrains, excessive tillage and frequent downpours that led to serious wind and water erosion. This became even more serious each year from the 1950s to the 1970s.

No-tilling corn and soybeans as early as possible gives us 15-40 additional days to no-till a second summer crop of corn or soybeans. The winter crop is typically barley and we seed annual ryegrass as a cover crop. In many fields, we grow three crops in a year's time.

Between 1997 and 2010, corn and soybean yields doubled or tripled with no-till. Average yields rose to 195 bushels for corn and 66 bushels per acre for soybeans. Besides the no-till cropping operation, we milk 500 cows and have 400 sows in a farrow-to-finish operation.

Why No-Till Matters: It offers both economic and soil sustainability and allows us to grow three crops in a year's time in many fields. ❋

SOUTH AFRICA

EARLY NO-TILL WORK WITH PARAQUAT

Thomas M. Borland,
Builth Wells, Wales,
United Kingdom

I started with ICI Plant Protection in 1965 on the research and development of paraquat and diquat in the vineyards and orchards of South Africa. The aim was to replace plowing and discing with herbicides.

It soon became clear that perennial weeds posed a serious problem, one that wasn't fully solved until glyphosate (Roundup) came along years later.

I moved to Rhodesia in 1969 to work for Ciba-Geigy before becoming a weed research officer there. In

A NEW ERA. Paraquat and diquat replaced plowing and discing in South African fields.

1974, I became the senior extension specialist for weed control in Rhodesia. My main responsibilities were the dissemination of weed and tillage research results within the farming community.

After 1978, I started farming, went into agribusiness banking and lecturing. I retired in 2005 and moved back to the United Kingdom in 2008, away from the chaos going on in Zimbabwe.

Why No-Till Matters: Our early work in South Africa showed intensive tillage could be replaced with herbicides. ❋

INDIA

NO-TILL DOUBLE-CROPPING PAYS OFF ON INDIA FARMS

Vijay Vardhan Vasireddy
Kolkata, West Bengal State, India

I work for ITC Ltd., and we are actively engaged with farmers and work closely with communities on natural resources management and the promotion of sustainable agricultural practices.

In the plains of Uttar Pradesh,

MORE INCOME, LESS COST. No-till wheat yields in India have increased by as much as 40%.

Bihar and West Bengal, farmers mainly cultivate paddy rice and wheat. After rice harvest, these farmers traditionally uproot and burn the crop stubble, plow and seed wheat, often by broadcasting the seed.

This traditional cropping practice has many unfavorable issues:

1 Loss of soil moisture due to opening up the top crust of soil.

2 Higher wheat production costs due to multiple land preparation trips, sowing and extra seed expenditures.

3 Delays in sowing, leading to wheat missing the critical cold period that's needed during the flowering and grain filling stages.

4 Paddy rice stubble being burnt and adding to the unfavorable atmospheric carbon situation.

To overcome these issues, we looked at various solutions and zeroed in on no-till. We promote no-till after paddy rice crop harvest by directly seeding wheat into these plots.

Within 2 years, no-till grew to 111,200 acres. We worked with eight grassroots groups in nine districts, trained 10,000 farmers through 400 farmer field schools and put out 12,000 demonstration plots. Through Agri Business Centres, we offered the use of 170 no-

NO-TILLED WHEAT. Interest has grown in no-tilling small grains in many areas of India.

till implements along with promoting locally available no-till tools.

Third party evaluations and farmer interviews demonstrated notable benefits:

◆ Drilling wheat instead of broadcasting led to a 67% reduction in seeding rates.

◆ There were savings of $27 per acre in the cost of field preparation and seeding costs.

◆ Dropping to only two essential irrigations due to improved retention of surface moisture.

◆ There are also long-term benefits related to improvement in organic carbon, soil texture and soil structure.

Why No-Till Matters: Wheat yields were improved by 16%-40% with the use of no-till. ✺

ITALY

EARTHWORMS DO OUR TILLAGE

Francesco da Schio,
Vicenza, Italy

No-till has allowed us to preserve soil from erosion, save water and farm more efficiently. I started to no-till 9 years ago because my soil used to

10 YEARS OF NO-TILL. Francesco da Schio no-tills to avoid costly soil and water erosion.

blow away with the wind and wash away with traditional tillage.

With no-tillage, we keep the soil on our own land and handle the work with smaller horsepower tractors. Earthworms are doing our tillage work.

Why No-Till Matters: We're saving soil, saving water and boosting yields. ✺

REPUBLIC OF THE CONGO

NO-TILL NEEDED IN LOW INCOME COUNTRIES

Terry Ellard,
Seattle, Wash.

The first of my two Peace Corps tours took place in 1972-74 in what is now the Democratic Republic of the Congo. Using my bachelor's degree in soil science and a metal working background, I taught agriculture at a rural school, started an oxen-powered project and began to design tools that could be locally made. One tool was a moldboard plow constructed of wood and sheet metal salvaged from oil drums.

My second tour in 2011-13 was in Lesotho, a small country completely surrounded by South Africa. Oxen power in agriculture was well known here and the moldboard plow originally introduced in the 1800s by French missionaries and later imported from China was used almost exclusively in clean tilled fields.

I started a project to locally manufacture tools other than plows. I'd become convinced that no-till farming is the only way to save the severely degraded soils in Lesotho. Some 80% of Lesotho's people depend on farming to survive and unemployment is nearly 50%.

Yields have declined by as much as 50% in the last few decades.

Due in part to climate change, droughts are much more common. When rains come, they are often intense downpours and erosion can be a serious problem.

I'm now back in the states and attempting to learn as much as I can about no-till so I can help design no-till tools that can be produced in African countries.

Why No-Till Matters: It's the only way to save the severely degraded soils found in many of Africa's smaller countries. ✺

BIG TIME SAVER. This corn sheller works 16 times faster than shelling with bare hands.

FRANCE

NO "BIG IRON" IN THE SOIL. Sarah Singla's fields have not been plowed in over 35 years.

NEIGHBORS THOUGHT MY DAD WAS CRAZY

Sarah Singla,
Canet de Salars, France

Back in 1980, my father was among the first farmers in France to switch to no-till to reduce the costs of production and to reduce erosion. He bought a no-till drill at a farm show and sold the moldboard plow at the same time, so there was no turning back if our reduced tillage system didn't work out.

He was criticized by neighbors who thought he was crazy. It was difficult as the results were not good during the first 2 years. But he was stubborn and committed to succeed when he sold the plow. At that time, there was no Internet or network of no-till farmers to call on for help for those who wanted to try this new way of farming. It was much different than today when there are many new communication tools and no-till farmer networks in many countries.

My father died in 1990 and my grandfather took over the farm again. He decided to keep no-tilling even when many folks said he would eventually have to plow the land. Many ag researchers at the time thought no-till wouldn't work in the long term.

My grandfather told me that if he had to return to using the plow, the farm would probably have been sold. We would have again had to invest in tillage equipment and big horsepower tractors. It would have also meant more dollars spent for fuel, more tractor hours, more labor to collect stones from the fields and more erosion.

He told me, "I'm going to keep this system and we'll see if one day we will need to plow again." He never saw any reason to plow again as we were able to no-till with very low production costs.

When I took over the farm in 2010, he told me that if I ever thought I needed to buy a plow in order to farm like the neighbors, he would offer me a trip to Brazil so I could fully understand why there is no reason to plow. But it was not my idea to touch the soil with iron as my goal is to replace iron with roots and fuel by photosynthesis.

I'm blessed to run the family farm and have the chance to grow crops on soils that have been untouched by iron for more than 35 years.

Farmers who want to switch to no-till need to understand no-till is not a goal, but just a tool within a system to help you be more efficient and to improve the economical, environmental and social aspects of your operation. Once you start no-tilling, the road is not finished and there are still many things to improve.

The critical questions that farmers need to ask are these:

◆ What do you want your farm to be in 100 years time?

◆ How can you add more value to the system with the crops you sell and the people you work with?

◆ How will you personally fit into a no-till system?

In the future, we must go beyond no-till and talk about regeneration agriculture, as we have a duty to leave the soil in better shape for future generations. We don't inherit the land from our parents and then borrow it from our children.

The future of agriculture is bright and we will see major changes over the next few decades. It is not an evolution, but rather a revolution if we look at what we are doing at the moment. Beyond agronomy, we must think about sharing ideas, networking, happiness, human life, healthy food and healthy people.

No-till goes much further than just the agronomy itself as the word "soil" comes from the Latin name "humus," which includes "humility" and "humanity."

Why No-Till Matters: If we had to return to using the plow, the farm would probably have been sold. ✳

ALGERIA

AMONG THE VERY FIRST GROWERS TO NO-TILL

Benkhelifa Mohammed,
Mostaganem, Algeria

We are the first to utilize the results of a non-tillage research trial in our semi-arid climate in Algeria.

Why No-Till Matters: With limited water and little mechanization, no-till will revolutionize small-scale crop production in many poor countries. ✳

NO-TILL AROUND THE WORLD...

Percent of Acres No-Tilled by Continent

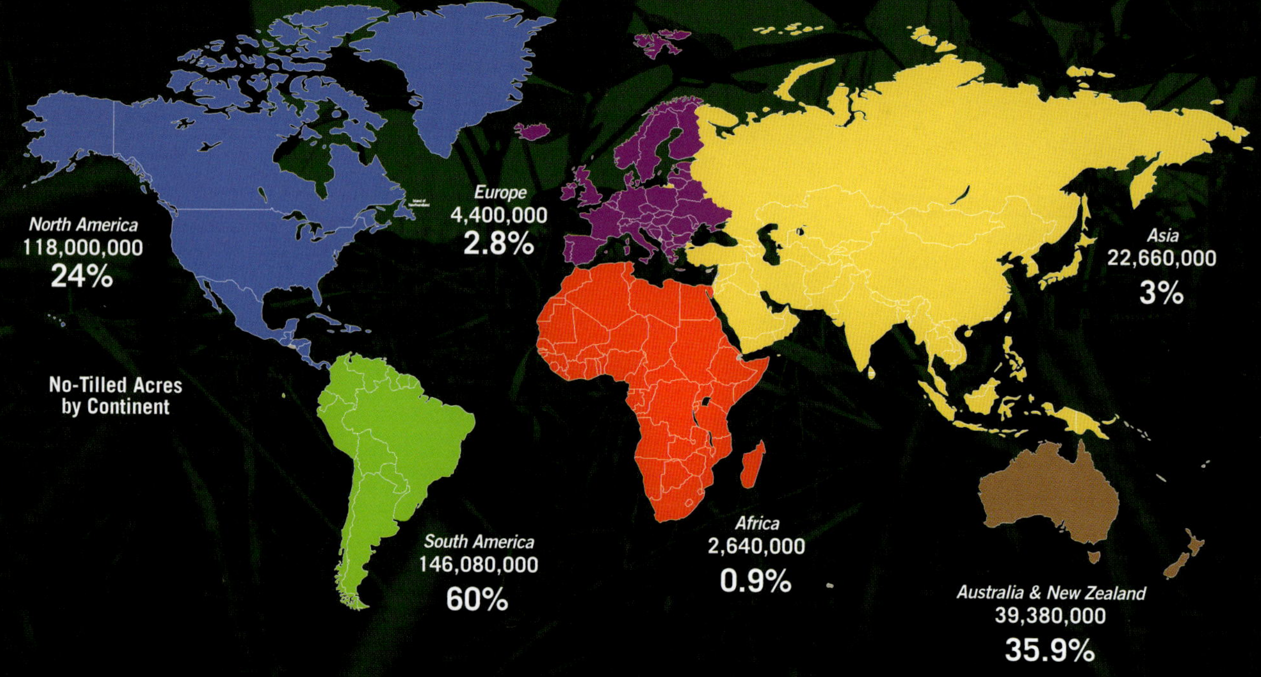

North America
118,000,000
24%

Europe
4,400,000
2.8%

Asia
22,660,000
3%

No-Tilled Acres by Continent

South America
146,080,000
60%

Africa
2,640,000
0.9%

Australia & New Zealand
39,380,000
35.9%

World's Continents	No-Tilled Acres	No-Till Amount of Continent Cropland	No-Till Amount of Worldwide Cropland
South America	146,080,000	60%	42.3%
North America	118,000,000	24%	34.4%
Australia & New Zealand	39,380,000	35.9%	11.4%
Asia	22,660,000	3%	6.6%
Europe	4,400,000	2.8%	1.3%
Africa	2,640,000	0.9%	0.8%
World's no-tilled acres	345,356,000	10.9%	—

— *2013 acreage data, Food and Agriculture Organization of the United Nations*

ZAMBIA

ONLY WAY TO FARM. No-till and fire lanes have helped overcome concerns with brush fires.

BRUSH FIRES SHOW HUGE VALUE FOR NO-TILLING

Adrian Bignell
Zambia

Among the biggest cropping concerns experienced in central Africa are the yield losses due to unexpected brush fires. Even with firebreaks around our fields, a brush fire managed to jump my fireguard and burned a corner of a field.

Soybeans were planted in that area and the results were startling. The retarded soybean growth in the burned corner was clearly visible during one of the hottest and driest growing seasons on record in our area of Africa.

At harvest, there was a 30 bushel per acre difference between the burned corner and the remainder of the field, with the unburned area yielding 48 bushels per acre.

This dramatic result demonstrated the many benefits of no-tilling and the importance of crop cover in effectively utilizing available moisture.

Why No-Till Matters: A loss in residue cover and less effective use of limited moisture clearly demon-

strated the value of no-tilling in some of the driest areas of the world. ✳

PARAGUAY

CRITICAL STEPS FOR NO-TILL SUCCESS

Rolph Derpsch,
Asuncion, Paraguay

For a grower shifting over to no-till, here are a few essential items for success under South American cropping conditions:

1. Behave like an agronomist, whether you are a landowner, landlord or manager of the soil.
2. Expand your no-till knowledge before shifting away from more intensive tillage.
3. Analyze your soils and aim for balanced nutrient and pH levels.
4. Avoid starting out with no-till on soils with poor drainage and level the soil surface in your fields.
5. Eliminate soil compaction before you start no-tilling.
6. Produce the largest amount of biomass possible.
7. Buy a straw harrow and no-till planter or drill.
8. Start no-tilling with less than 10% of your acres.
9. Follow a crop rotation and use green manure cover crops.
10. Continue to learn more about no-till and seek out advisors who understand this reduced tillage concept.

Why No-Till Matters: With nearly 350 million acres worldwide, innovative farmers have switched to no-till to do less work, make more money, control erosion more effectively, make farming more environmentally friendly and improve their quality of life. ✳

SOUTH AFRICA

CHEAP GRAZING. No-till residue and cover crops provide steers with low-cost forage.

ABOVE AVERAGE YIELDS, BELOW AVERAGE SOILS

Egon Zunckel,
Bergville, Kwa Zulu Natal, South Africa

South Africa has erratic weather patterns, at times going from howling, hot winds to icy, rainy conditions within hours. The average seasonal rainfall on our farm is 36 inches per year, but having four months of no rain in the winter is not uncommon.

With more extensive tillage, I used to deal with soil runoff, poor aeration, poor infiltration, erosion, low organic carbon, low microbial populations, little earthworm activity, high fuel bills, high maintenance, labor and machinery costs, time restraints, silted dams, dust storms and the like.

Since drastic measures were needed, I went 100% to no-till 20 years ago and have never looked back. After 10 years, the carbon content of our soils doubled and the problems listed above have disappeared. Today, I harvest above average yields on below average soils and can afford to buy decent tractors, planters and sprayers.

Cover crops, such as oats and radishes, became a permanent part of our crop rotation after we discovered their importance in combating nematodes and dealing with other problems. They are direct-seeded immediately after soybean harvest on one-third of our acres. Later, beef cattle utilize the vast quantities of cover crop and corn residue.

Why No-Till Matters: I'm confident this is the closet method of crop farming to what God intended at the time of the world's creation. ✽

IRAN

AVOID COSTLY BURNING. No-till helps Iran growers avoid traditional burning of residue.

MOISTURE SAVED BY NOT BURNING STUBBLE

Mohammad Esmaeil Asadi,
Gorgan, Golestan Province, Iran

In the province of Golestan, which is in the northern part of Iran, wheat and barley are grown on more than 740,000 acres. After harvest, farmers typically burn the stubble prior to conventionally tilling the ground before planting

a second crop such as soybeans, corn or grain sorghum.

These practices cause soil disturbance, compaction and soil deterioration. Conventional tillage has led to a sharp decline in crop yields and profitability.

In the past, nearly 300,000 acres of wheat and barley stubble in this area were burned each year. Thanks to our no-till activities, only 12,000 acres were burned in 2016.

Why No-Till Matters: Not burning crop stubble has led to healthier soils and higher yields. ✽

TURKEY

NO-TILL REPLACES THE NEED TO BURN STUBBLE

Muzaffer Avci,
Central Research Institute for Field Crops, Ankara, Turkey

Some 35 million acres in Turkey are farmed without any conservation measures, including 76% of the land where severe, very severe or medium class erosion occurs. The ground farmed with extensive tillage is much more vulnerable to reduced soil productivity due to soil and water erosion, frequent tillage, intensive crop production and residue removal. All these concerns are eliminated with no-till.

Farmers in Turkey normally burn wheat stubble for easier seeding of the following crop or on ground that is summerfallowed.

Both minimum tillage and no-tillage, which retain crop residue, play a critical role in water conservation, preventing soil loss, increasing yields and sustaining long-term crop production.

When wheat stubble is left on the soil surface in a wheat and fallow rotation, there's 36% less water runoff, 29% less soil loss and 23% higher wheat yield than when the residue is burned.

Why No-Till Matters: Both Mediterranean and worldwide experiences shows no-till and other reduced tillage systems are a pre-requisite for sustainable agriculture. Efforts to adopt these systems should be intensified in Turkey. ✽

SOUTH AFRICA

STRIP-TILL CONTRIBUTES TO SOUTH AFRICAN SUCCESSES

Anne Marie Joubert,
Northern Cape, South Africa

In 2012, George Wheeler of the GWK-branch at Douglas in the Northern Cape of South Africa, and some of the more prominent farmers in the district gathered around a kitchen table to discuss solutions to intensive tillage problems.

They realized they couldn't continue to do things the same way their fathers did 15 years ago and maintain profitability and sustainability. That meant trimming input costs and protecting the soil.

They decided to curb costs by utilizing whatever means they had at their disposal. Their starting point was John Deere's 1750 planter as almost every farmer in the area already had one. The idea was not to replace what they had, but to add to it.

Their local Deere representatives took up the crusade and found the answer in America's High Plains region in eastern Colorado and western Nebraska.

As in the Northern Cape, these farmers are also subject to water issues, windy conditions, sandy soils, the need for conservation tillage and labor-saving technology.

Their solution was to use a combination of the XDR Combo planter, Deere's tractor and planter technology and Orthman's strip-tillage system to prepare the soil and plant in one smooth operation.

At a farmers' day on the Cilliers brothers farm between Douglas and Prieska, the XDR Combo, the Orthman 1tRIPr and John Deere 1750 planter with an adjusted frame was put to the test. The unit was demonstrated on land covered with slightly moist barley residue under typical area conditions.

Immediately after harvesting barley or wheat, the brothers plant corn through the thick, wet plant material. There's too much plant material left to work it all into the soil as part of a no-till system. If left on the ground, the wind blows most of the residue away and the soil dries out. There was also a desire to avoid burning the residue

Strip-tillage offers the logical solution. The 1tRIPr removes the residue in the row area, yet leaves residue between the rows to conserve moisture.

The system adheres to Orthman's three basic principles for precision tillage.

The first principle is the placement of fertilizer at the desired depth. This is made possible by the parallel linkage wrap-around design of the implement, which helps keep it steady and level in all types of conditions.

The crop and soil determines the ideal seeding depth.

The second principle is creating the perfect root zone environment by shattering the soil. Using waffle-shaped coulters to perform vertical tillage encourages root development.

Third, the seedbed is prepared with no soil voids so the planter can move smoothly and evenly across the fields.

The Cilliers had previously used an Orthman 1tRipr, but planted in a separate trip. Now, the saving in fuel and time alone makes the combination worthwhile, with the added benefit of fewer trips and less soil compaction. The system has proven ideal with their soils and irrigated farming, but would be just as useful in dryland conditions.

Why Strip-Till Matters: The saving in fuel and time alone makes the one-trip strip-till and planter combination worthwhile. ✳

CANADA

CARBON SEQUESTRATION HUGE NO-TILL BENEFIT

Wayne Lindwall,
Agriculture and Agri-Food Canada, Lethbridge, Alberta

I dedicated most of my 40-year career to no-till as a scientist, engineer and senior executive with Agriculture and Agri-Food Canada. Papers for my two graduate degrees from Iowa State University were based on no-till research. Eventually, I published more than 55 peer-reviewed papers and book chapters dealing with conservation tillage and no-till.

Despite the significant benefits of conservation tillage and no-tillage

NO-TILL RESEARCHER HONORED. In 2008, Wayne Lindwall, left, received the highest honor in public service from the Prime Minister and Governor General of Canada. This was for his career-long contributions to no-till, the linkages to carbon-sequestration and for serving as the carbon sinks expert for agriculture and the government of Canada in multi-country negotiations.

that were well documented 40 years ago, development of appropriate seeding equipment was slow. It was due to the large mainline equipment manufacturers having a vested interest in conventional tillage equipment and high horsepower tractors and, as a result, seeing limited market potential for no-till.

Innovative producers and a few shortline manufacturers working with government and university researchers were the pioneers in no-till equipment development.

Although, there were some early and important success stories in the 1970s and 1980s with high-clearance hoe drills and a few specialized no-till drills, widespread adoption of direct seeding was made possible in the 1990s with the development of air seeders that could provide effective seed and fertilizer placement under a wide range of soil and crop residue situations.

Why No-Till Matters: By 2010, the use of air seeders dominated the Canada prairies. ✳

AUSTRALIA

NO-TILL PIONEERS. The Hine family farm in Western Australia was among the first growers in the country to try no-tilling, applying Gramoxone to pasture land in 1963. The 1927 Chevrolet in this 1964 photo shows Bill Hine driving and siblings Louise, Russell and Elaine seated in the back on their way to meeting the morning school bus.

AMONG AUSTRALIA'S VERY FIRST NO-TILLERS

Garry Hine,
Wellstead, Western Australia

Our original family operation included 420 acres of owned ground and 300 acres of rented land along with sheep and beef cattle.

In 1963, we seeded perennial ryegrass for the first time. The reason we went to no-till was due to the fact that tillage encouraged the germination of annual ryegrass, which is considered a weed when it comes to pure seed production. By spraying our fields with paraquat (Gramoxone), no-till was tried to see if we could sow perennial ryegrass into pasture land while keeping annual ryegrass from germinating.

Thanks to no-till, we eliminated the more expensive and time-consuming option of having to graze

EARLY DAY DIRECT SEEDING. The British chemical company ICI, which developed paraquat, promoted SpraySeed, also known as direct seeding, in the mid-1970s in western Australia. Tined seeders pulled with 100-hp tractors seeded 1-3 days after the ground had been sprayed with Gramoxone and diquat.

these fields for up to 3 years to "choke-out" the annual ryegrass.

Why No-Till Matters: It offered us a big savings in both time and money. ❊

THERE MUST BE A BETTER WAY TO FARM

Bill Crabtree,
Yalgoo Western Australia

In 1960 at the age of 12, I helped clear land on the family farm in Gardiner, Australia. Even then, I thought there must be a better way to farm than doing all of this hard work.

In the mid-1970s, the United Kingdom company ICI encouraged western Australian farmers to buy SpraySeed herbicide, which was a mix of diquat and paraquat. This herbicide was sprayed on newly emerged weeds in May since western Australia has a Mediterranean climate that includes wet winters and extremely hot and dry summers.

After spraying, farmers would fertilize and seed with tined seeders that did a limited amount of tillage. Pulled by 100 hp tractors, these drills were used to seed various crops 1-3 days after the weeds had been sprayed.

Known as SpraySeeding or

direct drilling, the concept was the forerunner of no-till in western Australia. There was nearly 100% soil disturbance with the tined seeder. By comparison, no-till in Australia today is defined as having less than 20% soil disturbance at seeding time.

I've devoted much of my career to expanding the no-tilled acres in western Australia and have worked with many of the country's no-till pioneers.

No-till has long been a passion of mine. My efforts to build and improve the practice in Western Australia earned me the nickname "No-Till Bill."

When I started working with the western Australian government, no-till wasn't practiced in western Aus-

NO-TILL CHAMPION. A long-term promoter of no-till in western Australia, Rick Madin was Bill Crabtree's early-day mentor on the critical principles of conservation tillage.

tralia, and even the governmental agencies supposedly backing it weren't entirely on board.

We've overcome many obstacles since that time, and adoption rates have topped 90% in western Australia — with 100% adoption in my home county.

A great benefit to no-till was that when new land in Western Australia was made available for production, it was mostly the brave, innovative farmers with fresh and open minds that saw the opportunity and took on the challenge.

After years of helping countless farmers make a lot of money, I finally decided to try and make a go of it myself. In 2007 I bought 7,000 acres of farm ground in northeast Morawa, Australia, and am now officially both a no-till farmer and consultant. Some 95% of my cropping program includes continuous wheat, but I also no-till canola, triticale and lupins.

I've been in countless no-till fields and seen the benefits of the practice, but nothing compares to experiencing it on your own farm. It's been wonderful to feel my own soil, crumble it and know that just a few years ago I couldn't crack through it with a screwdriver. With no-till, I can break my soils apart with my bare hands in the middle of summer, with no rain.

I'm good friends with many of the South American no-till champions and Australia's likely first no-tiller is a long-term friend of mine. Now in his 70s, Garry Hine first no-tilled in 1963 by spraying weeds with Gramoxone.

Why No-Till Matters: It offers a much better way to farm, makes for healthier soils and eliminates the hard work involved with intensive tillage. ✷

THAILAND

LONG DISTANCE LEARNING. Nuanlaong Srichumpon and Asavapat Asavapongsopon traveled 8,500 miles to attend the National Strip-Tillage Conference. Shown with them is *Strip-Till Farmer* Managing Editor Jack Zemlicka at right.

SEEKING NEW NO-TILL, STRIP-TILL IDEAS

Asavapat Asavapongsopon,
Bangkok, Thailand

Traveling nearly 8,500 miles from Thailand, my cousin, Nuanlaong Srichumpon and I attended the 2016 National Strip-Tillage Conference to learn more about North American conservation tillage practices.

We don't use no-till, strip-till or cover crops in our area. We came to the U.S. to meet and talk with farmers and university scholars to better understand the machines, technology and conservation concepts American farmers are using that we could use in our country.

Why No-Till Matters: The adoption of some of America's conservation tillage practices will help our farmers find new ways to reduce tillage, labor, erosion and other costs. ✷

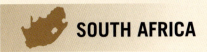

SOUTH AFRICA

NO-TILLED GROUND IS MUCH MORE VALUABLE

John Jackson,
Bergville, South Africa

Before new ground is purchased in South Africa, a soil pit is dug at least 36 inches deep and soil samples are taken at different intervals for analysis. The profile is checked for compaction layers and soil depth.

Why No-Till Matters: This lets a grower decide to pay for the farm's yield potential and not for the soil's looks if they expect to no-till the new ground. ✷

CANADA

NEEDED SPRAYER CONSISTENCY, SAFETY AND EFFICIENCY

Dale Wiens,
Beechy, Saskatchewan

Saskatchewan is where much of the Canadian air seeder and air drill technology was developed and refined. Without this equipment, no-till would not have not happened here and we would still be doing extensive tillage and having to summerfallow our black soils.

Why No-Till Matters: The air seeder and air drill technology is huge in western Canada and has dramatically increased our farm income and profit. ✷

SLOVAKIA

WHY GROWERS SHOULD NO-TILL

Alejandro Figueroa,
Prievidza, Slovakia

There are four major areas in regard to no-tilling that we are discussing with farmers in Slovakia:

BOOSTING ORGANIC MATTER. No-till has improved soil health for many Slovakia growers.

1 With a short growing season, no-till can offer grow good yields while providing time to spend with family and friends, take a second job off the farm or have more hours to enjoy your hobbies.

2 With no-till, you need fewer man hours, less fertilizer, fewer long-term pesticide concerns and dramatic savings due to less machinery, depreciation and reduced fuel costs.

3 No-till offers better water infiltration, reduces evaporation and improves water quality.

4 No-till will increase the organic matter of your soils. It will help you to store more nutrients in the soil, improve soil structure, boost the cation exchange rate, buffer and filter water storage and improve carbon sequestration.

Why No-Till Matters: The key is to utilize no-till crop rotations that include both winter and spring crops, make sure you include cover crops and feed the soil so it can feed the plants. ✺

CANADA

NO-TILL MAKES ME CONFIDENT OF AG'S FUTURE

Jocelyn Michon,
La Presentation, Quebec, Canada

Sometime in the mid-1970s, I saw a photo in a Case IH catalog of a no-till drill being used in a corn field. I was blown away and thought it was what I wanted to do. But it took almost 20 years to realize my dream.

After a few years of experimenting with reduced tillage, I abandoned the moldboard plow in 1986 and replaced it with offset discs. The discs were used for 8 years before I bought a Great Plains 1500 no-till drill in 1994.

My first no-till trials with spring wheat and soybeans were successful and I've never regretted my decision to transition to no-till, since I didn't experience any yield drag. The primary

SOIL BIODIVERSITY. No-till and cover crops helped Jocelyn Michon improve soil stability.

reason for this successful transition was because no-till favors the rehabilitation of the biological and physical properties of the soil.

I started using cover crops in 2003, which improves soil biodiversity and exploits the full nutritional potential of the soil. All my land is no-tilled into a living cover crop.

No-till allowed me to save significant time and money, and I took the opportunity to acquire high-performance equipment. Now that I'm over 60, my son is involved and I'm confident in the future of farming.

Why No-Till Matters: It has resulted in increased organic matter, better structural stability and greater soil resilience. ✺

NEW ZEALAND

NO-TILLERS MUST OVERCOME CARBON, WATER VAPOR LOSSES

John Baker,
Cross Slot Tillage Systems, Feilding, New Zealand

No-tillers in semi-arid environments need to take a serious look at how carbon and water can impact yields:

◆ While water vapor can't be seen in undisturbed soils, it's a vital part of plant and soil microbe growth. Tillage causes this valuable moisture to evaporate, where it's unavailable to plants and soil life.

Even when it's dry, the humidity in no-till residue can germinate seeds.

◆ There's a dramatic net gain of sequestered carbon in the soil from residue decomposition due to decaying organic matter when

RETAINING RESIDUE. John Baker predicts no-till will become world's #1 cropping system.

compared to a loss when burying surface organic matter with tillage.

Low-disturbance no-till helps soil gain up to 500 pounds of carbon per acre, while plowing and disturbance from drills can lead to an annual loss of 1,800 pounds.

Local and national government support for no-till can help growers feed the world's growing population while reducing carbon in the atmosphere. The type of support is important, but no-till's rapid growth demonstrates financial and environmental benefits.

Yet in countries where no-till adoption has been slower and government influence plays a bigger role in farming systems, maybe a not-so-subtle nudge is needed.

If it takes 500 years to rebuild 1 inch of topsoil at the rate we're eroding the world's arable land, time is running out to reverse the damage.

The world's food production industry must rekindle its interest in rebuilding soil health through residue retention if farmers are to continue to feed an increasing number of hungry mouths. We're not there yet, but there's little doubt that low disturbance no-till will eventually get us there.

Why No-Till Matters: The science and practice of no-till has passed the "point of no return." It is within sight of becoming the most common method on the planet for seeding arable crops.

SAUDI ARABIA

SOILS ARE MUCH MORE STABLE WITH NO-TILL

Michael Dunn,
Holsworthy, Devon, United Kingdom

I started no-tilling in Saudi Arabia in 2000 with a modified drill in the corner of a large center-pivot irrigated field. We subsoiled the area before direct drilling and the yields were consistent with conventional tillage.

No-till helped these soils retain their stability and minimizes the need for soil cultivation. This has allowed growing plants to root more effectively, has reduced soil runoff and effectively utilizes more of the plant residue.

Why No-Till Matters: It dramatically reduces the need for water under these desert conditions.

NO-TILL LOOSE ENDS...

NO-TILL'S A GROWING TREND

March 1984

In selecting the top 10 farming trends for 1983, *Successful Farming* editors and readers had some things to say about no-tillage.

Seeing no-till double-cropping as a way to earn more profit from the same assets, the editors highlighted farmers who no-tilled double-crop beans directly behind the combine at wheat harvest time.

Another trend was the growing use of post-emergence herbicides, which is important to no-tillers. This weed control option is of extreme importance to no-tillers.

No-tillage itself got contradictory views in the survey. Some farmers called no-till a major ag trend while others felt it was one of the biggest disappointments of the decade.

As the *Successful Farmer* editors pointed out, both groups may be right. But newly-developed planters and drills, along with better chemicals, have made no-till much easier than 10 years earlier.

Frank Lessiter

A First-Person Knowledge Bank from
Nearly 200 No-Till Veterans

> **"**This series of no-till articles revealed **more passion for farming than any other group of agriculturists** I've covered as an ag journalist in over 45 years...**"** — *Ron Ross*

When you think about the typical series of articles appearing in magazines, it will normally include three, four or five stories written about a specific topic. Seldom does a series run over more than for a few consecutive issues.

Yet that's what makes the "What I've Learned About No-Tilling" series of thought-provoking articles in *No-Till Farmer* so unique. Zeroing in on the very best cropping ideas from leading no-tillers, this series has been running non-stop since August of 2002.

Since that time, we've featured more than 185 leading no-tillers, educators, suppliers and researchers from 26 states in this on-going series. While we've concentrated the majority of the content in the U.S., we've also included a few articles from no-tillers in Australia, Canada, Italy, Paraguay and New Zealand over the years.

In 2007, we published a 172-page book that features the cropping secrets from 58 highly successful no-tillers who had been spotlighted in the early days of this on-going series.

What I've Learned From No-Tilling

Cropping Secrets From 58 Highly Successful No-Tillers
By Ron Ross

WHAT I'VE LEARNED FROM NO-TILLING. Published by *No-Till Farmer,* this 172-page book featured hundreds of cropping secrets from 58 of the highly successful no-tillers featured in the first 5 years of this on-going educational series.

Over the years, the late Minnesota farm writer Ron Ross and Montana ag writer Martha Mintz have interviewed and prepared the text and photos for this series of on-going articles as contributing writers. Each article is told from the personal vantage point of the featured no-till celebrity — in their own words.

"It's astonishing that American farmers have been so slow to adopt no-till with their 40-year head start…"

No one has captured the essence of this sought-after and unique approach than Ron Ross himself, who died in 2017. Here are his own words about this series that was a truly passionate project for Ron.

What the Series is All About

By Ron Ross

RON ROSS

I often tell my sources for these no-till articles that they reveal more passion for farming in their responses than any other group of agriculturists I've ever covered as an ag journalist in over 40-plus years. By "passion" I mean (taking it straight from my Merriam-Webster dictionary), "an intense, driving feeling of conviction."

This has been true whether I was interviewing strip-till corn and no-till soybean growers in Iowa, no-till innovators with breakthrough cover crop programs in Missouri, no-till vegetable farmers in Maine or Pennsylvania, no-till soybean and wheat growers in Alabama or pioneers with direct seeding of small grains, lentils and peas in arid areas of Washington, Montana and Saskatchewan.

They all have seen the potential of a new kind of farming. The longer they are involved with no-till, observing dramatic changes in soil biology and crop production — and enjoying improved farm profitability — the more they realize they are doing something "special." Yet most of them are too humble to admit they represent the cutting edge of American agriculture.

World-Class No-Till Ambassadors

Interviews done for this book also included conversations with world-class scientists like Paraguay's Rolf Derpsch, South Dakota's Dwayne Beck and Alberta's Jill Clapperton. Interviewing Extension, NRCS specialists and consultants like Paul Jasa, Barry Fisher, Joel Meyers, Bud Davis, Dan Towery and others also revealed an uncommon passion for spreading the no-till story.

Derpsch's observation about U.S. growers is an admonishment of sorts that should be read repeatedly by "doubting Thomases."

"It is astonishing that American farmers have been so slow to adopt no-tillage, given that you have 40

WHAT I'VE LEARNED FROM NO-TILLING. John Aeschliman has been no-tilling wheat and other grains successfully since 1977 on slopes as steep as 60% in the Palouse area at Colfax, Wash.

"Every farmer who regularly attends this mid-winter gathering of the **no-till clan has a valuable story to share...**"

Frank Lessiter

WHAT I'VE LEARNED FROM NO-TILLING. No-tilling since 1970, Mike, Bob and Jim Ellis have relied on double-cropping to harvest as many as 8,352 acres of wheat, corn and soybeans from 6,300 acres located on 31 farms near Eminence, Ky. They added precision technology to no-till in 1996.

years of proof that no-till sharply reduces erosion, slashes farming time, cuts fuel costs and often ups your per-acre net income by 130% or more," he says.

During those times when I made farm visits or visited with my subjects at no-till events, it confirmed to me that a successful no-till effort boils down to adapting a specific system for an individual farm. Just basing decisions on generalities simply doesn't cut it in today's agriculture.

I had fun interviewing many of the spotlighted no-tillers who attended the National No-Till Conferences. Often, they might say, "I probably don't have much of a story to tell." It often surprised them (and even me at times) that we would still be "passionately" talking about their no-till operations after filling a couple of 45-minute audio tapes.

I really believe that literally every farmer who regularly attends this mid-winter gathering of the no-till clan has a valuable no-till story to share.

As for the future of no-till, there were many barriers to expanded acreage that came up in these conversations. Certainly, every successful no-tiller has had times when he or she was sorely tempted to revert to steel, and there are still plenty of "excuses" that slow no-till adoption.

But within each of the following areas of rationale for staying with conventional or minimum tillage, I've found innovators who have turned the negatives into positives by switching to no-till:

1 **"I already have a full line of tillage equipment."** This is a poor excuse, as there's still a good market for used tillage tools. Several growers I talked to for this book got enough dollars at auction for used tillage tools to buy a new no-till planter.
2 **"Spring soils are too wet and cold in my area."** Many strip-tillers would disagree. New corn hybrids, cold-ratings, disease re-

sistance, planter designs, available add-ons, etc., are keys to success in problem soils.
3 **"I don't want to invest in rented land."** On the other hand, good no-tillers seldom lose their leases. Landlords love improved soil quality, cleaner fields, reduced costs and higher yields.
4 **"I'm afraid of no-till's yield drag, especially when grain prices are low."** Long-term no-tillers repeatedly report consistent yield increases over time.
5 **"No-till costs too much."** The chances are good that future farm bills will offer unprecedented cost sharing for conservation practices. Meanwhile reduced fuel and labor costs are easy to document.

Great Sense Of Humor

I've enjoyed the good sense of humor shared by this group of amazing no-till leaders. The Good Lord knows you better have a sense of humor if you're gutsy enough to shift from a 100-year precedent of farming with

NO-TILL-AGE®
What Equipment Do You Own?

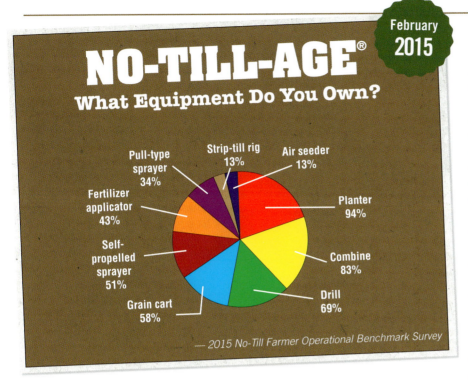

Pull-type sprayer 34%

Strip-till rig 13%

Air seeder 13%

Fertilizer applicator 43%

Planter 94%

Self-propelled sprayer 51%

Combine 83%

Grain cart 58%

Drill 69%

— 2015 No-Till Farmer Operational Benchmark Survey

and have told me over and over is that successful no-tilling depends on developing a system within which every component — soil health, residue management, equipment design and modification, seed selection, cover cropping, time of planting, harvest methods, etc. — are all interdependent.

As always happens with this type of venture (and adventure), there are many names missing from the series that should be included. But, given time, with *No-Till Farmer's* continued support — and a little help from nature (I intend to keep writing as long as I can get to the computer!) The next volume of "What I've Learned From No-Tilling" … can't be too far down the road.

— Ron Ross

Editor's note: Up until he passed away in the spring of 2017, Ron was always extremely proud of having established this on-going series that has continued to be an important part of the No-Till Farmer content offering for many years.

a plow to doing it all with a tractor, planter/drill, sprayer and combine, as many of these no-tillers have done.

The different dialects I've encountered on the audio tapes I've made have been fun — and challenging at times. Let's see, was that "aboot," "aboat" or "about"?

And it seems that north of I-70 and I-64 "fertilize" (meaning to apply) is a verb. Further south "fertilize" (meaning the product) becomes a noun. And a Minnesotan like me really needs his ears on when transcribing the comments of a Maine-born and bred grower like Andy Williamson.

Awesome No-Till Knowledge

While we didn't have any real "criteria" for selecting sources, I believe we're creating a "No-Till Hall of Fame" of sorts. This series is an awesome bank of no-till knowledge to turn to. With the broad range of growers, scientists and other experts, geography and no-till systems covered in this series, there is a "hands-on" answer somewhere within these

"first-person" stories for just about any question you may have on practical conservation agriculture.

These experiences are worth noting because literally every piece of advice is based on solid farming economics. What these veteran no-tillers and experts have learned

NO-TILL-AGE®
What's Straw, Chaff Worth?
(Based on net return price for selling wheat residue)

Residue	Spring Wheat
Straw	$15.60
Chaff	$03.17
Total return	$18.77

Figures are based on baling, loading and transportation costs of $20.87 per ton, a selling price of $40 per ton for straw delivered to the manufacturing plant and a cost of $4.81 per ton in the form of nutrients lost in the straw. This leaves a net value of $15.32 per ton of straw.

— Alberta, Canada, AgriPulp Workshop

WHAT I'VE LEARNED FROM NO-TILLING

Introduced by veteran farm writer Ron Ross in 2002, this series has featured many of the top minds in no-till, as they shared their extensive knowledge with our readers. The insights of these no-till veterans represents a gold mine of wisdom for others regardless of whether you're just starting out with no-till or have over 20 years of experience.

Pat and Chris Breen, Seneca, S.D.

Scott Day, Deloraine, Manitoba

Jim Hershey, Elizabethtown, Pa.

Doug Harford, Mazon, Ill.

Eddie Hoff, Boonville, Mo.

MEET THESE 'FOUNTAIN OF KNOWLEDGE' NO-TILLERS

Since 2002, *No-Till Farmer* has featured nearly 200 leading no-tillers, educators, suppliers and researchers from 26 states and several foreign countries in the "What I've Learned From No-Tilling" series.

Here's the list of the folks who have shared their no-till secrets in this on-going series.

Paul Ackley, Bedford, Iowa
Robert Adamic, Silverwood, Mich.
John Aeschliman, Colfax, Wash.
James Andrew, Jefferson, Iowa
Roy Applequist, Salina, Kan.
Joel Armistead, Adairville, Ky.
Jim Arnaud, Cole, Mo.
Don Bahe, Stanley, Iowa
Steve & Bill Bailey, Chatfield, Minn.
John Baker, Feilding, New Zealand
Ron Baurth, Alpena, S.D.
Dwayne Beck, Pierre, S.D.

Mike Beer, Keldron, S.D.
Mike Belan, Oil Springs, Ontario
Steve Berger, Wellman, Iowa
Allen Berry, Nauvoo, Ill.
Roger Bindl, Spring Green, Wis.
Ross Bishop, Jackson, Wis.
Larry Bonnell, Pittsford, Mich.
Bob & Monte Bottens, Cambridge, Ill.
Chip Bowling, Newburg, Md.
David Brandt, Carroll, Ohio
Donn Branton, Leroy, N.Y.

Pat Breen, Seneca, S.D.
Joe Breker, Havana, N.D.
Allan Brooks, Markesan, Wis.
Dan & Lana Buerkle, Plevna, Mont.
Howard Buffet, Decatur, Ill.
Billy & Sadie Burns, Hooper, Colo.
Cade Bushnell, Stillman Valley, Ill.
Cecil & Steele Byrum, Windsor, Va.
Mark Chapman, Bowling Green, Ky.
Brad Choquette, Upland, Neb.
Chad Christianson, Hooper, Neb.
Jill Clapperton, Lethbridge, Alberta

Mauro Collovati, Udine, Italy
Bill & Nina Cowan, Hartney, Manitoba
Bill Crabtree, Perth, Australia
Jerry Crew, Webb, Iowa
Lucas & William Criswell, Lewisburg, Pa.
Michael & Adam Crowell, Turlock, Calif.
Kip Cullers, Purdy, Mo.
Terry Dahmer, Marion, Ill.
Bill Darrington, Persia, Iowa

Ron Lampe, Havelock, Iowa

Doug Goehring, Menoken, N.D.

Hans Kok, Carmel, Ind.

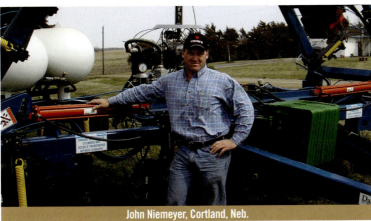

John Niemeyer, Cortland, Neb.

Howard Buffet, Decatur, Ill.

Cade Bushnell, Stillman Valley, Ill.

Wendte Farms, Shumway, Ill.

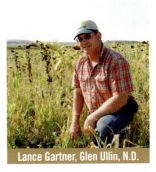

Lance Gartner, Glen Ullin, N.D.

Bud Davis, Salina, Kan.
Scott Day, Deloraine, Manitoba
Jodi DeJong-Hughes, Marshall, Minn.
John Derhman, Faribault, Minn.
Rolf Derpsch, Asuncion, Paraguay
Dan DeSutter, Attica, Ind.
Mike Dick, Munich, N.D.
Jerry Doan, McKenzie, N. D.
Joel Douglass, Martell, Neb.
David Dukes, Bedford, Iowa
David Dum, Lititz, Pa.
Jerry Eitle, Attica, Ohio

Mike Ellis, Eminence, Ky.
Tracey Eriksen, St. John, Wash.
Bob Featheringill, Attica, Ohio
Dean Fehl, LaPorte City, Iowa
Barry Fisher, Greencastle, Ind.
Wayne Fredericks, Osage, Iowa
Curtis Furr, Albemarle, N.C.
Ray Gaesser, Corning, Iowa
Gordon Gallup, Ririe, Idaho
Lance Gartner, Glen Ullin, N.D.
Josh, Justin & Jeremy Gerwin,
    Gibsonburg, Ohio

Dan Gillespie, Meadow Grove, Neb.
Don & Brian Glenn, Hillsboro, Ala.
Doug Goehring, Menoken, N.D.
Steve Groff, Holtwood, Pa.
Doug Gronau, Deniston, Iowa
Nick Guetterman, Bucyrus, Kan.
Don Halcomb, Logan County, Ky.
James Harbach, Loganton, Pa.
Doug Harford, Mazon, Ill.
Roger Harrington, Ollie, Iowa
Dennis Haugen, Hannaford, N.D.
Jim Hershey, Elizabethtown, Pa.

Keavin Hill, Orient, Ohio, and
    Maysville, Ky.
Spencer, Lynne, Dane, Sterling,
    Liana, Gordon & Viola Hilton,
    Strathmore, Alberta
Eddie Hoff, Boonville, Mo.
Dale Hollan, Suffolk, Va.
Ralph Holzwarth, Gettysburg, S.D.
Jim House, Port Stanley, Ontario
Larry Huffmeyer, Osgood, Ind.
David Hula, Charles City, Va.
Terry Huss, Faulkton, S.D.

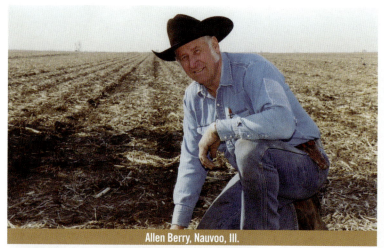
Allen Berry, Nauvoo, Ill.

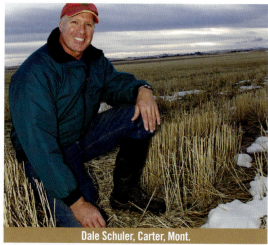
Dale Schuler, Carter, Mont.

Wittman Farms, Lapwai, Idaho

Jerry Crew, Webb, Iowa

Ellis McFadden, Fort Wayne, Ind.

Guy Williams, Iowa City, Iowa

Todd Intermill, Colman, S.D.
Paul Jasa, Lincoln, Neb.
John C. Johnson, Stewartstown, Pa.
Bryan Jorgensen, Ideal, S.D.
Paul Kelly, Midland, Ohio
John Kemmeren, Bainbridge, N.Y.
Keith Kemp, West Manchester, Ohio
Floyd Koerner, Laingsburg, Mich.
Hans Kok, Carmel, Ind.
Ron Lampe, Havelock, Iowa
Duane Lange, Ord, Neb.
Eugene Lapp, Kinzers, Pa.

Bill Lehmkuhl, Minster, Ohio
Bob Linnell, Weyburn,
    Saskatchewan
Rich & Marcy Little, Eaton, Ohio
Don Lobb, Caledon, Ontario
Fred Luhnow, Kewanna, Ind.
Dennis Lundy, Fontanelle, Iowa
Warren Macemon, Glencoe, Minn.
Jack Maloney, Brownsburg, Ind.
Frank Martin, Hallsville, Mo.
Howard Martin, Elkton, Ky.
Jeff Martin, Mt. Pulaski, Ill.

Brad Mathson, Whitehall, Wis.
Carl Mattson, Chester, Mont.
Joel McClure, Hugoton, Kan.
Ray McCormick, Vincennes, Ind.
Jason McCunn, Red Oak, Iowa
Ellis McFadden, Fort Wayne, Ind.
Maury McLean, Lancaster, Wis.
Russell McLucas, McConnellsburg,
    Pa.
John McNabb, Inkom, Idaho.
Tim Melville, Enterprise, Ore.
John Merchant, Cass City, Mich.

Perry Miller, Bozeman, Mont.
Clay Mitchell, Buckingham, Iowa
Jeff Mitchell, Davis, Calif.
Dave Moeller, Keota, Iowa
Tom Muller, Windom, Minn.
Joel Myers, Spring Mills, Pa.
Ed Neesen, Rogers, Neb.
Joe Nester, Bryan, Ohio
John Niemeyer, Cortland, Neb.
Chris & Dick Nissen, Vermillion, S.D.
Randy Norby, Osage, Iowa
Malcolm Oatts, Hopkinsville, Ky.

Steve and Dennis Berger, Wellman, Iowa

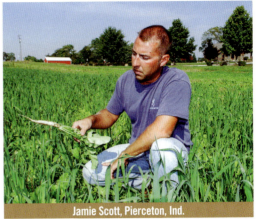
Jamie Scott, Pierceton, Ind.

Mike Starkey, Jack Maloney, Brownsburg, Ind.

Bill Darrington, Persia, Iowa

Brad Choquette, Upland, Neb.

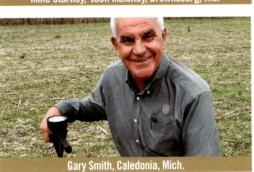
Gary Smith, Caledonia, Mich.

Tom Oswald, Cleghorn, Iowa
Jerry Peery, Clinton, Ky.
Patrick Petersen, Gordon, Neb.
Jim & Charles Pilling, Mediapolis, Iowa
Phil Pinch, Brandon, Wis.
Mike Plumer, Carbondale, Ill.
Allen Postlethwaite, St. Arnaud, Victoria, Australia
Bill Richards, Circleville, Ohio
Jack Rigby, Blenheim, Ontario
Leslie Roy, Moxee, Wash.
Rodney Rulon, Arcadia, Ind.
Paul Schaffert, Indianola, Neb.
Keith Schlapkohl, Stockton, Iowa
Rod Schmidt, Clinton, Iowa

John & Hans Schnekloth, Eldridge, Iowa
Dale Schuler, Carter, Mont.
Ed & Wayne Scott, Jacksonville, Ill.
Jamie Scott, Pierceton, Ind.
Pat Sheridan, Caro, Mich.
Gary Smith, Caledonia, Mich.
Read Smith, St. John, Wash.
Gary Sommers, Clinton, Wis.
Mike Starkey, Brownsburg, Ind.
Craig Stehly, Mitchell, S.D.
John Stigge, Washington, Kan.
Dan Stokes, Omro, Wis.
Gordon Stoner, Outlook, Mont.
Kent Stuckey, Bucyrus, Ohio
Ray Styer, Reidsville, N.C.

Alan Sundermeier, Bowling Green, Ohio
Donald Sutherland, Leroy, Kan.
Gary Sweet, Columbia Station, Ohio
Sam Swinford, Flat Rock, Ind.
Curtis Strand & Gabe Sybesma, Platte, S.D.
Lyle IV Tabb, Kearneysville, W.Va.
Terry Taylor, Geff, Ill.
Jimmy Thomas, Timberlake, N.C.
Keith Thompson, Osage City, Kan.
Kelly Tobin, New Market, Iowa
Dan Towery, West Lafayette, Ind.
Grant Troop, Oxford, Pa.
Mark Turner, Livermore, Ky.
Ralph Upton Jr., Springerton, Ill.

Bryan Von Holten, Cole Camp, Mo.
Allen Verell, Jackson, Tenn.
Jacob Wagers, Woodrow, Colo.
Eugene Wahling, Shelby, Iowa
Wendte Brothers, Shumway, Ill.
Roger Wenning, Greensburg, Ind.
Guy Williams, Iowa City, Iowa
Andy Williamson, Jefferson, Maine
Ed Winkle, Martinsville, Ohio
Danny Wipf, Lake Andes, S.D.
Dick Wittman, Lapwai, Idaho
R.D. Wolheter, Wolcottville, Ind.
Mike Wolpert, Buffalo, W.Va.
John & Alex Young, Herndon, Ky.
Russ Zenner, Genesee, Idaho
Chuck Zumbrun, Churubusco, Ind.

More No-Till Needed to Feed
World's 10 Million Mouths

If no-till is going to help to feed an ever-expanding worldwide population, here are **19 'crystal ball' predictions that may change the way you farm in the decade ahead.**

With the world needing to produce twice as much food to keep 10 billion people around the world from going hungry just 30 years from now, we must ratchet up crop yields. If we can't do that, the only option is to start cropping millions of acres around the world that are better suited for wildlife and recreation.

Over the next few decades, nothing will have a bigger impact than the continued growth of no-till and its efficiencies when it comes to feeding the world. Simply combining no-till with new innovations in the adoption of heat-and-drought-tolerant seeds, improved nitrogen-use efficiency and precision agriculture over the next three decades could reduce the number of malnourished children around the world by more than 1 million annually. ✻

As we moved into the new century nearly two decades ago, the *No-Till Farmer* editors came up with 14 major changes that we felt could occur with no-till by 2015. For this chapter, we've taken another look at those earlier predictions and added a few more that should take place over the next decade.

1 **NO-TILL'S FUTURE.** Over the next several decades, no-till may become the world's "conventional" farming practice. It offers terrific soil protection with continuous residue cover, minimum soil disturbance, more diverse rotations and the expanded use of cover crops. Keeping the soil covered with living plant roots all year long helps control erosion, increases economic returns, improves soil health and will help overcome climate change concerns.

Further adoption of no-till can make worldwide food production truly sustainable while regenerating soil health, increasing productivity from the world's remaining cropland and improving air and water quality.

2 **GLOBAL WARMING, CLIMATE CHANGE.** "Pessimism" and "disaster" are two words found extensively in numerous negative climate change reports when it comes to the many changes facing agriculture. Even so, a careful analysis of these reports offers bright spots for both no-tilling and cover cropping.

Unfortunately, scientists are scaring folks with predictions of killer heat waves, melting ice that will raise ocean levels, higher night-time temperatures, more drought, increased humidity, stronger storms, changing rainfall patterns, warmer winters and new threats from weeds, pests and diseases.

If the predictions of many clima-tologists come true, we'll see gradual warming around the world that will lead to more soil evaporation and severe droughts. Climate change could result in the northern movement of weeds, growing weed and insect resistance concerns and greater disease pressures due to warmer winters and earlier springs.

While yearly precipitation may not undergo much change, more moisture will likely fall during the winter with less rain during the growing season. Increased evaporation and higher transpiration rates could occur, boosting the need for no-till and irrigation alike.

3 **WATER USAGE CRITICAL.** More water is stored in no-tilled soils than in fields worked with intensive tillage, which leads to costly evaporation. Many groundwater aquifers will see a dramatic decline unless tighter water allocations are placed on wells.

Research indicates extensive use of no-till in developing countries could boost corn yields by 20% over the next 36 years. More irrigation could push up yields even more dramatically. But with municipalities needing additional water, farmers must learn to produce crops with less water.

4 **CURBING EROSION.** Soil, wind and water erosion will continue unless more reduced tillage is adopted at a faster rate around the world. With thin and depleted topsoil layers in many areas, no-till is the best way to overcome losses of our valuable soils.

5 **NO-TILLING ORGANICALLY.** Growers in Paraguay have developed an organic system that relies on no-tilling green manure crops as soon as possible after harvesting no-till cash crops. With this system, sunflowers and sunn hemp are knocked down with a knife roller, which lets farmers no-till into the dead mulch without needing herbicides for weed control.

With cover crops controlling the weeds, these no-till farmers boosted returns by $15 per acre with more mulch, higher organic matter in the soil and less nutrient leaching.

U.S. growers don't yet have the kinks worked out for making a no-till organic farming system practical, but refinements are coming.

6 **SLICING HERBICIDE RATES.** While most no-till crops won't be herbicide-free, we're seeing reductions among no-tillers. Manitoba researchers are developing a program to avoid using herbicides in as many as three of five crops with an expanded no-till rotation.

7 **MORE PRECISE PRECISION.** Between now and 2030, precision farming tool usage will increase rapidly. We'll see remote-driven machinery, more accurate placement of fertilizers and pesticides, infrared spot spraying of weeds, more variable rate farming and robotic seeders in no-tilled fields. Plus, there will be other innovative precision tools that nobody has even dreamed about today.

8 **SHARING NO-TILL IDEAS.** When no-till was introduced into several South American countries, the fact that farmers, farm groups, educators, shortline equipment manufacturers, consultants and government agencies immediately all jumped on the no-till bandwagon is one of the major reasons for its quick adoption and success.

For example, the rapid acceptance of no-till allowed Argentina's growers to expand double-cropping, conserve water, boost soil organic matter, trim fertilizer needs and decrease soil erosion. No-till helped Argentina's growers trim soybean production costs by 50% while bringing more marginal land into efficient crop production and boosting yields.

9 IMPROVED NO-TILL RESIDUE MANAGEMENT.
To overcome concerns with wet and cold soils, several no-till spinoffs — including vertical tillage, zone tillage and strip-tillage — have been developed.

Expect still more no-till modifications as growers adopt systems to specific field needs.

10 DIVERSIFIED NO-TILL ROTATIONS.
Adding diversified rotations that include cover crops help no-tillers deal with compaction, reduced yields, drought, insect, disease and weed problems and cash in on value-added crop markets.

Yet expanding no-till rotations is tough in the Corn Belt when you can't find crops with the same earning power as corn and soybeans. Look for oilseeds to gain popularity among Midwestern no-tillers.

11 MORE COVER CROPS.
South American no-tillers don't understand why North American growers don't fully recognize the many benefits of cover crops.

Even so, U.S. no-tillers are ahead of the game when it comes to cover crop acceptance as 83% seed cover crops compared with fewer than 5% of all U.S. farmers.

As more growers see the many soil health benefits of no-till, they'll add cover crops to control erosion, boost soil organic matter, improve overall soil quality, suppress weeds, provide low-cost nutrients and add grazing options.

12 SEQUESTERING CARBON.
In 2000, we predicted no-tillers would soon be earning carbon credit payments. It hasn't happened, as there's still plenty of governmental maneuvering to be done around the world before this idea is fully accepted.

Since no-till slows the loss of greenhouse gas emissions, it will play a key role in capturing carbon and providing long-term storage of atmospheric carbon in the soil. In fact, no-till alone could offset 20% of the world's annual carbon dioxide emissions.

Combine no-till with expanded crop rotations, rotational grazing, nitrogen-fixing crops and cover crops and we'll see a dramatic reduction in carbon losses into the atmosphere.

13 EARLIER NO-TILLING.
Three concepts dealing with earlier planting haven't materialized to the extent that was predicted nearly two decades ago. The idea

WHAT'S COMING FOR NO-TILL, STRIP-TILL?

The popularity of these two conservation tillage systems will continue to grow, as they have too many crop production, economic and environmental benefits for that not to happen.

◆ **No-Till Acres.** Serving since 1972 as the *No-Till Farmer* editor, I've had a ringside seat in watching the U.S. no-till acreage grow from 3.2 million acres in 1972 to 50.8 million acres in 2000 and to an anticipated 107 million acres by 2019.

In 2011, the U.S. Department of Agriculture reported that no-till and strip-till together accounted for 31% of the corn acres, 46% of the soybean acres, 43% of wheat acres and 39% of the cotton acreage. While it's likely that these percentages have increased in recent years, nobody knows by how much.

Our "best crystal ball guess" for 2030 is that no-till will be used on 145 million acres in the U.S. We expect the worldwide no-till acreage to reach 466 million acres in 2030, an increase from 388 million worldwide no-till acres in 2013.

◆ **Strip-Till Acres.** It's difficult to predict the anticipated increase in strip-tilled acres since neither the government or anyone else collects data indicating how many acres are farmed with this system today.

In 2007, we estimated there were 3.6 million acres of strip-tilled ground, but it's grown dramatically since that time.

Our guess is that an estimated 10 million acres in the U.S. will be strip-tilled by 2030. ❈

What's Bt Corn Worth?

(Estimated Value of YieldGard Corn for European Corn Borer Infestation Levels)

Crop Income Per Acre	Probability of European Corn Borer Infestation		
	20%	30%	40%
$250	$2.96	$4.59	$6.32
$300	$3.27	$5.50	$7.59
$350	$4.15	$6.42	$8.85
$400	$4.74	$7.34	$10.12
$450	$5.33	$8.25	$11.38
$500	$5.92	$9.17	$12.65

—Purdue University

behind polymer-coated seed was that corn and soybeans could be no-tilled in early April into cold, moist ground.

Thanks to the protective coating, germination would be delayed for a few weeks until the soil reaches the right temperature. This earlier planting concept offered a way to spread out a farmer's critical labor needs during the critical spring season.

Canadian growers have successfully seeded polymer-coated canola in the fall when temperatures are as low as 10 degrees. This technology has helped western Canadian no-tillers shift to fall-seeded crops that previously couldn't survive severe winter weather.

Another prediction earlier in the decade anticipated a big boost with no-till double cropping. By 2010, Purdue University agronomist Ray Bressan predicted that farmers could no-till corn or soybeans in March, harvest an early-maturing crop in July, immediately no-till another crop in those same fields and harvest a second round of crops by November.

The key to making this concept work was developing new seed vari-eties to overcome frost, hot weather, drought and stress tolerance concerns. So far, commercialization of the concept has not happened.

Polymer seed coatings idea could also be used with relay intercropping soybeans into standing wheat. While a few growers have no-tilled coated soybeans into standing wheat 6 weeks ahead of small grain harvest, we still need to determine how bean maturities, seed vigor, plant populations and other traits affect yield.

None of these three concepts have gained much momentum over the past 15 years. However, new research and technology breakthroughs could have a huge impact on earlier planting.

14 GMO SITUATION.

While GMO corn, soybeans and cotton have caught on rapidly with North American no-til-lers, we'll see still more specialized genetic traits incorporated into many more crops in the future. This is particularly true in areas of the world where the GMO technology has not yet been accepted by farmers, governments or environmental and consumer groups.

The use of biotech-created seeds will continue to grow as more countries get on board and overcome their food safety and environmental fears.

15 800-BUSHEL CORN YIELDS.

In 2017, David Hula achieved what many seemed was the impossible. The Jamestown, Va., no-tiller harvested 542 bushels per acre that represented a new corn yield record.

He says the key to turning out these extraordinary high yields involves help from Mother Nature, irrigation, seed selection and nurturing the crop throughout the entire growing season.

Hula believes some of today's corn hybrids offer the potential for 800-850-bushel yields. But to get these kind of yields, Hula maintains you must protect the full yield potential throughout the growing season.

16 DESIGNING NO-TILL VARIETIES, HYBRIDS.

By choosing corn hybrids, soybean varieties and wheat varieties specifically for no-till conditions, seed suppliers will start doing a better job of identifying lines with strong emergence and enhanced disease resistance to perform well under tough no-till conditions.

More seed suppliers need to perform the necessary cold tolerance tests to determine which corn hybrids are ideal for no-till.

17 REDESIGNED GENETICS.

Since yields must increase dramatically to meet the growing food demand, we need genetics that can handle higher plant populations and likely new plant redesigns. As an example, no-tillers will need improved genetics to boost corn populations to 60,000-75,000 plants per acre.

We'll also need drought-tolerant seeds, along with lots of fertilizer and careful management to keep nitrogen from leaching into streams.

18 STILL NARROWER ROWS.
The Harry Stine family at Adel, Iowa, is developing new genetics for growing corn in a 12-by-12-inch row pattern.

Regardless of the crop, look for even narrower rows to take full advantage of new genetics, improved weed control with earlier shading and other improvements that will pump up yields.

19 MORE EFFICIENT NITROGEN USAGE.
As was the case 15 years ago, the over-application of fertilizer is still a major concern. Around the world, improved nitrogen-use efficiency could boost rice yields by 22%. Adding irrigation will boost yields by another 21%, which will provide much of the nutrition for the world's growing population over the next few decades. ✺

NO-TILL-AGE®
March 2017

What Motivates Growers to Use Cover Crops?

Cover Crop Benefits	% of Growers
Increase overall soil health	87%
Increase soil organic matter	83%
Reduce soil compaction	76%
Reduce soil erosion	76%
Scavenge nitrogen	59%
Provide another nitrogen source	57%
Choose diverse rooting systems	51%
Increase cash crop yields	47%
Improve weed control	43%
Attract pollinators	29%
Increase economic return	25%
Decrease production costs	25%
Diversify termination methods	22%
Reduce cash crop disease potential	20%
Improve winter hardiness	16%
Improve insect control	13%
Other	9%

— 2015-16 Cover Crop Survey from the Conservation Technology Information Center and Sustainable Agriculture Research & Education

MAJOR CONCERNS IMPACTING AGRICULTURE'S FUTURE

An ever-expanding population, less fresh water and warmer worldwide temperatures will have a major influence on whether we can grow enough food to feed the world over the next few decades.

◆ By 2030, the United Nations predicts the world's population will reach 8.5 billion, compared with 7.3 billion people today. By 2050, the world expects to have 10 billion mouths to feed.

◆ Since 1900, the earth's average surface temperature has increased by 1.3 F. Some researchers expect another 7.2 F temperature gain by the end of this century.

◆ By 2030, the world will need 30% more fresh water than what's used today. Increased emphasis must be placed on irrigation to boost yields in order to feed the world. ✺

WHY NO-TILL WORKS IN ASIA

Peter Hobbs is convinced no-till is the solution for feeding Asia's increasing population. With growers in Bangladesh, India, Nepal and Pakistan leading the way, the continent's no-tilled acres are growing at an amazing pace.

"Without the conservation of water through no-till, this region will become dependent on imported food — which no one can afford," Hobbs says.

The natural resource agronomist in Nepal with the International Maize and Wheat Improvement Center says that forcing the Asian population to rely on imported food will create inevitable food shortages and severe malnutrition. As an example of the seriousness of the situation, Hobbs says 40% of the Nepal population in 2002 was living on less than $2 per day.

Hobbs sees 5 major benefits for increasing no-tilled acres throughout Asia.

1 **Saving Water.** As much as 50% of the available water found under Asia's extremely dry ground can be saved by no-tilling.

2 **Higher Yields.** No-tilling wheat immediately after rice harvest allows a 3-4-week earlier seeding than with plowed ground. This allows the wheat to fully mature before hot, dry weather sets in, increasing both yields and income.

3 **Reducing Costs.** No-till with wheat dramatically reduces land preparation costs while boosting yields. As farmers work toward no-tilling rice, both their income and cost-cutting benefits will improve.

4 **Fewer Weeds.** By leaving the soil undisturbed with no-till, fewer weeds will germinate while earlier wheat seeding leads to shading out weed growth. With up to two-thirds fewer weeds, herbicide needs are reduced.

5 **Reducing Greenhouse Gas Emissions.** By shifting to no-till and reducing the need for tractor and water pump power, as much as 75% less fuel is burned each year. Some 800,000 fewer tons of carbon dioxide will be emitted as greenhouse gases into the atmosphere once no-till is widely adopted in the region.

WHAT'S HOLDING BACK NO-TILL?

John Hebblethwaite, former executive director of the Conservation Technology Information Center in West Lafayette, Ind., lists seven reasons why no-till has not caught on faster among U.S. farmers.

1 Conflicting agendas among farmers, educators, government officials and suppliers about the value of no-tilling.

2 Concern about the economic payback from no-tilling.

3 Acceptance of the 25%-30% labor-saving benefit of no-till is much higher among growers in Latin America than in North America.

4 Soybeans are more favorable for successful no-till than corn.

5 Site-specific solutions are needed to gain acceptance with no-till corn due to cool, wet planting conditions in many areas of the Corn Belt.

6 More intensive and profitable rotations need to be combined with no-till in the Great Plains.

7 There must be faster adoption of cover crops in the southern U.S. before no-till will become widely adapted.

No-Till Success
Spawns 'Spin Offs'

"Purists hate hearing these tools are considered part of a no-till practice; consider it no-till with an asterisk..."— *Phillip Brown*

Just as reducing tillage had gained acceptance during the 1960s, several variations of no-till began to take shape in the last few decades.

◆ The "zone tillage" concept was developed in Michigan by no-till corn and soybean grower Ray Rawson.

◆ An offshoot that gained even greater acceptance was strip-till, where growers plant in a narrow, blackened strip of bare soil while leaving the rest of the field undisturbed. Its success also led to "bio strip-till," which is a similar system for seeding cover crops in strips.

◆ Still another approach that evolved later was the use of "vertical tillage." Running fast and shallow, these tools feature discs or coulters set at unaggressive angles to manage tough Bt corn stalks, incorporate residue, fertilizer or cover crops, improve seedbed conditions and even fight weeds.

As these "spin offs" have matured, the role of these tools in a no-till system hasn't been without controversy, particularly with the growth of vertical tillage. ❁

WARMER SOILS, DEEPER ROOTS WITH ZONE TILLAGE

Many variations of tillage systems have been used around the globe, and one that grew in prominence beginning in the 1970s was the zone-tillage concept developed by Ray Rawson of Farwell, Mich. He's known as the "Father of Zone Tillage."

Over the years, Rawson developed several zone tillage tools to go along with a cropping system that he felt blended the best of both no-till and conventional tillage. The purpose of zone tillage was to expand the size and depth of the root zone while preparing an effective seedbed. The system also allowed accurate placement of liquid or dry fertilizer in a warm, moist area of soil to promote rapid and expanded root development.

Overcoming Poor Soil Structure

Rawson advised growers to use zone tillage in situations where the soil structure wasn't friendly and when the ground had been overtilled. In these cases, he believed zone tillage offered improved soil-to-seed

SUMMER BREAK. With the family vacationing for a number of summers in mid-Michigan, I'd always visit Ray Rawson to see latest planter ideas.

Frank Lessiter

contact and helped warm up the soil in cooler areas of the country.

Rawson pioneered a combination of three wavy coulters in a zone tillage system, with the center or lead coulter clearing away the residue. Two parallel coulters running behind the lead coulter provided in-row tillage to promote soil-to-seed contact and help accurately place dry or liquid fertilizer.

Digging Deeper

The system can also include a deep zone tillage tool to open a slot up to 22 inches deep that allows air, water and roots to move deeper into the soil. Rawson preferred using a straight leg with a narrow point at the bottom to run deep to take out the natural soil pan.

Sometimes mounted behind the shank, covering coulters were followed by a rolling basket to build a "zone" that offers a 6-8-inch clean berm of soil.

"By doing this, we're driving microlife deeper into the soil and we begin to see changes in the way soybeans nodulate, the nitrification

4 DAYS OF NON-STOP LEARNING. Zone till pioneer Ray Rawson, (center) shares data on 80,000-200,000 per acre no-till soybean populations at a 2003 National No-Tillage Conference Roundtable.

Frank Lessiter

"Zone tillage **improves seed-to-soil contact and helps warm up** cooler wet soils..."

gets deeper and so on," says Rawson.

While Rawson normally performed deep zone tillage in the fall, he often made another trip with the same zone-tilling machine in the spring. Running through the same zones established in the fall, the tool only runs 6 to 9 inches deep. By planting directly into the slots established the previous fall, growers could get a day or two jump on emergence and root mass growth.

Rawson was constantly trying new ideas to perfect the zone-till system. For a number of years, he sold his planting rig each summer and would build one with new innovations over the winter months.

True No-Till or Not?

In a 1992 postcard survey, 23% of *No-Till Farmer* readers were doing zone tillage while nearly 60% had thought about trying it. Some 44% of readers believed zone tillage represented true no-till. Along with its detractors, the system also had its defenders, especially in fields having heavy, poorly drained soils.

Zone tillage helped Paul Beere of Burlington, Wis., reduce hairpinning and remove residue for better depth control, He found this led to quicker germination with faster warming of soil in the row.

Three coulters working an 8- to 9-inch wide strip makes a perfect corn seedbed for Richard Chamberlin of Allenton, Mich. The 2-inch wavy coulters on his Blu-Jet coulter cart help warm the soil while placing 28% nitrogen on both sides of the row.

"If it weren't for a Rawson triple-coulter setup, plus no-till coulters on my John Deere row units, I wouldn't be no-tilling," adds Rex Ulrich of Sand Creek, Mich.

Zone-Till Goes Mainstream

Equipment manufacturers like Brillion and Unverferth soon introduced equipment built around Rawson's zone-tillage concept. Launched in 1992, Unver-

NEW YEAR, NEW PLANTER. For a number of years, each summer Ray Rawson sold off his current zone tillage rig and would build a new one over the winter with his latest ideas for use the following spring.

ferth's Zone-Builder subsoiler was billed as the first tool offering an automatic reset and a 3/4-inch-thick shank to reduce soil disturbance in no-till situations.

"The tool was developed as a minimum-disturbance subsoiler that would work by itself for conservation compliance to alleviate compaction or in conjunction with a zone-tillage cart or zone-till attachment," says Unverferth's Jerry Ecklund. "When used in zone tillage, the tool went in between the old corn rows. Growers would place next year's row right on top of where the Zone-Builder had passed in the fall."

Over the winter, water flows down into the shank's path. The freeze-thaw cycle eliminates compaction faster while filling in the vertical slice created by the shank.

Current Usage

Zone tillage is still used on many farms in the U.S. and globally, although its popularity began to decline with the advent of much larger planters. The practice has been revived to some extent with GPS-guided "strip-till" systems that feature coulters, knives or openers that create a favorable seedbed for crops while minimizing soil disturbance to preserve soil health benefits. ✹

STRIP-TILL OFFERS SOLUTION FOR COLD, WET SOILS

BUILDING RIDGES. Unhappy with no-till corn yields on their cold. poorly drained soils, some growers were quick to jump on the strip-till bandwagon in the early 90s. A modified form of zone tillage and ridge-till, strip-till features 6-8 inch tall berms that can be built and fertilized in the fall or spring.

While growers were moving toward no-tilled soybeans in a big way during the 1990s, no-tilled corn acres remained somewhat flat. Many farmers feared problems when they tried to no-till corn into cold, wet ground covered with residue.

As a result, strip-till, which is a modified form of zone tillage, soon emerged as a compromise. These machines move residue away from the narrow row area in the fall or spring and create a strip of bare soil. The system combines the soil drying and warming benefits of conventional tillage with the soil-protecting advantages of no-till while only disturbing soil in the row area.

Strip-till usually meant working the soil 6-to-8-inches deep and allowed deep placement of phosphorus and potassium in areas where roots could easily access them. In some cases, row cleaners moved residue away from the row area where the 6-8-inch-tall berm was being built. A coulter could also be used to slice through residue, and a shank or mole knife sliced a narrow slot through the soil.

Rich Follmer, the retired president of Progressive

> ## "It helped overcome the concerns with no-till..."

Farm Products in Hudson, Ill., recalls a day in 1995, when *No-Till Farmer* Editor Frank Lessiter asked him to give a 30-minute presentation on strip-till at the National No-Tillage Conference in Indianapolis.

"After wrapping up my talk, I spent the next 3 1/2 hours in the hotel hallway fielding questions about strip-till from attendees," Follmer recalls. "The response was unbelievable — I'd never seen so much interest in strip-till."

Follmer, who was a No-Till Innovator in 2008, wasn't the first strip-till speaker at the conference. During the first annual mid-winter event in 1993, Cliff Roberts of Kentland, Ind., and a strip-till pioneer since 1987, shared his successes with strip-tilling corn.

Many growers in those days still weren't getting a favorable yield response with no-till corn, mainly because of the heavy residue. Combines weren't spreading residue far enough, which led to cold and wet seedbeds.

When no-tilling corn into soybean residue, Follmer says the soil would be so damp and cold that you could squeeze the furrow shut. Or it would open back up after the soil dried out and you could count the corn kernels you had planted 2 days earlier.

Serious About Strip-Till

In the late 1990s, the Illinois Department of Agriculture and the state's McLean County Soil and Water Conservation District were concerned about the future of no-tilled corn. They unveiled a statewide effort to increase conservation-tilled acres with a program that included research on the benefits of fall strip-tilling to overcome concerns with the cool, wet soils often found with no-till.

The department hired the Lexington, Ill., father and son team of Jim and Brian Kinsella to help plan ways to increase the adoption of high-residue management systems over current levels by 25% by 2005 and some 50% by 2010.

Grants were provided to Soil and Water Conservation Districts to help farmers establish 10 tillage research and demonstration plots in highly visible areas and in a variety of soil types throughout Illinois. The plots compared strip-till, no-till and full-width tillage in corn and soybeans.

Many of strip-till's benefits were outlined in a 2001 article in *No-Till Farmer* where Ohio State University ag engineer Randall Reeder indicated soil temperatures in a strip-tilled seed zone could be 6-10 F warmer on sunny spring afternoons than untilled areas between the strips.

The areas between the strips were left untilled and covered with residue, which conserved soil moisture and reduced erosion. Eliminating full-width tillage increased soil organic matter and soil structure.

More Uniform Emergence

In 2001, Ohio State University ag agent Alan Sundermeier analyzed 3 years of strip-till data collected from research and private farms in northwestern Ohio. He found strip-till provided a slightly quicker, more uniform plant emergence than other reduced tillage systems, although final stand counts and yields showed little or no difference.

At the Ohio State research farm in northwest Ohio, Sundermeier and colleagues compared chisel plowing, strip-till and no-till. They buried recording thermometers to get a precise picture of soil temperature and seed emergence.

They analyzed data for a 24-hour period on a clear day in late April, a key time for no-tilling corn in that part of Ohio. They found the strip-till seed zone was cooler at night, while the no-till and chisel-plow systems with 30%-50% residue were 5 F warmer.

"You lose some energy with strip-till during the dark hours, but when the sun comes out, strip-till shoots right back up and bypasses the other systems," Sundermeier said. "By mid afternoon, we saw an 8-10 F temperature difference."

Kinsella, who has no-tilled and

NO-TILL-AGE®

October 2008

Zone-Till vs. Conventional Tillage
(17-year study)

Tillage System	Harvest Population	Broken Stalks	Corn Yield
Zone-till	29,125	1.3%	182.8 bu.
Conventional tillage	27,438	4.2%	183.5 bu.

For zone-till, a 7-inch width was tilled with three coulters.
For conventional tillage, a V-ripper and S-tine tillage tool was used.
A plant population of 30,000 plants per acre was planted following soybeans.
Zone-tilled corn produced higher yields in 8 of the past 12 years.

—*2007 Practical Farm Research, Beck's Hybrids*

MOLE KNIFE. Illinois manufacturer Herb Stamm developed the mole knife, which allowed strip-tillers to place fertilizer deep in fall or spring built berms.

Frank Lessiter

strip-tilled for many years, considered strip-till to be a good insurance policy in cold and wet weather. As an example, Kinsella's 1996 strip-tilled corn yielded 15-22 bushels per acre more than no-tilled corn following a cold, wet spring.

Around the same time, University of Minnesota soil scientist Gyles Randall conducted a 3-year study comparing strip-till, no-till and chisel plow systems with corn. Strip-till averaged 4-6 bushels per acre more than no-till, and 1 bushel more than chisel plowing.

As it turns out, strip-till quickly caught on, new strip-till units were introduced at farm shows and the acreage increased. Some of the popularity was likely due to overcoming the psychological divide some conventional farmers faced in making the switch to no-till.

In 2007, *No-Till Farmer* surveyed NRCS state agronomists on strip-till corn acreage estimations. From this survey data, it was estimated that 3.6 million corn acres were strip-tilled in 2007. The biggest concentration of strip-till was in the Corn Belt with 1.7 million acres, and in the Northern Plains at 823,500 acres.

Among growers who strip-till, the average number of strip-tilled acres has increased from 877 in 2013 to 1,112 acres in 2017, according to the fourth annual Strip-Till Operational Benchmark Study. Some 99% of the respondents strip-till corn, compared with 56% who strip-till soybeans.

In the 10th annual No-Till Operational Benchmark

What Tillage Systems Do You Use?

Practices Used in 2018 by No-Till Farmer Subscribers

Tillage System	% of Growers
No-tillage	97%
Minimum tillage	25%
Vertical tillage	19%
Strip-till	17%
Moldboard plow	1%

— *2018 No-Till Farmer Operational Benchmark Study data*

STRIP-TILL IS GROWING

Since 2014, the *No-Till Farmer* editors have conducted an annual Strip-Till Operational Benchmark study. This comprehensive annual research project is the industry's only study that analyzes strip-till practices among growers.

This comparison demonstrates how strip-till has changed and grown during a recent 4-year period.

Practice	2014	2017
Acres strip-tilled...	877	1,112
Corn	707	877
Soybean	444	430
When strips are built...		
Fall	42%	46%
Spring	29%	32%
Fall & spring	29%	22%
Row unit setup...		
Shank knife	44%	43%
Mole knife	34%	38%
Coulter	38%	37%
Rows with strip-till rigs...		
6 rows	16%	8%
8 rows	11%	11%
12 rows	43%	43%
16 rows	26%	31%
24 rows	4%	7%
Fertilizer placement...		
Banded below berm	73%	68%
Mixed into berm	27%	35%
Between berms	5%	4%
Seed cover crops in strip-tilled fields	44%	61%
Corn yields...		
Strip-till	184 bu.	199 bu.
No-till	161	171
Soybean yields...		
Strip-till	50 bu.	62 bu.
No-till	49	58

— *2014 and 2018 Strip-Till Operational Benchmark Study data*

"Bio strip-till offers the best of both worlds with cover crops and no-till..."

RADISH-WHEAT COMBO. Seeding radishes and wheat cover crops in separate rows is becoming a popular bio strip-till practice. The radishes improve drainage and help the soil warm up faster.

Study, strip-till corn once again yielded the best among survey participants, reaching an average 203 bushels per acre in 2017. This was 22 bushels higher than the average no-till corn yield of 181 bushels per acre.

Even as no-tilled acres increased across the U.S., strip-till also continued to gain legitimacy. *No-Till Farmer's* parent company, Lessiter Media, convened the first-ever National Strip-Till Conference in Cedar Rapids, Iowa, in 2014, which was an instant success and is repeated annually as the world's largest gathering of strip-till minds.

Strip-till systems continued to mature, buoyed by increasing options available for GPS guidance and more sophisticated strip-till rigs designed to reduce soil loss.

Bio Strip-Till Debuts

An interesting offshoot of strip-till developed during the 2000s, as growers started to seed cover crops in strips. This let plant roots do the work of creating a seedbed for the following crop rather than relying on a shank, mole knife or coulter to build the strips.

Dubbed as bio strip-till, the practice represents an advanced way of using cover crops to improve the soil environment within the row, says Joel Gruver, an agronomist at Western Illinois University in Macomb, Ill.

Seeding cover crops with bio strip-till started with cover-crop pioneer Steve Groff of Holtwood, Pa., who started using the concept with tillage radishes.

Ironically, it was Lisa Stocking — a University of Maryland graduate student who later married Gruver — who came up with the idea. Groff and Stocking blocked off holes in a no-till drill with 7 1/2-inch spacings to seed tillage radishes in 30-inch rows. The following spring, corn was planted into strips of alternated winter-killed radishes and Austrian winter peas.

Soon after, David Brandt was among the first farmers to do bio strip-tilling with planters used to no-till soybeans in 15-inch row soybeans on his Carroll, Ohio, farm.

Other farmers filled half of the planter boxes with radish seeds and the other half with Austrian winter peas or another cover crop species. Some planted twice, splitting the 30-inch row to create a 15-inch row spacing with RTK guidance.

"Precision planting lets you cut seeding rates in half, due to precise spacing of the seed vs. drills," Groff says. "Going over a field twice with a planter that has a 30-inch row spacing can be economical.

"The seed spacing with planters is generally 3-4 inches with most cover crops. We have the disc or plate-part numbers for most planters, so we can help farmers choose the right plate and settings for different size cover crop seed."

In addition to Brandt, another grower who pioneered this idea was Joe Breker of Havana, N.D. He sowed a combination of high-residue cover crops such as turnips, radishes and flax into wheat stubble with a disc drill. He also drilled peas between what would become next year's corn rows.

The cover mixes grew for 75 days before winter-killing, and corn was no-tilled into the residue the following May. He uses flax and sunflowers to support the growth of mycorrhizal fungi to help make phosphorus tied up in the soil more available for plants.

The growing organic content of Breker's soils proved a boon for more effective fertilizer management. Given the long-term no-till practiced on his land and a diverse rotation, Breker believes he's receiving an nitrogen credit of 50 pounds per acre.

Gruver credits the bio strip-tilling of radishes with three distinct benefits:

1 Planting radishes in the same rows that will be planted to a subsequent crop may move old crop residue away from the row, which can help when no-tilling the following spring.

2 Radish growth may accelerate the decomposition of other crop residue.

3 Since little radish residue remains, using a burn-down herbicide is easy if the tops don't winter-kill.

"Seeding radish as a cover crop tends to improve drainage and helps the soil warm up faster," Gruver says. "Radishes winter-kill and decompose very quickly. They leave a zone of nearly bare soil, but it's soil with better aggregation and structural stability and it's not likely to crust in the spring."

Works with Manure

Decatur, Ind., strip-tiller Gene Witte has used bio-strip-tilling after wheat harvest since 2009 to grow oilseed radishes, annual ryegrass and clover. He finds spreading manure helps create a better strip-till seedbed and reduces soil compaction from liquid-manure tankers and dry-manure spreaders.

The radishes winter-killed in mid-December. The following spring, the annual ryegrass took off aided by the hog manure and got out of hand because of wet weather plaguing the eastern Corn Belt. So Witte mowed and baled the annual ryegrass and sold the hay.

"Having the cover crop get away from you in wet weather is one of the risks you take," Witte says. "We got the annual ryegrass baling done on June 14 and

IMPROVED SEEDBED. Building strip-till berms in the fall in an annual ryegrass cover crop leads to a much better seedbed in the spring, says Matt Van Tilburg, a Celina, Ohio, bio strip-tilller. The opening photo in this chapter shows how he seeds covers.

no-tilled Roundup Ready corn on June 16. The new growth of annual ryegrass had just started as the corn emerged."

After harvesting wheat, Witte has frost-seeded a mix of medium red clover and alsike clover before strip-tilling the field.

"I don't kill the clover until spring, when I apply a herbicide before no-tilling corn," Witte said at the time. "With RTK, I can find the fall strips as I'm planting corn. By seeding the clover, I'm increasing the nitrogen in the soil, which the wheat crop had depleted."

Tried Several Options

In 2011, Matt, Luke and Kyle Van Tilburg solid seeded annual ryegrass in the fall before strip-tilling. In 2009 and 2010, the Celina, Ohio, brothers had seeded annual ryegrass into soybeans using a high-boy seeder with drop tubes. They returned with a 60-foot-wide Wil-Rich toolbar holding 24 Soil Warrior strip-till units on 30-inch row spacings.

The Soil Warrior coulters killed the annual ryegrass growing in the strips. For corn following soybeans, the Van Tilburgs didn't apply phosphate or potash because enough of these nutrients are available in the poultry manure applied to the fields.

In the spring, the brothers make another shallow pass with the strip-till rig, then no-tilled corn. Several years earlier before they had started strip-tilling, the roots were so thick and tough that the "V" furrow wouldn't close. Thanks to strip-till, they capture the benefits of annual ryegrass and no-till into that narrow-tilled strip in the spring and control wheel traffic.

The brothers say this is the best of both worlds with cover crops and no-till. They want something growing on their soil year-round.

While it's hard to quantify how much cover crops improve bio-strip-till yields, their best corn yield from the dry 2010 growing season occurred in fields where annual ryegrass had been strip-tilled. ✺

STRIP-TILL ALLOWED A STEP AWAY FROM TILLAGE

The Jepson Family Farm in Orlinda, Tenn., was almost completely 100% no-till except for one crop — tobacco.

While only taking up 100 acres, Willis Jepson and his father, William, felt they were taking a big step back by tilling their tobacco ground. They had tried no-tilling, but saw too big of a yield drag.

So, in 2014, the family decided to try strip-tilling. Using a 4-row "ripper stripper" coulter tool by Kelley Mfg. Co., they created a 2-foot-deep row that creates a navigable path for water and nutrients to easily reach plant roots. Rollers on the back of each row unit break down clods and smooth the strips. They run a Multivator strip-cultivator to finish off a 16-inch strip where the tobacco plants are set.

The Jepsons have seen positive results from making the switch to strip-till. Tobacco yields are typically between 2,500 and 3,500 pounds per acre, and their yield for 2015 clocked in at 3,000 pounds.

"I don't know why we didn't do it sooner," Willis says. "With the wet year we had in 2015, our soil erosion was very little compared to what it would've been if we had been planting conventionally." ✳

BANDING OR BROADCASTING DEBATE. Many strip-tillers maintain banding phosphorus and potash deep in the berm is the most efficient way to fertilize high-yielding corn and soybeans. Yet other strip-tillers have found broadcasting these nutrients works just as well. University of Illinois agronomists say banding these nutrients reduces the fixation of applied phosphate and potash into soil minerals and clay layers.

WHEN RIDGE-TILL LOST OUT TO STRIP-TILLING

December 2013

For years, ridge-tillers had complained about how no-tillers continually bad-mouthed this system. The biggest complaint was that ridge-tillers relied on cultivation for weed control while no-tillers preferred to apply herbicides. No-tillers saw ridge-till requiring more labor — one or two cultivations when hay should be harvested — and not being an effective choice for acreage expansion.

After strip-till got going in the late 1980s and early 1990s, ridge-tillers began to wonder why no-tillers were so quick to embrace many of the ideas they hadn't liked about ridge-tilling. Here are a few examples.

Controlled traffic. Ridge-tillers ran tractor, planter and combine tires in the same paths in the field year after year to reduce compaction. Tractors and combines were often equipped with extended axles to line up similar tire paths. Mechanical guidance devices kept the planter row units running in the center of the ridge.

Then GPS and auto-steer came along and made controlling traffic and precise fertilizer and plant placement in strip-tilled berms much easier.

Banding fertilizer. Ridge-tillers banded nutrients in the ridge or close to it for maximum usage by corn and soybean plants.

Strip-tillers soon started banding phosphorus, potassium and micronutrients in the berms. This led to more effective use of fall-applied nutrients.

Building ridges (or berms). Strip-tillers rely on berms to help heavy soils dry out and warm up faster for spring planting. Whether built in the fall or spring, the berms allow growers to band and place nutrients where they can earn the biggest return. On the other hand, ridge-tillers relied on one or two late spring cultivations to build ridges for planting 10 or 11 months later.

Deep banding fertilizer. Just like ridge-tillers, strip-tillers deep band nutrients to reduce costly nutrient runoff and make nutrients more readily available to growing plants.

Continuous corn. Ridge-tillers for years had seen the benefits of moving heavy corn residue away from next year's row area. When building berms, strip-tillers do the same with residue.

There's no doubt that ridge-till served as the foundation for strip-till. Yet veteran ridge-tillers still remember when no-tillers had little respect for many of the ideas considered essential in today's strip-till operations. ❋

RIDGE TILLING QUICKLY LOST GROUND. In the 1970s and 1980s, many Corn Belt growers overcame yield concerns with cold and poorly drained soils by moving to a ridge-till system. However, the system was very labor intensive with two cultivations usually required to build ridges for next year's crops.

NRCS

"Vertical tillage is a way to wean farmers off plowing..."

Great Plains Mfg.

VERTICAL TILLAGE. Vertical finishers used with no-till are set to only work as deep as the seedbed to size residue, loosen problem soils and create a warm, firm seedbed for accurate seed placement. However, many die-hard no-tillers insist that vertical tillage does not have a place in a no-till system.

VERTICAL TILLAGE INTEREST GROWS, STIRS CONTROVERSY

Another tool that emerged during the first decade of the new century was vertical tillage — or what some industry observers call "vertical finish." The concept features a specialized implement with disc gangs set at a non-aggressive angle that runs shallow and fast over no-tilled fields to deal with heavy residue concerns.

As corn hybrids continued to improve and yields kept increasing — especially with the advent of tough *Bt* hybrids — many no-tillers expressed frustration that residue wasn't decomposing fast enough in continuous corn fields, causing problems with seed placement and hairpinning. Even heavy wheat residue was posing problems when no-tilling.

Vertical tillage tools cut and size residue, improve the seedbed, bring residue in contact with the soil for faster breakdown, incorporate fertilizer or cover crop seed, dry out wet soils or even kill weeds.

Many implements fall under the vertical-tillage umbrella. But the most common were "vertical finishers," featuring a gang of straight, wavy or fluted coulters followed by a harrow and rolling basket that worked the top 1-3 inches of soil.

While some units have disc gangs that run nearly straight up and down, other tools feature gang angles that are fairly aggressive and move more soil. These often draw the ire from no-till traditionalists and NRCS staffers who consider this full-width tillage.

Manufacturers like AerWay introduced equipment that was slightly different, using rotating knives to slice and fracture the soil without creating as much soil disturbance.

Not for Everyone

In 2010, 12% of growers responding to the *No-Till Farmer's* annual No-Till Operational Benchmark Practices Survey were using vertical tillage. This had increased to 19% by 2018.

In 2010, Ohio State University ag engineer Randall Reeder indicated that while vertical-tillage tools could lead some farmers away from pure no-till, they could have the opposite effect, too.

"These are great tools for transitioning to conservation tillage — sort of a way to wean farmers off plowing," Reeder says. "I would hope that in the long run, these producers would possibly transition to no-till."

NO-TILL-AGE®

Who's Still Cultivating?

(% acres cultivated with various tillage systems from 1990-1993)

Tillage System	Number of Cultivations			
	0	1	2	3 plus
Soybeans...				
Conventional with Plow	14%	50%	32%	4%
Conventional without plow	21%	56%	21%	2%
Mulch till	19%	57%	23%	1%
No-till	78%	18%	4%	0%
Ridge till	4%	55%	41%	0%
Corn...				
Conventional with Plow	24%	53%	19%	4%
Conventional without plow	23%	61%	15%	1%
Mulch till	22%	56%	20%	2%
No-till	68%	26%	5%	1%
Ridge till	0%	22%	77%	1%

— USDA's Agriculture Resources

A big advantage with vertical-tillage tools is that they don't create compaction layers like is the case with many more intensive tillage tools.

"With a moldboard plow, you get a smear layer 6-8 inches deep across the entire field," Reeder says. "With vertical-tillage, even if there's some smearing at the bottom of the coulter, it would only be 1/4-inch wide every 10 inches or so, which shouldn't create any restrictions for root growth."

Transition Tool

At one time, Chris Hudson and his father, Curt, used vertical tillage on every acre in making a transition from twin-row corn to 20-inch-row corn. They now use the tool more on a prescriptive basis on their Crawfordsville, Ind., farm, where they no-till 2,800 acres of corn and soybeans.

"We use our Great Plains 4000 Turbo-Till unit on 5%-15% of our acres each year," Chris says. "It's a good tool to work fields for maybe 3 years where we're changing the cropping direction or switching from 30-inch to 20-inch rows to get the field reoriented to the new direction or in areas prone to ponding."

The Hudsons ran the tool up to 2 inches deep to help manage residue on corn-on-corn ground. While it's a good tool, it does violate their definition of no-till. If they used it across more acres, they feel it's definitely recreational tillage. And as these tools chop and size residue, it can create bad feelings with non-farm neighbors when residue starts blowing around.

Rethinking Practices

When corn prices were $6 a bushel or more, better economic returns from continuous corn over a 50-50 corn-soybean rotation prompted Phillip Brown and his sons, Josh and Jared, to rethink their cropping practices.

"We tried to plant 5 inches to the side of the previous row, but the depth wheels were hitting last season's root balls, and we weren't getting the seed placement we wanted," says Phillip.

The Browns ran a 30-foot Kuhn Krause Excelerator vertical-tillage rig pulled with a 320-horsepower tractor at up to 7 mph. They ran it 3 inches deep in the fall over fields going back to corn the next spring.

"Purists don't consider vertical tillage to be a no-till practice, but it's reality," he says. "I consider it no-till with an asterisk, as it's the only way to handle heavy corn residue to prepare a uniform seedbed for the next crop."

Tackling Residue Issues

Planting corn hybrids with stronger stalks at higher populations put Mike Reichart in a bind when no-tilling into corn residue. The stalks wrapped up on the planter's row cleaners, chains and sprockets and led to uneven planting depth and emergence, says Reichart, who strip-tills 690 acres of corn and no-tills soybeans near Tallula, Ill.

He found making a vertical-tillage pass in the fall to chop the stalks and incorporate residue into the soil led to a better seedbed. He also uses the tool to incorporate cereal rye after application with a PTO-driven spinner box in erosion-prone areas.

Although he's a proponent of no-tilling, Reichart thinks vertical tillage is a good compromise for managing residue.

"This possibly violates the basic theory of no-till and strip-till, but I'm only cutting a wavy slit in the ground and not totally disturbing the soil like a disc does," he says. "If you could use a chopping corn head, you probably wouldn't need this tool."

Incorporating Covers

Donn and Chad Branton, who no-till and strip-till 1,300 acres near LeRoy, N.Y., started their vertical-tillage journey in 2008 with a Case IH 330 Turbo Disk. They pulled the 25-foot rig at 6-7 mph with a 175- or 250-horsepower tractor. Since speed is important, Donn says you'll need more horsepower in hilly terrain.

Based on soil moisture and the amount of residue, the fluted coulters are set at a slight angle to only move the top 2-3 inches of soil as it's mixed with residue. They've used the tool mostly in the fall, but also in the spring.

"We don't feel we're violating the basic principles of no-till or strip-till," Donn says. "We're not doing deep tillage and we're leaving a mat of residue on the surface. We think these implements need to be reclassified by government agencies as a conservation-type tool."

Sounding Off

Not every no-tiller is convinced of these vertical-till advantages, and *No-Till Farmer* shared their concerns in an "equal time" article.

"Growers that can spend $40,000 on one of these machines should spend it on a bigger no-till drill and seed a cover crop," said Lewisburg, Pa., no-tiller Lucas Criswell.

"The notion that anyone who uses vertical tillage can still be considered a no-tiller shows why the current concept is unsustainable," says Gabe Brown of Bismarck, N.D. "I've been on thousands of farms in the U.S. and around the world and have never seen an operation that uses vertical tillage or strip-till that has healthy, regenerating soils."

Critics of vertical tillage seemed to win some points in 2015 after *No-Till Farmer* was contacted by farmers armed with photos of washed-out stalks covering roads and filling up ditches in Ohio. This raised the question of how vertical tillage could be used without ruining no-till objectives.

No-tillers enrolled in federal conservation programs should make sure using such a machine won't violate the government rules on the use of full-width tillage.

Fall vs. Spring Work

The time of year when growers make a vertical tillage pass impacts the potential for residue movement. Purdue University agronomist Tony Vyn sees less opportunity for residue to wash off fields with spring vertical tillage since residue decaying takes place. Corn after

WORK THE ZONE. Numerous strip-tillers find the ability to place nutrients both shallow and deep within the berm can maximize strip-till corn yields. They've learned shallow-placed fertilizer is a big help in providing essential nutrients to corn plants that get off to a slow start during the spring.

STRIP-TILL TWINNING WORKS. Planting corn in twin rows on 20-inch centers is helping some growers capture as much sunlight and nutrients as possible while giving each corn plant needed room to grow. Each of the twin rows is normally planted 1-inch away from each edge of the 10-inch wide strips.

corn yields were also better with a single vertical tillage pass in the spring.

How a grower runs a vertical-tillage tool in regard to the old crop rows can also impact residue movement. Running across the old rows, or at an angle, brings more soil in contact with the residue. This cracks and crushes the corn stalks so moisture and bacteria can speed up the digestion process. Unfortunately, running at an angle can loosen up small pieces of residue that can blow and float into ditches or off fields.

Tim Harrigan says farmers in Michigan have had good luck broadcasting cover crop seed ahead of vertical tillage as it improves seed-to-soil contact. While it's not as effective as drilling, it's better than broadcasting cover crop seed alone, says the Michigan State University ag engineer.

Universities Weigh In

Several universities began studying the effects of this relatively new tillage practice that seemed to lend a perspective to the emotional debate. Their work didn't paint a dire prediction from using vertical tillage tools, nor did it say the machines were miracle workers.

While initial research shows vertical tillage can be used with minimal impact on soil conservation measures, University of Wisconsin agronomist Kevin Klingberg says it's important to use the tools effectively. If they are used more aggressively or with multiple passes, growers may lose many of the soil and water conservation benefits.

Kansas State University soil specialist DeAnn Presley spearheaded research to assess the effects of vertical-tillage practices on residue coverage and mass, disease incidence and severity, soil-and residue-borne pathogens, soil moisture, bulk density and yields.

In 2011, a Kansas study of vertical tillage with

continuous corn showed no-till left 96% residue cover, compared to 90% with a vertical-tillage pass and 68% with a modified "spader" vertical-tillage implement. In nine of 10 trial locations, Presley found little or no yield difference with vertical tillage.

Her vertical tillage research also found:

◆ Soil moisture was reduced by vertical tillage in the upper 2 inches of the soil profile when compared to no-tilled fields.

◆ Compaction could not be detected with any of the vertical-tillage passes, although running the tools in wet conditions affected the soil more than if fields are dry.

◆ No difference in soil structure or bulk density was found at most sites.

In Michigan, Harrigan conducted research on vertical tillage with 10 onfarm soybean trials from 2012-2014. He measured bean emergence, final plant population and yield with five vertical tillage tools.

Regardless of whether done in the fall or spring, he found vertical tillage of corn residue generally led to more rapid soybean emergence, but didn't increase final plant population or yield. Because corn stalks were tough in the fall, slicing and sizing didn't reduce the amount of residue left at planting time. He found little benefit from making two passes compared to making only one pass in either the fall or spring.

Since farmers can't control the weather, Harrington says vertical tillage can offer a slight edge in avoiding planting delays. Some growers report being able to plant up to 1 mph faster following vertical tillage, which ties into the efficiency issue and risk management factors related to short planting windows with large acreages.

Summing up, Harrigan says the key with using any vertical tillage tool is to know exactly what you want to accomplish. ✺

Defining and Developing
a Strip-Till Niche

Here are a few takeaways from the National Strip-Tillage Conference, sired in 2014 by *No-Till Farmer* and the National No-Tillage Conference to meet a knowledge void.

Strip-till is an imperfect farming practice and some variables can't be conquered. But strip-tillers can curb the impact Mother Nature has on their crops with a flexible, progressive system, which may include a customized cover-cropping strategy, well-timed fall or spring strip building and an adaptive nutrient management program.

While there are many variations, strip-tillers plant in a narrow, raised

blackened strip of soil while leaving the rest of the field undisturbed. For growers not happy with no-till results in problem soils, it's a way to dramatically trim tillage without having to move back to more extensive tillage.

Seeing the need to create a unique learning and networking environment, the National Strip-Tillage Conference was launched in 2014. Building on the established model created more than 20 years earlier at the National

No-Tillage Conference, the strip-till event filled a face-to-face void in the industry to advance and expand adoption of this unique farming practice.

Attracting attendees from 30 states and 8 foreign countries, the conference continues to evolve as an annual hub for strip-tillers, researchers and supporting companies to compare cropping strategies. Here you'll find a summary of a few of the most memorable moments from past strip-till conferences. ✿

Co-sponsors, along with *No-Till Farmer* and *Strip-Till Farmer* ... Dawn Equipment | Environmental Tillage Systems | Kuhn Krause | Ingersoll Tillage Group | Montag | Orthman Mfg. | Raven Industries Thurston Mfg./Blu-Jet | Yetter Mfg. Co.

2014

'Superb Strip-Tillage Learning'

July 30-31, Cedar Rapids, Iowa • 446 attendees

NSTC Insights...

Strip-Till 101

Set up your strip-till system for success by paying close attention to equipment setup, says University of Minnesota Extension educator Jodi DeJong-Hughes.

She says soil type, rotation and moisture will determine how aggressive your strip-till rig should be. Also know whether you want to chop or spread your corn residue. Since crop residue is the lifeline of your soil. Some strip-tillers say 75% of residue should remain on the soil surface.

Dialing Down Applications

Keith Schlapkohl, a strip-tiller in Stockton, Iowa, has made incremental improvements to his system that are inching his corn yields toward the 300-bushel mark. He explains that it's not just one piece of equipment, strategy or input in his arsenal that is edging up yields, but rather a mix of methods.

Schlapkohl has been on a quest to reduce N application for 20 years. When a stalk nitrate test showed Schlapkohl was applying too much N, he decided to cut in half his 21-1-0 foliar application from 2 gallons to only 1 gallon per acre.

Co-sponsors, along with *No-Till Farmer* and *Strip-Till Farmer* ... Copperhead Ag Products | Dawn Equipment | Environmental Tillage Systems | Ingersoll Tillage Group | Kuhn Krause | Montag | Nufarm | Orthman Mfg. | Raven Industries | The Andersons | Thurston Mfg./Blu-Jet | Yetter Mfg. Co.

2015

'Charting A New Strip-Till Course'

Aug 5-6, Iowa City, Iowa • 408 attendees

NSTC Insights...

Digging Into Strips

In several presentations, Kevin Kimberley stressed the importance of letting the soil tell you what it needs. For this, he suggested that a strip-tiller's best friend is his spade.

When digging up a freshly built strip, the consultant from Maxwell, Iowa, is on the lookout for sidewall compaction fractures and voids left along the seedbed. His preferred spading depth is about 30 inches.

Kimberley stresses the importance of knowing where the hardpan is. A simple test is running a knife into a wall of a dug hole at its lowest point and pulling it slowly toward the surface. When resistance is felt on the knife, you can gauge how thick, deep and hard the compaction layers are.

Kimberley suggests using the same knife each time, as a different blade may cut easier and skew results. A tapered blade, like what is found with a putty knife, works best.

Sweeten up the Strips

Bill Darrington, who's been strip-tilling and no-tilling in western Iowa since 1987, focuses on a "ground up" perspective. While he sees many salient points about soil health, the Persia, Iowa, grower says that if attendees could walk away with only one idea from this conference, it'd be to add sugar to their fertility mix.

Darrington says that every time he goes across the field, no matter the reason, he applies 1 pound of sugar per acre. He says that for 75 cents an acre, it's a bargain for the increased nutrient uptake and energy source it provides for microorganisms.

To further emphasize the point, he shared an anecdote about a time when he ran out of sugar. In the pinch, he bought out the local grocery store's supply of RC Cola and added that to his mix.

"It worked great," he says. "Think about what's in soda — carbon dioxide, sugars, phosphoric acid and caffeine."

Co-sponsors, along with *No-Till Farmer* and *Strip-Till Farmer* ... Calmer Corn Heads | Copperhead Ag Products | Dawn Equipment | Environmental Tillage Systems | Ingersoll Tillage Group | Kuhn Krause | Montag | Orthman Mfg. | The Andersons | Thurston Mfg./Blu-Jet | Vulcan | Yetter Mfg. Co.

2016

"I prefer to build fall strips, plant a mix of oats, barley, cereal rye and peas **and rest the ground until planting**."

— *Dustin Mulock, Woodville, Ontario*

'Comparing Strip-Till Strategies'

Aug 3-4, Springfield, Ill. • 361 attendees

NSTC Insights...

Covering Up

One of the hottest topics at the conference was the practical use of cover crops in a strip-till system.

"Contemplate when you want to do your strip-till pass," says Woodville, Ont., strip-tiller Dustin Mulock. "I see a lot of farmers making a pass after the cover crop is growing late in the fall. It's my belief that at that point, we've already used the energy from the cover crop to push the roots down, to create the warm channels and the root channels."

'Bio' Reacting to Water Quality

With more emphasis on reducing fertilizer and water runoff, Eagle Grove, Iowa, strip-tiller Tim Smith shared how a combination of strip-till and enrollment in the Mississippi River Basin Initiative have reduced nutrient loads that filter into streams leading to the Mississippi and the Gulf of Mexico.

His tile water flows through a wood-chip-filled bioreactor that serves as a carbon source for bacteria that denitrify the water. Before installing the bioreactor, tile water nitrate levels were 13-14 parts per million. Nitrate levels are now only 5-6 parts per million.

Co-sponsors, along with *No-Till Farmer* and *Strip-Till Farmer* ... Agro Liquid | Copperhead Ag Products | Dawn Equipment | Environmental Tillage Systems | Ingersoll Tillage Group | Kuhn Krause | Montag | Schlagel Mfg. | The Andersons | Thurston Mfg./Blu-Jet | Vulcan | Yetter Mfg. Co..

2017

"If you're using a shank or mole knife, and the manufacturer says you need 18 horsepower per row, add an extra 10%..."

— *Joey Hanson, Elk Point, S.D.*

'Finding New Strip-Till Opportunities'

Aug 3-4, Omaha • 423 attendees

NSTC Insights...

Row Unit Roulette

Knife, shank or coulter? They definitely all have their fit and Elk Point, S.D., strip-tiller Joey Hanson has run them all while custom strip-tilling more than 20,000 acres on 25 different soil types.

"I run a mole knife in the fall and a mini mole knife in the spring," he says. "If you're going with a shank or mole knife setup, and the manufacturer says you need 18 horsepower per row, add at least 10% more to that total. The way I have my machine set up, I like to run at about 6½ mph. The results change drastically when I slowed down or sped up."

Hanson also ran a series of coulters and wanted a little bit more aggressive action in front. Those worked really well in a dry fall. However, really wet conditions a few years ago made it difficult to get strips built on 160 acres in heavy clay soils, and his rig paid the price over a 3-day period.

"The strips ended up turning out fine, but what I learned is that when I'd leave one location for another 3 miles down the road, the soil conditions changed, but I didn't account for that with my rig," he says.

Reaping the Benefits

Attendees represented a diverse range of climates, soil types and crop rotations, and success stories are always abundant throughout the conferences. Aaron Wickstrom, a converter to strip-till after running a conventional tillage operation for years in Hilmar, Calif., saw substantial strip-till economic advantages.

Believing 100% no-till was an unrealistic fit for his finer-texture sandy soils, Wickstrom experimented with a mostly strip-till and part no-till combination. He was impressed with both yield and economic benefits.

"We're running about $164 per acre for a full-year, three-crop cycle, which saves us $153 per acre," he told attendees. "That was nothing to sneeze at as yields also went up."

Co-sponsors, along with No-Till Farmer and Strip-Till Farmer ... Agro Liquid | Copperhead Ag Products | Dawn Equipment | Environmental Tillage Systems | Ingersoll Tillage Group | Kuhn Krause | Montag | Schlagel Mfg. | Thurston Mfg. (Blu-Jet) | Topcon Agriculture | Vulcan | Yetter Mfg. Co.

2018

"**I only have trouble with slugs** when interseeded cereal rye residue touches the corn row..."

— *Laurent Van Arkel, Dresden, Ontario*

'Let's Get Strip-Tilling'

July 26-27, Iowa City, Iowa • 364 attendees

NSTC Insights...

Pricing Cover Crop Seed

When selecting specific varieties for a cover crop tankmix, Jack Boyer recommends relying on seeds-per-acre rather than pounds-per-acre. The Reinbeck, Iowa, strip-tiller seeds cover crops on 100% of his corn and soybean acres.

Boyer tool the time to count the seeds in a sample of two separate purchases of cereal rye seed —a bushel with no stated variety and a bushel of Elbon seed. He found 35 pounds of Elbon seed contained the same number of cereal rye seeds as 56 pounds in the no variety stated seed sample.

"Going to seeds per acre has a tremendous benefit to our cost of production and has given us as good or better crop coverage," he says.

Make Variable Rate Pay

Josh McGrath reminds strip-tillers that there's plenty of uncertainty when it comes to interpreting soil test results for variable-rate fertilization. The University of Kentucky soil scientist adds that maximum yield does not equal maximum profit and that over fertilization rates may often have an impact on the environment.

While he recommends yearly soil testing, he suggests shifting soil sampling grids by a quarter-acre every year, since taking samples at quarter-acre intervals for more accurate variable-rate computer analysis is both cost prohibitive and labor intensive.

OUR COMMITMENT TO STRIP-TILL

LOTS OF ONE-ON-ONE NETWORKING TIME.

STRIP-TILL ROUNDTABLES OFFER SOLUTIONS TO KEY CONCERNS.

Jeff Lazewski

I don't recall when we wrote our first *No-Till Farmer* article on strip-tilling, but it was soon after Cliff Roberts first used this no-till spin-off on his Kentland, Ind., farm in 1987. We've followed with dozens of strip-till articles in *No-Till Farmer* over the past 3 decades and also had Roberts share his strip-tilling experiences at the first National No-Tillage Conference in 1993.

In 2007, we'd estimated there were 3.6 million acres of corn being strip-tilled in the U.S. As strip-till gained momentum, we saw a need for carving out more extensive learning opportunities for strip-tillers in addition to the coverage in *No-Till Farmer* and at the National No-Tillage Conference.

Here's how our ag publishing company has served the growing strip-till market:

2010: Launched the *Strip-Till Strategies* electronic newsletter that has evolved over the years into a daily electronic strip-till update, a podcast series, special reports on specific strip-till topics and webinars.

2014: Launched the mid-summer National Strip-Tillage Conference that offers a unique farmer-to-farmer learning experience.

2017: *Launched Strip-Till Farmer* as a print magazine.

I'm sure there's much more to come as we continue to develop new educational opportunities for growers as the strip-till acres continue to expand. ✿

The National No-Tillage Conference —
No-Till Learning at its Best

Here are a few year-by-year memories from this annual midwinter event that got its start in 1993.

If you were to footnote all the reduced tillage learning opportunities shared over the years at the National No-Tillage Conference, the numbers would be astronomical.

Produced each year by *No-Till Farmer* and co-sponsored by leading industry suppliers, these annual gatherings have offered unprecedented learning opportunities over the years for growers hailing from 48 states and 24 foreign countries.

As you can see from the accompanying chart, our National No-Tillage Conference numbers are impressive in bringing farmers together on all aspects of no-tilling.

This mid-winter gathering of the no-till clan has always served as a great learning experience for no-tillers. The educational mix of General Sessions, No-Till Classrooms, No-Till Roundtables and hallway networking still works as well today as it did in 1993.

It's also been gratifying to have attendees tell us they've increased their annual income by as much as $95,000 after attending just one of the conferences.

On the following pages you'll find a brief summary of all the National No-Tillage Conferences held to date. Each year's conferences is chronicled with a few facts, figures and a couple of no-tilling tips from that year's event

Here's hoping that you enjoy taking this look back at the National No-Tillage Conference. ❀

Jeff Lazewski

THE NATIONAL NO-TILLAGE CONFERENCE
By the Numbers
1993-2018

26 Mid-Winter Events	**851** Speakers	**456** No-Till Classrooms	**48** States Represented	**7** Midwestern Locations
21,597 Total Attendance	**169** No-Till General sessions	**997** No-Till Roundtables	**24** Foreign Countries Represented	**1,201** Hours of No-Till Learning

Co-sponsors, along with *No-Till Farmer*... American Cyanamid | DowElanco | DuPont Ag Products | ICI Agricultural Products (Syngenta) | Monsanto

1993

"Folks in the eastern Corn Belt no-till to get rid of excessive moisture **while out west we no-till to save every drop...**"

— *Dwayne Beck, Dakota Lakes Research Farm, Pierre, S.D.*

'No-Tilling For The Future'
Jan. 12-14, Indianapolis • 814 attendees

NNTC Insights...

An Introduction to Strip-Tilling

Building strips in the fall leaves a mound of soil in the row area that remains 5-7 degrees warmer than the surrounding soil, says Cliff Roberts of Kentland, Ind. Fall stripping allows ideal fertilization placement and eliminates compaction in the row area. Having already strip-tilled for a half dozen years, Roberts indicated he had done it all wrong the first 3.

Grid Testing Pays When No-Tilling

It's a time-consuming practice, but Tom O'Dell says grid soil testing is relatively cheap if you have access to a computer. The veteran no-tiller from Losantville, Ind., urges no-tillers to soil test more frequently and pay attention to the acidity levels in the top 3 inches. He also told attendees the capacity for cation exchange in no-tilled soils is as critical to success as the actual presence of nitrogen or phosphorous.

TALK IT OVER. Some 50 No-Till Roundtables at the first event gave 814 attendees a chance to ask very specific no-till questions of other farmers from numerous states and several foreign countries.

381

Co-sponsors, along with *No-Till Farmer*... American Cyanamid | DowElanco | DuPont Ag Products | Monsanto | Zeneca Ag Products (Syngenta)

1994

'Ride The No-Till Success Express'
Jan. 11-13, St. Louis • 865 attendees

"My head is still swimming, as **it was impossible to absorb everything presented over the last 3 days**..."

— *Richard Rice, Philo, Ill.*

NNTC Insights...

Huge Fuel Savings with No-Till

In an 11-year experiment at the University of Tennessee's Milan Experiment Station, John Bradley found the fuel costs of no-tilled corn amounted to $4.90 per acre less than conventional tillage. The long-time no-till researcher found corn yields were also higher with no-till.

For no-till corn success, he recommends equipping a no-till planter with plenty of weight, a sturdy toolbar, 3/4-to-1-inch ripple coulters, heavy down-pressure springs and cast-iron press wheels.

No-Till Adds Up to Extra Profit

Ray Brownfield has data to show landowners that no-till adds up to additional profits when compared with other tillage systems. The head of Capital Agricultural Property Services in Oakbrook, Ill., says operating costs for an Illinois farm averaged only $33 per acre for no-till vs. $55 per acre on comparable farms doing conventional tillage.

Machinery costs were $142 per acre for no-till while equipment costs for conventional tillage averaged $208 per acre, he says.

Co-sponsors, along with *No-Till Farmer*... American Cyanamid | DowElanco | DuPont Ag Products | Monsanto | Zeneca Ag Products (Syngenta)

1995

'Racing For Bigger No-Till Profits'
Jan. 19-21, Indianapolis • 822 attendees

"The long-term threat to human existence is **soil erosion**..."

— *Dennis Avery, Center for Global Food Issues, Churchville, Va.*

NNTC Insights...

Avoid Hairpinning with No-Till

Todd Intermill equipped a 6-row John Deere no-till planter with six Case IH offset disc openers. The Colman, S.D., no-tiller took one disc off each row, alternating them left and right with the new discs so the leading discs would not pull the planter to the side.

He removed all but one of the shims for the row units so the new opener would serve as the leading disc and sharpened the older discs. Intermill says the result is no hairpinning and he saved half the cost of replacing the disc openers.

Roundup Problems Often Due to Hard Water

Roundup would be an even more effective herbicide if it offered better plant penetration, says Lloyd Wax, a USDA weed scientist at the University of Illinois. However, it's difficult to get enough glyphosate into the leaf where it can work.

Wax says the calcium, magnesium and sodium found in the water causes most of Roundup's ineffectiveness. He says the results are much better when ammonium sulfate is added to Roundup, as it removes the metal solvents, calcium, iron or whatever else is tying up the Roundup0

LOYAL ATTENDEE. Bryan Von Holten of Cole Camp, Mo., shares ideas with a conference speaker. He's one of six no-tillers who have attended all of these events.

Co-sponsors, along with *No-Till Farmer...* American Cyanamid | DowElanco | DuPont Ag Products | Monsanto | Zeneca Ag Products (Syngenta)

1996

'Meeting No-Till's New Challenges'
Jan. 11-13, St. Louis • 874 attendees

"If you're in 30-inch rows and I'm in 20-inch rows on my 2,400 acres, **I'll earn $50,000 more income than you will**..."

— *Dick Gremel, Sebewaing, Mich.*

NNTC Insights...

Balance Diversity, Intensity in No-Till Rotations

Include at least three different types of crops in your no-till rotations to control weeds and reduce disease problems, says Dwayne Beck. The director of the Dakota Lakes Research Farm in Pierre, S.D., says crop intensity and diversity need to be balanced for no-till profitability.

Beck says the best way to handle both wet and dry years is by diversifying and expanding your no-till rotations.

No-Till Keeps Brazilians Farming

Thanks to no-till, Brazilian growers like Franke Dijkstra often grow three crops a year on the same ground. The president of the Cooperativa Agro-Pecuaria Batavo, Ltd., (700 members that are 100% no-till on 330,000 acres) says there's no way other than no-till to produce crops on much of the highly erodible land found in Brazil's tropical areas.

Without keeping these soils covered and avoiding tillage, Dijkstra says Brazilian growers will lose both their soils and farming operations.

Co-sponsors, along with *No-Till Farmer...* American Cyanamid | DuPont Ag Products | IMC-Agrico Co. | Monsanto | Zeneca Ag Products (Syngenta)

1997

'More Profitable No-Till Strategies'
Jan. 16-18, Des Moines • 680 attendees

"You want to match **crop genetics to environmental needs** to maximize your no-till yields..."

— *Bill Preller, Advanced Technical Information Services, Goodfield, Ill.*

NNTC Insights...

Deep-Banding Starch Pays with No-Till

John Walker told attendees that deep-banding starch has the potential to increase yields with almost any no-till crop. The agronomist at Ricks College in Rexburg, Idaho, told attendees to use granulated starch rather than powdered starch, mix it with starter fertilizer and deep band the mixture with the planter at a rate of at least 20 pounds of granulated starch per acre.

No-Tilling the Worst Ground on the Planet

Once Carlos Crovetto started no-tilling in 1978, the quality of his farming life greatly improved. Over nearly two decades of no-tilling, the Conception, Chile, grower has used less fertilizer while increasing soil quality and yields.

He maintains no-till must be permanent, which means always leaving 100% of the residue on the soil surface. Yields with continuous no-till corn have been as high as 316 bushels per acre along with 170-bushel no-till wheat yields.

TARGETED KNOWLEDGE. No-Till Classrooms let attendees zero in on specific topics. Howard Martin (L) of Martin Industries in Elkton, Ky., and John Bradley of the Milan Experiment Station in Milan, Tenn., stress the critical role of row cleaners in no-till operations.

Co-sponsors, along with *No-Till Farmer*... American Cyanamid | Agrotain | DuPont Ag Products | Monsanto | Zeneca Ag Products (Syngenta)

1998

'No-Till ... The Future Is Now'
Jan. 22-24, Indianapolis • 660 attendees

"With precision farming, it's up to no-tillers to manage the variability and fill in the missing pieces..."

— Doug Harford, Mazon, Ill.

NNTC Insights...

Less Compaction with Higher Organic Matter Soils

With long-term no-till, Lloyd Murdock says growers can increase organic matter levels to a point where soils are very difficult to compact. The University of Kentucky soil scientist urges no-tillers not to resort to rotational tillage after a few years when problems arise. He predicts a number of benefits for no-till that won't happen for 12-15 years down the road.

If there's a barometer to know how well you're actually doing with no-till, Murdock says it's being able to increase your soil organ-ic matter. Over 20 years of Kentucky studies have shown organic matter doubled in the top 2 inches of no-tilled soils.

No-Till Works with Cold, Wet Soils

For 7 years, Ray Rauenhorst successfully no-tilled corn under less than ideal conditions in his Easton, Minn., operation. He's careful to keep a detailed list of what works and what doesn't each year — all the way from pre-planting through post-harvest.

When strip-tilling wet soils, Rauenhorst moves the row cleaners 1 1/2-to-2-inches to the side of the row to work in drier soil.

Co-sponsors, along with *No-Till Farmer*... American Cyanamid | IMC-Agrico Co. | Monsanto | Zeneca Ag Products (Syngenta)

1999

'No-Till ... Your Gateway To The Future'
Jan. 21-23, St. Louis • 660 attendees

"If we band phosphorus in our no-tilled fields, we can get by with half as much..."

— Duane Lange, Ord, Neb.

NNTC Insights...

Non-Uniform Nitrogen Application a Concern

Jim Andrew says corn prices may be so low that the cost of applying extra nitrogen may not be covered by the additional yields. As a result, the Jefferson, Iowa, no-tiller says nitrate tests are critical to avoid runoff into the nation's water supplies and to protect the environment while earning the best possible returns.

Andrew urges growers to check fertilizer hoses and application units to catch serious nutrient problems before they impact corn stands.

Try Earlier Planting with No-Tilled Soybeans

For attendees interested in earlier soybean planting dates, Dave Savage says he hasn't seen any major yield swings or significant visual differences in plant health.

The Farley, Iowa, no-tiller says growers in his area should target April for best yields, be prepared for early weed control concerns, be ready to no-till when conditions are right, consider GMO seed and proceed with caution.

AIM HIGHER. Wayne Humphreys challenged fellow no-tillers to appreciate farming and aspire to be the best. "We're at this conference so we can be smarter," says the no-tiller from Columbus Junction, Iowa.

Co-sponsors, along with *No-Till Farmer...* American Cyanamid | Aventis Crop Science | Monsanto | Zeneca Ag Products (Syngenta)

2000

"No-tilled soils have up to **250% more carbon and nitrogen** than conventionally tilled fields..."

— *Dean Martens,*
National
Soil Tilth
Laboratory,
Ames, Iowa

'Tough Times ... Tough Decisions ... Prospering With No-Till'

Jan. 12-15, Des Moines • 715 attendees

NNTC Insights...

Heavy No-Tilled Soils Need More Field Scouting

With the heavy soils in northwestern Ohio, Joe Nester likes to no-till with what he calls a defensive soybean program. The crop consultant from Antwerp, Ohio, says Roundup Ready soybean fields that are no-tilled need to be scouted more often than fields planted to traditional varieties. Just because these fields are free of weeds today doesn't mean they will be next week.

Nester is recommending no-tilling more *Bt* corn hybrids than in previous years on these heavier soils. He's convinced hybrids with this trait will perform better under cold and wet soil conditions.

Aussies Concerned with Salinity Content of Their Soils

By 2025, Bill Crabtree says Australia farmers expect salt to be a major soil concern on 30% of their ground. The no-till crop consultant from Northam, Western Australia, says 30% of fields may have major salinity concerns. Just 40 yards under the soil surface, there may be as much as 2,000 tons of salt per acre, so there is a huge problem with salinity.

With 60% of the Australian cropland being no-tilled, Crabtree says precision seed placement and improving the ability to tolerate drought have been major benefits of no-till.

Co-sponsors, along with *No-Till Farmer...* Agrotain International | Aventis CropScience | Monsanto | Landec Ag | Na-Churs Alpine Solutions | Zeneca Ag Products (Syngenta)

2001

"No-tillers are the **best environmental stewards the nation has,** and it offers you the greatest opportunity for profitable farming..."

— *Jay Lehr,*
Ostrander, Ohio

'Expanding The No-Till Toolbox'

Jan. 10-13, Cincinnati • 720 attendees

NNTC Insights...

No-Till Requires Different Equipment, Management Skills

No-till seeding equipment must be able to handle wet soils, heavy residue and soils that have an existing structure built around intensive tillage to get uniform and vigorous stands, says Matt Hagny.

The no-till crop consultant from Salina, Kan., says growers must set aside both outdated equipment and traditional habits that were part of the long history of planting into tilled seedbeds.

Take Full Advantage of Nutrients in Manure

Jim Arnaud says his top cost-saving idea is fully capitalizing on the valuable nutrients in poultry manure. The veteran no-tiller from Monetta, Mo., says manure provides cheap fertilizer and also boosts soil health.

Since the poultry manure decomposes and continues to feed the no-tilled corn crop throughout the growing season, Arnaud can often avoid the expensive cost of sidedressing nitrogen.

ASK AWAY. Finding an answer to every question any attendee could ask about any aspect of no-tilling has been an essential part of each of these mid-winter events.

Co-sponsors, along with *No-Till Farmer*...Agrotain International | Aventis Crop Science | Landec Ag | Monsanto | Na-Churs Alpine Solutions | Syngenta Crop Protection

2002

'Bigger, Better, Bolder'
Jan. 9-12, St. Louis • 682 attendees

"Brazilian research indicates no-till soybean yields **can be increased as much as 60% by adding black oats as a cover crop**..."

— *Rolf Derpsch, Asuncion, Paraguay*

NNTC Insights...

No-Tillers Wasting Dollars with High Soybean Plant Populations

Ten years of results from Marion Calmer's extensive onfarm research studies didn't show any economic advantage for soybeans that were planted in late April or early May with plant populations of over 175,000 seeds per acre.

The no-tiller from Alpha, Ill., says this was true when 15-inch-row beans were either drilled or planted with either no-till or conventional tillage.

Properly Sizing Residue Essential When No-Tilling

Kevin Anderson says it's essential to properly size your crop residue at harvest. The veteran no-tiller and shortline equipment manufacturer from Andover, S.D., says you must take the time to adjust your combine properly in order to effectively slice residue at the proper length.

Anderson says correctly sizing residue will improve seed-to-soil contact when no-tilling your crops the following spring, lead to improved microbiological soil activity and provide you with much more efficient fertilizer usage.

Co-sponsors, along with *No-Till Farmer*... Bayer Crop Science | Dow AgroSciences | Landec Ag | Monsanto | Na-Churs Alpine Solutions | Syngenta Crop Protection

2003

'No-Tilling For A More Prosperous Tomorrow'
Jan. 8-11, Indianapolis • 666 attendees

"Over a **10-year period, we increased soil organic matter from 0.6% to 1.2%** with no-till..."

— *Joe Bradford, Agricultural Research Service, Weslaco, Texas*

NNTC Insights...

Slow Down at No-Till Planting Time

Try to keep your no-till planting speed at less than 5 mph, suggests Scott Davidson. The no-tiller from Bethany, Ill. (an attendee at all of the National No-Tillage Conferences), says most planters are designed to drop a 30,000 population at 4 1/2 mph.

Davidson also says to avoid calibrating your no-till planter at 4 1/2 mph when you will be planting at 7 mph.

What You Need to Know About Carbon Sequestration

Recognize that heavy clay soils respond slower to management change, says Dan Towery of the Conservation Technology In-

formation Center in West Lafayette, Ind. Since adding no-till soybeans to a rotation results in less carbon, he suggests seeding a cover crop following soybean harvest.

Towery says a single pass with a chisel plow can erase much of the potential for carbon accumulation in no-tilled fields.

NONSTOP LEARNING. Well after her general session presentation concluded, soil microbiologist Jill Clapperton continues to field questions from no-tillers in hotel hallways.

Co-sponsors, along with *No-Till Farmer*... Bayer Crop Science | Dow AgroSciences | Landec Ag | Monsanto | Na-Churs Alpine Solutions | Syngenta Crop Protection

2004

'No-Till Knowledge Is Power'

Jan. 7-10, Des Moines • 680 attendees

"We've seen a significant response in no-tilled fields **from the application of sulfur**..."

— Brad Mathson, Whitehall, Wis.

NNTC Insights...

Keep a Skilled Driver on the Tractor Seat

A tractor driver's talents can be extremely critical to the success or failure of strip-tilling, maintains Bill Rohrs. The head of the Conservation Action Project in Findlay, Ohio, says an efficient tractor driver with strip-till must be able to build a mound that will still have an adequate height in the spring, leaving a strip that will be drier and preparing an air pocket-free seeding area to increase plant germination.

No-Till Trims Machinery Costs

By no-tilling, British grower Jim Bullock has trimmed machinery costs by 30%. The no-tiller from Worcestershire, England, relies on a pair of 150-horsepower tractors and rents another tractor in the same horsepower range at harvest time on his 850-acre no-till operation. Before switching to no-till, a pair of 200-horsepower tractors were necessary for tillage work.

With no-till, Bullock says parts needs have been reduced, fuel consumption was cut in half and he and his brother no longer need to hire extra labor at harvest time.

Co-sponsors, along with *No-Till Farmer*... Bayer Crop Science | Dow AgroSciences | IntelliCoat | Na-Churs Alpine Solutions | Oregon Ryegrass Commission | Syngenta Crop Protection

2005

'Charting A New No-Till Course'

Jan. 12-15, Cincinnati • 635 attendees

"Only three things can bring soils back to their natural organic matter level: **permanent forestland, permanent pasture or continuous no-tilling**..."

— Carlos Crovetto, Conception, Chile

NNTC Insights...

More Continuous No-Till Needed

Dan Towery called on growers to shoot for higher profitability by transitioning to continuous no-till. The former staffer at the Conservation Tillage Information Center estimated only 10-15% of U.S. cropland has been continuously no-tilled for over 5 years.

Towery says any yield drag with continuous no-till that occurs is likely due to poor management decisions during the transition.

Strip-Tilling Still Another Option

Jim Kinsella, a widely known no-tiller, advocated the use of strip-tilling to take full advantage of the warm, dry seedbed of conventional tilling while also reaping the environmental benefits of no-tilling.

CHALLENGE THE EXPERTS. Out in the hotel hallways, conference speakers tackle specific questions asked by attendees after their scheduled presentations.

The Lexington, Ill., grower suggests fall strip-tilling ahead of corn provides low-cost insurance against the slow growth that can occur in a cold and wet spring.

Co-sponsors, along with *No-Till Farmer*... Bayer Crop Science | Equipment Technologies (Apache) | Oregon Ryegrass Growers Seed Commission | Syngenta Crop Protection

2006

NNTC Insights...

'No-Till – Your Best Chance For Success'

Jan. 11-14, St. Louis
• 702 attendees

TAKING THE FLOOR. Every attendee has a chance to address the speakers and ask questions as they seek solutions to their own no-till challenges.

"Consider spraying fungicides to increase soybean yields, even if Asian soybean rust never reaches your fields..."

— *Wayne Pedersen, University of Illinois, Urbana, Ill.*

No-Tilling Corn Into a Living Cover Crop

Frank Martin shared his ideas on getting into his fields extra early by looking at the possibility of no-tilling corn into living cover crops. The Hallsville, Mo., grower will burn down the cover crop 2-4 weeks after planting.

Admitting that the plan is untried and ambitious, Martin says his list of potential advantages — for making better use of available moisture and extending the growing season — outweigh his list of potential disadvantages.

Green Bridge is an Unknown Yield Thief

Root diseases may have more of an impact on yields than you think, says Jim Cook, a retired Washington State University soil sci-

entist. The cause may be the so-called green bridge dilemma where pathogens live on the roots of growing volunteer crops and weeds between the harvest of one crop and planting the next crop. As the new crop emerges, pathogens move to these roots, weakening the crop and reducing yields.

With green bridge being a serious concern in the Pacific Northwest, the remedy for no-tillers is to establish a green-free period during which no plants, including weeds and volunteer crops, are allowed to grow in a field.

Co-sponsors, along with *No-Till Farmer*... Agro-Culture Liquid Fertilizers (AgroLiquid) | Bayer Crop Science | Case IH/DMI | Cross Slot No-Tillage Systems | Equipment Technologies (Apache) | Exactrix | Goodyear Farm Tires (Titan) | Great Plains Mfg. | Greenleaf Technologies | Oregon Ryegrass Growers Seed Commission | SFP (Verdesian Life Sciences) | Syngenta

2007

NNTC Insights...

'Fueling The No-Till Revolution'

Jan. 10-13, Des Moines • 743 attendees

"Waterhemp in Roundup Ready soybeans in Missouri wasn't killed with the application of 176 fluid ounces per acre of Roundup..."

— *Ian Heap, International Survey of Herbicide-Resistant Weeds, Corvallis, Ore.*

Do-It-Yourself No-Till Plot Studies

Putting out your own no-till test plots is definitely a paying proposition, says Jim Leverich. The Sparta, Wis., no-tiller says most no-tillers already have the equipment to effectively evaluate different hybrids and varieties, row spacing, plant populations, variable-rate fertilizer and seed application, compaction concerns, starter fertilizer, nitrogen placement and much more.

"You can't effectively manage your no-till system without measurements that can help you make the proper judgment calls," Leverich says.

Fitting All the No-Till Pieces Together

Although there are numerous methods for successfully no-tilling, Paul Jasa suggests a systems approach that strives for uniformity. This includes emphasis on seed spacing, row spacing, emergence, crop height, herbicide application and residue cover.

The University of Nebraska ag engineer says most residue concerns result from uneven spreading of stalks or straw and from row cleaners that move residue while creating uneven distribution. Uneven residue leads to differing soil temperatures and moisture levels, leaving non-uniform conditions for seeds and uneven emergence.

Co-sponsors, along with *No-Till Farmer...* Agro-Culture Liquid Fertilizers (AgroLiquid) | Bayer Crop Science | Case IH | Cross Slot No-Tillage Systems | Equipment Technologies (Apache) | Exactrix | Goodyear Farm Tires (Titan) | Great Plains Mfg. | Oregon Ryegrass Growers Seed Commission | SFP (Verdesian Life Sciences) | Syngenta | Trimble Navigation

2008

'Gaining The No-Till Edge'
Jan. 9-12, Cincinnati • 766 attendees

"Save the good soil structure that's been developed by no-tilling ground coming out of the Conservation Reserve Program..."

— *John Baker,*
Cross Slot
Tillage Systems,
Feilding, New Zealand

NNTC Insights...

Less Nitrogen Needed with Long-Term No-Till

After 12-14 years of continuous no-till, Mark Alley says you'll be able to dramatically trim nitrogen costs. The retired Virginia Tech University agronomist has research data that indicates long-term no-tillers definitely won't need as much fertilizer as conventional-tilling neighbors.

Alley also pointed out that most growers can maximize their overall profitability by investing in starter fertilizers for quick plant growth when no-tilling corn.

Let Earthworms Do Your Soil Ripping

Stressing the value of huge earthworm populations in no-tilled soils, Jill Clapperton suggests doing everything you can to house them, feed them and protect them. The rhizosphere ecologist at Alberta's Lethbridge Research Center also cited the huge value that comes from all of your underground no-till partners. "Fungi form intimate relationships between the plant, soil and soil organisms," she says. "This illustrates the potential for using rhizosphere processes to dramatically improve both your no-till soil quality and productivity."

Co-sponsors, along with *No-Till Farmer...* Agro-Culture Liquid Fertilizers (AgroLiquid) | Agrotain International | Bayer Crop Science | Case IH | Equipment Technologies (Apache) | Exactrix | Goodyear Farm Tires (Titan) | Great Plains Mfg. | Oregon Ryegrass Growers Seed Commission | SFP (Verdesian Life Sciences) | Syngenta | Trimble Navigation

2009

'Charging Ahead With No-Till'
Jan. 14-17, Indianapolis • 899 attendees

"Since soil fertility is a moving **target**, soil sample more often than every 3 years."

— *Joe Nester,*
Nester Ag Crop
Consultants,
Bryan, Ohio

NNTC Insights...

Balance Your No-Till Soils

Along with providing nutrients, Neal Kinsey says no-tillers must maintain ideal balances of air, water, organic and mineral soil components to support the soil microbial life. The soil fertility specialist with Kinsey Ag Services in Charleston, Mo., says calcium and magnesium are critical for soil life since they impact the amount of available air and water in the soil.

"No soil will perform without the proper amount of air and water," he says. "Once nitrogen, phosphorus, potassium and sulfur are in line, then calcium and magnesium optimize plant quality, yields and profits."

For High No-Till Yields, Work Your Plan

National yield champ David Hula says the key to turning out profitable no-till corn yields is coming up with a good plan, following the plan and evaluating the success at the end of the year.

A long-time no-tiller from Charles City, Va., Hula holds the U.S. corn yield record of 542 bushels per acre set in 2017. He finds the keys to high no-till yields include seed treatments, residue management, controlled traffic patterns, calibrating seed meters, using seed sensing monitors, providing fertilizer when plants need it, evaluating nutrient needs all season long, frequent soil testing and season-long planning.

IT'S OFFICIAL. Many states grant pesticide recertification credits to farmers who attend National No-Tillage Conference sessions and sign the required logbooks.

Co-sponsors, along with No-Till Farmer... Agro-Culture Liquid Fertilizers (AgroLiquid) | Agrotain International | Bayer Crop Science | Case IH | Equipment Technologies (Apache) | Goodyear Farm Tires (Titan) | Great Plains Mfg. | Oregon Ryegrass Growers Seed Commission | SFP (Verdesian Life Sciences) | Syngenta | Thurston Mfg. (Blu-Jet) | Trimble Navigation

2010

"Minds are like parachutes, **as they only work when open**..."

— *Keith Schlapkohl, Stockton, Iowa*

'A Powerful New Decade For No-Till'

Jan. 13-16, Des Moines • 789 attendees

NNTC Insights...

Narrower Rows Needed For 300-Bushel No-Till Corn Yields

Allen Berry is convinced rows narrower than 30 inches are needed to achieve top no-till yields. The veteran no-tiller from Nauvoo, Ill., believes equidistant spacing of plants between and within the rows is the wave of the future.

Berry is convinced corn populations need to rise to around 45,000 plants per acre and that breeders need to breed corn hybrids that will routinely produce two ears per stalk. Berry envisions a future where corn is grown in 12-inch rows with corn plants 12 inches apart within the rows.

Cover Crops No Quick Fix for Poor Soils

Dale Mutch believes cover crops can help solve most soil problems and improve water quality. But the Michigan State University cover crops specialist cautions that they are a long-term investment and not a quick fix.

Mutch says cover crops continue to grow in popularity among no-tillers because they protect the soil after harvest, suppress weeds and enhance fertility.

"Each farming system is different," he says. "The success of cover crops is dependent on the year and the weather, as they need moisture and light to work."

Co-sponsors, along with *No-Till Farmer*... Agro-Culture Liquid Fertilizers (AgroLiquid) | Agrotain International | Bayer Crop Science | Case IH | Equipment Technologies (Apache) | Goodyear Farm Tires (Titan) | Great Plains Mfg. | Oregon Ryegrass Growers Seed Commission | SFP (Verdesian Life Sciences) | Syngenta | Thurston Mfg. (Blu-Jet) | Trimble Navigation

2011

"If you can **increase organic matter levels**, you can reduce your nitrogen usage..."

— *Jim Hoorman, Ohio State University educator, Lima, Ohio*

'No-Tilling Today For A Better Tomorrow'

Jan. 12-15, Cincinnati • 935 attendees

NNTC Insights...

No-Tillers Should Think of Themselves as Ecologists

Since you're managing a complex ecosystem that includes millions of organisms both above and below ground, Ray Weil says no-tillers need to add "ecologist' to their farming title. The soil scientist at the University of Maryland extols the virtues of earthworms, which incorporate organic matter and burrows that improve soil drainage, aeration and filtration.

"When your neighbors have water standing in their fields, they blame compaction," he says. "However, they probably don't have earthworms either."

Investing in Precision Pays Big-Time for No-Tillers

The benefit of investing in the latest fast-moving precision ag technologies is that these products offer no-tillers an economic edge where investment returns will continue for years, says Dietrich Kastens. The Herndon, Kan., no-tiller says no-till and precision ag technology follow similar adoption rates with the key questions being whether these investments will pay off and how fast the payback can be.

Kastens says no-tillers have accepted the obvious technology improvements such as auto-guidance and Roundup Ready crops. But you can also gain considerable profit from taking on less obvious technologies, such as variable-rate fertilizer application.

NEW IDEAS, NEW CONCEPTS. Natural Resources Conservation Services agronomist Ray Archuleta stresses how to improve soil health in your no-till rotations.

Co-sponsors, along with *No-Till Farmer...* Ag Leader Technology | Agro-Culture Liquid Fertilizers (AgroLiquid) | Agrotain International | Bayer Crop Science | Case IH | Cover Crop Solutions | Equipment Technologies (Apache) | Goodyear Farm Tires (Titan) | Needham Ag Technologies | SFP (Verdesian Life Sciences) | Syngenta | Thurston Mfg. (Blu-Jet)

2012

'Two Decades Of No-Till Know-How'
Jan. 11-14, St. Louis • 973 attendees

"Glyphosate is toxic to microorganisms, so it can affect nitrogen fixation..."

— *Don Huber, retired Purdue University educator, West Lafayette, Ind.*

NNTC Insights...

Spreading Residue Can Overcome Long-Term No-Till Concerns

Phil Needham finds uniform residue distribution at harvest is critical for no-till success, as it has numerous indirect effects on crop emergence and early plant growth. The owner of Needham Ag Technologies in Calhoun, Ky., says a drop off in combine speed going up a hill and dull chopper blades dramatically affect spreader performance and residue distribution.

Needham says many farmers in Europe have begun making combine-purchasing decisions based on the machine's ability to spread residue evenly across the entire width of the header.

Cover Crops Reduce Soybean Cyst Nematode Concerns

Research from southern Illinois indicates fields seeded with rapeseed prior to soybeans reduced soybean cyst nematode (SCN) egg counts compared to fields where no cover crop was seeded. Mike Plumer, a retired University of Illinois educator, says preliminary data suggests cereal rye and annual ryegrass may also have this impact on SCN egg counts, although more research is needed.

Co-sponsors, along with *No-Till Farmer...* Ag Leader Technology | Agro-Culture Liquid Fertilizers (AgroLiquid) | Agrotain International | Bayer Crop Science | Case IH | Cover Crop Solutions | Equipment Technologies (Apache) | Goodyear Farm Tires (Titan) | Needham Ag Technologies | SFP (Verdesian Life Sciences) | Syngenta | Thurston Mfg. (Blu-Jet)

2013

'Powering Up Your No-Till System'
Jan. 9-12, Indianapolis • 1,222 attendees

"Earthworm populations can increase by as much as 100% when tillage is reduced..."

— *Odette Menard, Agriculture and Agri-Food Canada, Saint-Hycainthe, Quebec*

NNTC Insights...

Better Water-Holding Capacity Another No-Till Benefit

The key to retaining moisture under no-till conditions depends on improving soil organic matter and leaving residue on the surface to reduce storm runoff, says Jerry Hatfield. The head of the National Soil Tilth Laboratory at Ames, Iowa, says saving moisture also reduces soil-water evaporation, decreases soil temperatures in the summer and allows soil biological activity to function more effectively.

To move from harvesting 200-to 300-bushel no-till corn, Hatfield says you will need to conserve 5 additional inches of water that need to be available to the crop. "You can't just depend on rainfall, as 300-bushel corn doesn't come free," he says.

No-Till, Cover Crops Overcome Drought Concerns

About 90% of Ryan Speer's acres in Sedwick, Kan., are seeded to cover crops or wheat during the winter, while cover crops or grain sorghum are seeded in the summer. Aerial photos show he's getting 100% cover in fields after harvest, which is his ultimate goal.

"If I can breakeven on the cost vs. the initial benefits of increased yields, reduced chemical costs and lower fertilizer costs, the long-term benefits of cover crops will pay off for many years to come," he says.

Co-sponsors, along with *No-Till Farmer...* Ag Leader Technology | Agro-Culture Liquid Fertilizers (AgroLiquid) | Agrotain International | Bayer Crop Science | Case IH | Cover Crop Solutions | Equipment Technologies (Apache) | Goodyear Farm Tires (Titan) | Needham Ag Technologies | SFP (Verdesian Life Sciences) | Syngenta | Thurston Mfg. (Blu-Jet)

2014

"It's rare for corn to take **up more than 50% of applied nitrogen**..."

— *Joel Gruver, Western Illinois University soils specialist, Macomb, Ill.*

'An Honest No-Till Education'

Jan. 15-18, Springfield, Ill. • 1,106 attendees

NNTC Insights...

Understand the Biology of Weeds, Then Exploit Their Weaknesses

Kevin Bradley urges no-tillers to fully understand the biology of each weed they're fighting and then find ways to deal with their weaknesses. As an example, the University of Missouri weed scientist says research studies indicate that introducing pigweed seed into previously pigweed-free soils could be brought under control within a couple of years.

"Pigweed seed doesn't emerge from lower soil depths," says Bradley. "By controlling pigweed early, the growers didn't let pigweed seed go back into the soil, and that's what we must do."

Making Useful Use of Data is Crucial

An important aspect of your no-till farm management program is making extensive use of precision-farming data, says Will Cannon. The Newton, Iowa, strip-tiller finds extensive data plays a key role in making important decisions on crop insurance, hybrid selection, fertility rates and more.

He relies on a field-to-farm office wireless cloud storage transfer system that automatically stores data for planting, spraying, harvesting and other field operations.

Cannon believes this technology could also be extremely useful when used with controlled traffic systems in no-till or strip-till systems.

Co-sponsors, along with *No-Till Farmer...* Ag Leader Technology | Agro-Culture Liquid Fertilizers (AgroLiquid) | Agrotain International | Bayer Crop Science | Case IH | Cover Crop Solutions | Equipment Technologies (Apache) | Goodyear Farm Tires (Titan) | Gypsoil | Needham Ag Technologies | Syngenta | Thurston Mfg. (Blu-Jet) | Verdesian Life Sciences | Yetter Farm Equipment

2015

"No-till offers **a 23% return on our overall investment** compared to a 19.8% return with deep tillage..."

— *Brian Watkins, Kenton, Ohio*

"Building Better No-Till Practices"

Jan. 14-17, Cincinnati • 935 attendees

NNTC Insights...

No-Tiller Grows Much of His Own Nitrogen Supply

After losing more than 80 pounds per acre of valuable nitrogen (N) from December to April, Cameron Mills found several ways to make sure cover crops were seeded on every acre. The Walton, Ind., no-tiller says soil testing showed the value of covering every acre over the winter.

He found using annual ryegrass as a cover crop scavenged 51 pounds of N, 62 pounds of potassium and 10 pounds of phosphorus per acre. That's a return of $66.98 per acre.

More to Automation than Replacing Labor

Scott Shearer predicts the no-till cropping system of the future will include several 60- to

70-horsepower autonomous tractors weighing less than 10,000 pounds each. The ag engineer at Ohio State University says these tractors will offer adjustable ground clearance with sensing technology that will let them work 24 hours a day, 7 days a week, without needing an operator in a tractor cab.

"For a long time, people thought of automation as a way to replace labor," says Shearer. "Going to automated equipment should also offer a gain in productivity."

IN-DEPTH NO-TILL LEARNING. Even with 1,000 National No-Till-age Conference attendees in recent years, 40% are normally first-timers.

Co-sponsors, along with *No-Till Farmer*... Ag Leader Technology | Agro-Culture Liquid Fertilizers (AgroLiquid) | Agrotain International | Bayer Crop Science | Case IH | Dawn Equipment | Equipment Technologies (Apache) | Goodyear Farm Tires (Titan) | KB Seed Solutions | Kinze Mfg. | Needham Ag Technologies | Nufarm | Syngenta | Verdesian Life Sciences | Yetter Farm Equipment

2016

'Pushing No-Till Beyond Barriers'

Jan. 6-9, Indianapolis • 1,104 attendees

"Earthworms give no-tillers **twice as much soil to draw water from**…"

— *Doral Kemper, National Soil Tilth Laboratory, Ames, Iowa*

NNTC Insights…

Need to Reduce Selection Pressure on Herbicides

No-tillers fighting herbicide resistance with waterhemp, Palmer amaranth or marestail in no-till soybean fields should consider using more soil residual herbicides, maintains Bryan Young. The Purdue University weed scientist says you need to match application timing and length of residual effectiveness to the period of peak emergence for these weeds.

Data from southwestern Illinois fields showed 90% of waterhemp emerged by June 30, while 90% of Palmer amaranth didn't emerge until sometime in August. As a result,

Young says one application of a residual herbicide may not last long enough to control all weed species in no-tilled fields.

Should You Soil Sample in the Fall or Spring?

Jimmy and Branson Howard take soil samples in the spring ahead of no-tilling corn ground. The father-and-son team from Mooresville, N.C., prefers to test for fertility when they feel soils are at their best.

They have found that this occurs after the heavy no-till residue has had an opportunity to break down over the winter and has released considerable amounts of valuable nutrients back into the soil.

Co-sponsors, along with No-Till Farmer … Ag Leader | Agrotain | Case IH | Dawn Biologic | Equipment Technologies (Apache) | Exapta Solutions, Goodyear Farm Tires (Titan) | KB Seed Solutions | Needham Ag Technologies | Nufarm | Pure Grade Liquid Fertilizer (The Andersons) | Syngenta | Verdesian Life Sciences | Yetter Mfg.

2017

'Quarter Century of No-Till Learning'

Jan.10-13, St. Louis • 950 attendees

"Before you can capitalize on data, **you need to understand how and why it's collected…**"

— *Jeremy Wilson, Effingham, Ill.*

NNTC Insights…

Modify Ways You Apply Nutrients

If you spread fertilizer evenly across a field, J.C. Cahill says each plant needs roots everywhere possible in order to effectively use the available nutrients. The plant ecologist at the University of Alberta says the result is that plants end up using too much energy to grow their root system.

But if a plant is able to find fertilizer in patches, it may be able to grow roots with less energy, which leads to higher no-till yields.

Cahill says that if no-tillers modify their fertilizer program to maximize the ability of plants to use their roots more effectively in seeking out nutrients, there's definitely higher yield potential without increased costs.

Cover Crops Boost Water Infiltration

Among the first benefits Johnny Hunter saw after combining cover crops with no-till was that water infiltration improved by an estimated 100%. The Dexter, Mo., no-tiller also learned seeding cereal rye as a cover crop had a dramatic impact on the control of pigweed.

When it comes to planting green, Hunter says no-tillers need to toss the traditional planting calendar out the window.

"All I care about is what's going on in each field," he says. "What is the cover crop telling me? Has my hairy vetch produced all the nitrogen I want it to produce? Is my cereal rye tall enough to form a weed barrier?"

If the answer to these kind of questions is "no," Hunter lets the cover crop continue to grow regardless of when the calendar is telling him to plant.

Co-sponsors, along with No-Till Farmer ... Ag Leader | Case IH | Equipment Technologies (Apache) | Exapta Solutions | Fennig Equipment | Goodyear Farm Tires (Titan) | KB Seed Solutions | Montag | Needham Ag Technologies | Nufarm | Pure Grade Liquid Fertilizer (The Andersons) | Syngenta | Verdesian Life Sciences | Yetter Mfg.

2018

"Gathering a global farming perspective helped me develop a much more effective no-till operation…"

— *Blake Vince, Merlin, Ontario*

'Racing Toward No-Till Success'

Jan.9-12, Louisville• 990 attendees

NNTC Insights...

Declaring War on Palmer Amaranth

If no-tillers can effectively use cover crops to keep palmer amaranth shaded in early spring, Larry Steckel says they have a good chance of keeping the troublesome weed from germinating. The University of Tennessee weed specialist indicates this means taking steps to keep heat, water and light away from these weeds during the early spring. By doing so, no-tillers can trim herbicide needs by as much as 50% while dealing with resistance concerns with this weed.

Steckel has seen no-tillers rely on cover crops to shade out resistant palmer amaranth when planting corn and soybeans "green" into covers and waiting to terminate them 2 weeks later with Roundup and dicamba.

Changes Taking Place with Soil Biology

John Macauley can't heap enough praise on the changes cover crops have made to his soils. The Groveland, N.Y., no-tiller says cover crops supply the equivalent of green manure in the family's 1,200-acre cropping operation.

Macauley has been able to hold cover crops seed costs to around $20 per acre with a six-way mix of radishes, Austrian winter peas, buckwheat, turnips, crimson clover and oats.

NO-TILL LOOSE ENDS...

March 2011

WHO WAS THE NATIONAL NO-TILLAGE CONFERENCE SPY?

I remember well the 2001 National No-Tillage Conference when lawyers attempted to tell one of our speakers what ideas he could and could not share with attendees. The situation dealt with the on-going rivalry between the Keeton seed firmers and the Schaffert seed firming devices.

This ridiculous situation eventually led the *No-Till Farmer* editors to go the generic language route and start referring to both devices as seeding attachments rather than the better known terminology of seed firmers or seed firming devices.

This was an irritating case of seeing lawyers hassle our staff, make threats to NNTC speakers and threaten to disrupt attendees over something that doesn't mean a hill of beans to no-tillers.

Here's What Happened

In the fall of 2000, I asked Paul Schaffert, a veteran no-tiller, a No-Till Innovator and the owner of Schaffert Manufacturing Co. in Indianola, Neb., to speak at the mid-January conference about seeding accuracy under no-till conditions. The inventor of the Rebounder Furrow V Closer planter attachment had worked with no-tillers around the country to find novel ways to seed more efficiently.

So I was surprised when he called to tell me about a letter received from an attorney stating that he'd better think twice about presenting the information that was listed in the conference program and that it would be monitored on their client's behalf.

No Laws Broken

Yet nobody representing the law firm, manufacturer or distributor paid the conference registration fee.

I checked the people who were attending this session to see if I could spot a spy. I've got a pretty good idea of how they monitored the session. And I'm sure their spy reported no trade secrets were violated.

Should I invoice the law firm for a conference registration since they were going to monitor the session?

Iron–Streak Attendees Reflect on
25 Years of No-Till Learning

At the **25th anniversary of the National No-Tillage Conference,** six attendees who've never missed a single event shared their path into no-till and how they leveraged the knowledge of their peers.

'CONTINUING YOUR NO-TILL EDUCATION IS CRITICAL'

Since its debut in Indianapolis in 1993 and through 2018, a half dozen individuals have attended every one of the National No-Tillage Conferences.

- ◆ **ALLEN BERRY,** *no-tiller, Nauvoo, Ill.*
- ◆ **ALLAN BROOKS,** *no-tiller, Markesan, Wis.*
- ◆ **SCOTT DAVIDSON,** *no-tiller, Dalton City, Ill.*
- ◆ **RANDALL REEDER,** *Ohio State University ag engineer, Columbus, Ohio*
- ◆ **BRYAN VON HOLTEN,** *no-tiller, Cole Camp, Mo.*
- ◆ **R.D. WOLHETER,** *no-tiller, Wolcottville, Ind.*

"CONVENTIONAL TILLERS WHO SEND US THEIR NUTRIENTS SAVE US FERTILIZER DOLLARS."

ALLEN BERRY,
no-tiller, Nauvoo, Ill.

At the National No-Tillage Conference's (NNTC) 25th anniversary in St. Louis, *No-Till Farmer* editor Frank Lessiter invited a half-dozen attendees to reflect back on the past 25 years of the conference and their experiences with no-tilling. Here is a recap of that discussion with six attendees who were part of each one of these mid-winter conferences starting in 1993.

Q: How did you get started in no-till?

Von Holten: I came to the first conference in 1993 and had just completed my full season of no-till. I was farming about 300 acres and was nearly 100% no-till corn and soybeans with a small acreage of wheat while holding down an off-farm job.

By '96, I'd grown the operation to 900 acres and the no-till concept was working out well enough to quit my day job and try full-time farming.

In the summer of 1997, I planted my first cover crop of hairy vetch and no-tilled corn into it. I had some struggles, as it stayed very wet after I no-tilled the corn. With my flat, poorly drained soils, the corn struggled and I didn't get the stand I wanted. I tried it again the next year, had the same problem and gave up planting cover crops in front of corn.

In the fall of 2002, I seeded cereal rye after corn and got along well. I soon switched to annual ryegrass for its deep rooting characteristics and got along well. While erosion was reduced with no-till, I still fought gully erosion in a few areas and decided I wanted some type of cover on every acre 100% of the time.

After listening to North Dakota no-tiller Gabe Brown, Ohio no-tiller Dave Brandt and other conference speakers talking about the need for more crop diversity, I switched to a 100% no-till corn, soybeans and wheat rotation. I'm no-tilling corn into a cereal rye and a few other species. This is followed by no-tilled soybeans and no-tilled wheat.

After harvesting the no-tilled wheat, we'll spread poultry litter and add a diverse cover crop to build soil organic matter. This spring I'll be no-tilling 3,100 acres.

Wolheter: I started no-tilling in '81 with corn. We farm in three counties in northeast Indiana. In a field area the size of this conference room, we can have two or three soil types. It changes that quick.

When I first tried no-till corn, we had 25 inches of rain in April, May and June. The corn was a success, even though we used bubble coulters. A few years later I figured out it was referred to as a compaction coulter since it seemed more suitable for conventional tillage or when no-tilling sandy soils to help compact the sidewalls. Fortunately, the first couple years bubble coulters worked well for me, and by '82, I was close to 100% no-till corn.

The first time I no-tilled soybeans was in '84 and it was a disaster. We used a 15-foot Marliss drill, which had a width good for driving down the road and getting through gates. However, the end wheels made it 18 feet wide and the drill proved too dangerous to take down some roads.

We didn't have much in the way of post-emerge herbicides at that time, but the no-till results were great where we had good weed control. A couple years later, I tried a Great Plains 10-foot drill and the beans yielded above average.

Brooks: I'm a contract vegetable grower and no-till has been more of an evolution for me because of the interest my father had as a professor at the University of Wisconsin. My dad and I first met *No-Till Farmer* editor Frank Lessiter in 1972, which was the year *No-Till Farmer* first came out.

We were trying low disturbance residue cropping, which was scandalous in the vegetable industry. Our vegetable production contracts had boxes you were expected to check for fall plowing and spring plowing and they expected to see the box checked for fall plowing.

We didn't plow one year and they were so upset that they sent a man out from headquarters who told us, "No more contracts for you unless you plow. Not just disturb the surface, but moldboard plowing."

My dad said to the vegetable grower field man, "Well, have you ever done no-till before?" The man answered no. Then Dad said, "Why don't you wait and see what happens and then make a decision?"

That harvest was a wet one and they learned they could harvest our no-till peas a few days earlier than with tilled fields. One time, we had as many as a half dozen pea combines working in our fields instead of sitting idle until other fields dried out.

We never heard any more about moldboard plowing from the canning company. We may disturb the surface while no-tilling, but the seed always goes into firm, undisturbed soil.

Berry: Some 25 years ago we were growing 1,000 acres of corn and soybeans, with a little no-till. In '88, I had a neighbor no-till some of my soybeans into standing corn stalks. In the early '90s, I bought a 20-foot no-till drill and outfitted it with a Rawson cart. Around '96, I tried my first 15-inch row corn.

Over the years, no-till helped us expand our operation and our son, who has a full-time crop consulting scouting business, came into the operation. We now no-till around 3,000 acres of corn and soybeans.

A dozen years ago, we rented a field in northeast Missouri. The Des Moines River runs on the north side of the field while the Mississippi River runs on the east side and sometimes water runs over the top of the field.

That field represents the best that Illinois and Iowa have to offer in the way of fertility. Thanks to the conventional tillers who send nutrients down the river, it saves us considerable on our fertilizer investment.

Reeder: I started at Ohio State

"WE WERE TOLD THERE WILL BE NO MORE VEGETABLE PRODUCTION CONTRACTS UNLESS WE PLOW!"

ALLAN BROOKS,
no-tiller, Markesan, Wis.

"MY BIGGEST NO-TILL MISTAKE WAS NOT RECOGNIZING WHEN IT'S TOO WET TO PLANT."

SCOTT DAVIDSON,
no-tiller, Dalton City, Ill.

University in 1979 and took over a research project that included ridge-till and plowing, to which we soon added no-till. Those plots are still going in northwest Ohio and cover crops were later added to the research.

Compaction plots were started in '87 with shallow chisel plowing to see if we could eliminate deep tillage with subsoiling. The compaction was studied with a 600-bushel grain cart with 20 tons of weight on the axle.

Subsoiling reduced the compaction in the deep layers of soil and boosted yields. In fact, this research has sold quite a few subsoilers in that area.

In 2002, we dumped chisel plowing to compare continuous no-till with subsoiling. NNTC attendees will love this result: Continuous no-tilling resisted compaction better than the plots that were subsoiled every 3 years.

Much of my job also included organizing the annual Ohio Conservation Tillage and Technology Conference that was started in the mid '80s.

Davidson: We no-till flat, black drummer soils with a corn and soy-bean rotation in central Illinois. We made the transition to no-till between '91 and '94 as a better way to take care of our renewable resource.

The main thing we had to learn was patience. If it's too wet to plant, then go fishing or spend some time with your family. When the other guys were out working the ground, we had to let the soil dry out. If you didn't wait, then you'd have to clean plenty of mud off the no-till planter.

At the end of the first no-till conference, I gave NNTC coordinator Alice Musser the registration dollars for the following year in St. Louis. I knew I needed to get back with this group — all who were on the same page as I was to learn how to take care of our soils better.

Q: What's the biggest impact no-till has had in your operation?

Berry: No-till is a lot less stressful, especially at planting time. When we did lots of tillage in the '80s and the early '90s, we had more tractors to fool around with and you had to keep the tillage tools running well ahead of the planter. There were many more things to deal with in those days.

Now, all we do is get on the tractor and plant. No-till is a lot more relaxed and easier to get the crop in with less stress at that real critical time.

Von Holten: No-till has allowed me to farm on my own terms. I've always been fussy about equipment — not liking my machinery to get bent, scraped or scratched. No-till allows me to do what I enjoy doing every day without the stress of managing people.

Our fields aren't as big as many attendees have, but I can take one tractor with a 40-foot no-till planter, 40-foot air seeder and a 120-foot sprayer and farm 3,000 acres on my own with only some part-time help moving grain away from the combine. With conventional or minimum tillage, that would be impossible.

No-till has allowed me to live my dream and do what I enjoy doing. I couldn't do this with a more intensive tillage system.

Brooks: It's learning how to think objectively. Throughout the years, I've been exposed to hundreds of no-till ideas at all of these events. Learning how to cope with numerous no-till ideas, to decide to try new ones and objectively analyze them is one of the biggest benefits.

Wolheter: No-till has enabled us to cover a lot more acres. There's less erosion in our hilly areas and now rocks are less of a concern.

Reeder: Besides eliminating erosion concerns with cover crops, no-till can help growers be more precise with injecting fertilizer. This is particularly true in northwestern Ohio due to phosphorus problems in Lake Erie. There are also concerns with algae buildup in the Gulf of Mexico where nitrogen flows down the Mississippi River from Corn Belt fields.

I intentionally used the term injecting. When many farmers hear the term incorporate, they think of plowing down residue. And today we have many machines that can precisely inject nutrients under no-till conditions.

Q: What's the biggest mistake you've made in no-till?

Davidson: When it's too wet, it's too wet. Plus, I remember a story Illinois no-tiller and strip-tiller Jim Kinsella shared in the '90s about what people do when the fire at a hot dog roast starts to die down. You take a stick and stoke the fire, so it gets hotter and burns up the wood quicker.

It's the same thing with tillage. When cropping problems exist, some farmers use more tillage, which burns up the organic matter. They think they're using intensive tillage to improve their soils, but in reality the extra tillage is destroying the soil.

Von Holten: It was giving up too early on cover crops. I planted my first cover crop in '97, had a stand failure and had to replant corn. The second year wasn't much better, so I gave up on cover cropping.

Looking back, I should have had the confidence to stick with protecting the land with a cover crop. If I'd stuck with it, I'd be over 20 years into cover crops at this point. I would probably be seeing the same incredible long-term results other no-tillers are seeing.

Wolheter: Most farmers underestimate the cost of running their equipment. More growers would shift to no-till if they knew how much it can reduce machinery costs.

We tried variable-rate fertilization with two companies each working with about 10% of our acreage. Over 3 years, one company wanted to build up the nutrient levels of our soils to much higher levels and wanted to use pelleted lime to correct the pH. While pelleted lime has its place, you don't need to apply it every year.

We have some pretty light soils that don't hold potash very well and I thought they were trying to build our soils up to a much higher fertility level than needed. The way I saw it, all they wanted to do was sell us more fertilizer.

The other company followed a 4-year program and it was the third year before they applied any lime.

"CONTINUOUS NO-TILLING RESISTED COMPACTION BETTER THAN PLOTS SUBSOILED EVERY 3 YEARS."

RANDALL REEDER,
Ohio State University ag engineer, Columbus, Ohio

> ## "NO-TILL ALLOWS ME TO DO WHAT I ENJOY DOING WITHOUT HAVING TO MANAGE ANY PEOPLE."
>
> **BRYAN VON HOLTEN,**
> *no-tiller, Cole Camp, Mo.*

They had too many acres to cover and we didn't appear to be a priority.

They were sampling on 3-acre grids while we do zone soil sampling. However, they just split a 21-acre rectangular field down the middle with half a grid on each side. That wasn't my idea of how to make variable rate fertilization work.

Brooks: Assuming weeds in a no-tilled field were dead when they weren't.

Berry: I was too slow at parking the disc and field cultivator. It took me a few years coming to the NNTC to figure out that I needed to park those tillage tools. Around 2000, we totally quit using those tillage tools. ❁

> ## "IT WAS A FEW YEARS BEFORE WE REALIZED BUBBLE COULTERS WERE ACTUALLY COMPACTION BUILDERS."
>
> **R.D. WOLHETER,**
> *no-tiller, Wolcottville, Ind.*

A Joint Leap of Faith: Reflections on the
Launch of the National No-Tillage Conference

Here are a few serious and not-so-serious things that occurred over the years at this annual mid-winter gathering of the no-till community.

During the early 1990s, we'd given serious discussion to whether *No-Till Farmer* should start an annual no-tillage conference. But we always seemed too busy to think about producing such an event and kept putting off a decision until the summer of 1992, when Monsanto announced they were going to start several regional no-till conferences the following winter.

That's when we decided we had to start the National No-Tillage Conference (NNTC) the following winter or get left behind in the Monsanto wake.

Before we made the decision , we asked *No-Till Farmer* readers via a postcard survey in July of 1972 whether they felt holding a non-commercial, in-depth and unbiased event devoted 100% to no-till was a good idea or not.

Results from the postcard survey produced an overwhelming vote in favor of *No-Till Farmer* offering an annual mid-winter meeting of the no-till community. In addition, these subscribers offered plenty of valuable opinions and program suggestions.

As a result, subscribers have always felt that they are part owners of this event, a key to success over the years. They helped planned the first year's program, the speakers, the location (51 suggested cities), month, week and even the specific days of the week for an exclusive conference that's devoted 100% to helping growers no-till more profitably.

That first year's program featured 45 nationally-acclaimed no-tillers and researchers who tackled 96 profit-building no-till topics.

659 MAKE-IT-HAPPEN VOTES. Event coordinator Alice Musser holds a portion of the postcard surreys returned by readers encouraging *No-Till Farmer* to launch the National No-Tillage Conference.

Along with highly valuable hallway networking, the 3 days of non-stop no-till learning was spread over 32 hours at the Hilton hotel in downtown Indianapolis.

Our goal was to attract at least 150 no-tillers.

150, 300 or 814 Attendees?

After an 8-page program was mailed in late October of 1972, 400 no-tillers quickly registered for the first upcoming mid-January event. After we signed up 600 no-tillers,

the downtown Indianapolis hotel told us that their meeting rooms couldn't handle any more. At that point, we started a waiting list.

By mid-December, the hotel had found a way to accommodate another 200 attendees. Those growers on the waiting list were delighted with the news, and 197 immediately signed up — the other three had made other plans during the waiting period.

By feeding attendees in two rooms on different floors, the hotel was somehow able to pack in 814 no-tillers. Even so, that first year there was still a waiting list of nearly 200 disappointed farmers who we simply could not accommodate.

We'd outgrown the Indianapolis hotel a month before the first NNTC was ever held!

Imagine our shock when 814 folks showed up from 21 states, Canada, Brazil and Argentina.

If someone wanted to learn more

about any aspect of no-till, these folks understood that this conference was the place to be. No-tillers had found "their people" — growers who were eager to share what they knew.

That's Crazy Pricing

After we announced the first year's registration price tag of $169, an Illinois friend called and asked if we were "off our rocker." He said growers could go to extension meetings on no-till for free or only spend $10-$15 fee for a no-till extension program that included lunch. He wondered what made us think we could get no-tillers to pay $169.

I explained that we were going to deliver an intensive 3-day unbiased program on all aspects of no-till that wasn't available anyplace else. And since we believed in what we could deliver, we were offering a money-back guarantee to any attendee who felt the event wasn't all it was

expected to be. We've offered that money-back guarantee every time for more than 25 years. (And oh, yeah, the Illinois caller became a frequent NNTC attendee.)

Co-Sponsors Had Faith

A key to our success that first year were the five co-sponsors who believed in our ability to produce an in-depth no-till conference. These firms included American Cyanamid, Dow Elanco, DuPont, ICI Americas and Monsanto.

Attendees told us they appreciated learning about new no-tilling ideas without getting pressure to buy from these five co-sponsors such as was taking place at local supplier meetings.

"This conference is fantastic," remarked an Illinois no-tiller. "I picked up many new no-till ideas and was able to compare the latest weed control programs offered by these

five firms. There were no high-pressure tactics to buy as happens back home when a supplier or ag chemical dealer holds a local meeting."

A No-Till Winner

It was encouraging to hear attendees rave about the '93 conference format and the outstanding speakers. Most felt it was a fantastic learning experience and indicated they had never been part of an event on any topic that had proved as valuable.

The opportunity to learn from 45 speakers discussing 96 no-till topics in 32 hours of intense no-till learning spread over 3 days meant there were no excuses for not learning.

Four hours into the conference, an Indiana no-tiller shared this insight:

"The handout materials you gave us are outstanding, he said. "If a no-tiller only read these materials and didn't sit in on any of the sessions, he would have more than received full value from attending this conference."

The fact that more than a dozen attendees paid on the spot for the following year's NNTC in St. Louis proved the event was on the right track.

Perfect Format

What's remarkable is the fact that the overall conference format of general sessions, No-Till Classrooms, No-Till Roundtables and including time for hallway networking has met the needs of attendees for over 25 years.

Among the things we've always been proud of is how, from the first year, attendees were sold on the benefits of networking with no-tillers from around the country. While many growers love to swap ideas with their neighbors, they welcomed the national no-till view gained from one-on-one hallway conversations with peers from all over North America. These conversations with others have led

to phone calls over the years, along with numerous summer visits to other areas of the country to see how others no-till.

A good example of the value NNTC attendees place on hallway networking took place after we'd held a dozen of these events. One August, an Iowa no-tiller called to register for the next winter's conference. I explained that we hadn't finalized the program yet, he recited the same message we'd shared in 1993.

He said the fact that we didn't have the program done didn't bother him, as he knew we would have another great one. But even if we offered what he felt was a "dud program," he said he'd still come because of the value he'd gained from networking with other no-till veterans. Meeting and talking with other no-tillers was worth the price all by itself, he said.

Monsanto Backed Us

Even though Monsanto was running their own no-till conferences in '93, they were NNTC co-sponsors right from the start. They continued for a number of years until they became more of a GMO-oriented seed company with less interest in selling herbicides and insecticides.

When the first NNTC was held in Indianapolis in '93, Purdue University ag economist Howard Doster presented three topics to the more than 800 attendees at this inaugural mid-winter event.

"Frank likes to remind me that I was worried that he could not make these conferences a long-term success," says Doster. "He was right that I was concerned, as that was the same year Monsanto — with their deep pockets — decided to hold several Roundup-oriented no-till conferences across the country."

Within a few years, Monsanto stopped holding these conferences while *No-Till Farmer* has produced more than 26 (and counting) of these January events that now often attract over 1,000.

Monsanto believed in what we were doing to expand no-till.

On two occasions, Bruno Alesii, the conservation tillage guru at Monsanto, asked me to immediately invoice them for the next year's conference sponsorship — 3 months ahead of when we usually did it. Both years, he knew he was going to get his budget slashed for the coming year and wanted to make sure Monsanto continued to be a NNTC sponsor.

No Hidden Agendas

One thing that set the NNTC program apart from the Monsanto program was "no company line" for speakers. During the summer, Monsanto would have their no-till conference speakers travel to St. Louis for training. An important part of this training was to get the speakers to deliver the same no-till message. Monsanto wanted the same no-till message presented as was being pitched in their company-wide marketing and ads.

After a couple of the NNTC events, one of our speakers suggested that we take a lesson from Monsanto. That is, to get the speakers together so their messages didn't conflict with other presentations.

After thinking it over, we said no. We were convinced NNTC attendees wanted to hear all sides of every idea without any built-in agendas or biases and reach their own decisions on what would work best in their farming operations. They did not want someone else deciding what they should hear or what one company thought was best.

This unbiased and independent

NNTC speaker philosophy has worked well. We like to say, "We can give you the ingredients for a successful no-till program, but you have to write your own recipe for success on your own farm."

While providing 21,597 attendees over 26 years with an in-depth no-till education at the NNTC, a few unusual, wild and crazy things have occurred. Here are a few of my favorites.

If Your Cell Phone Rings

Allen Berry recalls how the NNTC staff grew very strict on enforcing the event's rule against cell phones going off during the presentations. The no-till veteran from Nauvoo, Ill., recalled that the usual fine was $10.

"At one of the general sessions in Des Moines, Marion Calmer is making a presentation on reducing nitrogen rates for no-till corn," says Berry. "Near the end of Marion's presentation, Frank's cell phone goes off.

"Marion looks over at Frank and says, 'That will be $10.' As Frank and Marion banter back and forth, he hands Marion's daughter $10, and then he adds, 'By the way, that was Alice Musser calling to wake me up at the end of your talk.'"

Berry says it's not often that Marion is speechless, but it happened … and with a big roar from the crowd.

Not Best Way to Start

Among attendees at the very first NNTC was a grower from Michigan's "Thumb" area. Keen to learn all he could about no-till, he'd never no-tilled a single acre.

A few weeks after the conference, he called and told me he was so convinced that no-till would work that he was selling all of his tillage equipment and several high horse-power tractors at a late February auction. He was going 100% no-till on 1,200 acres of corn and soybeans that spring, based on what he had learned at that year's NNTC.

That was a scary situation for me, as we've always told growers to transition to no-till on a small scale to make sure they can make it work.

A couple years later, the Michigan grower told me he'd been successful in plunging whole-hog into no-till based on what he learned at that winter's NNTC. Thankfully, he'd made no-till work that first year.

Tough Questions

When sponsors were determining which company staffers would attend the NNTC, early-day co-sponsor American Cyanamid always gave it serious thought. They realized their area sales personnel weren't qualified to answer the highly technical weed questions asked by our attendees.

NATIONAL NO-TILLAGE CONFERENCE BRINGS COUPLES TOGETHER

Over the years, there have been at least three instances where NNTC attendance led to marriage proposals.

♦ Julie Coors and Bob Barker became engaged during the 1996 National No-Tillage Conference in St. Louis, Mo. A rumor at the time was going around that the Jeffers, Minn., couple was planning to honeymoon at the following year's event in Des Moines.

♦ Traveling 10,500 miles from Western Australia, no-tiller Tony White and his Australian girlfriend, Julie Symons, who traveled 4,100 miles to Des Moines from England, chose the 2000 NNTC in Des Moines to become engaged. *No-Till Farmer Editor* Frank Lessiter announced their engagement to 715 attendees at the Friday evening banquet.

♦ Sandy Cox, a computer software trainer from Missouri, spotted Denny Roth, a no-tiller from Indiana in the hotel bar, during a break in the sessions during the 1999 NNTC in St. Louis. She complimented his cowboy hat; he asked her to dance. It was the beginning of the rest of their lives together.

They stayed in touch for months and got together whenever they could, including at the annual NNTC gatherings. Denny eventually sold his farm, moved to Missouri and took a job in ag sales to be closer to Sandy. They eventually tied the knot.

"I would not have met the man of my dreams had it not been for the NNTC," she says. "The NNTC and your crew will always have a special place in our hearts." ❋

Because many weed control questions were so detailed and complicated, the company required the best technical specialists to be on hand.

How Do We Tell Dad?

We were in St. Louis in 1999 for the seventh edition of this mid-winter event. A no-tiller from Iowa who hadn't missed attending the conference didn't attend that year, but instead sent his two sons.

One of our staff members was in the elevator with the two boys around 11 a.m. on the last day of the conference and overhead the brothers talking. She heard one say, "How are we going to tell Dad all we did was party and that we never went to any of the NNTC sessions?"

Cruising' Downtown Cincinnati

The proud owner of a brand-new Buick, a farmer from LaCrosse, Wis.,

made the one-way 578-mile journey to Cincinnati for the 2001 event. These were practically the first miles he'd put on the new car.

Wanting to check his fuel mileage, the grower jotted down the mileage shown on his speedometer as he drove into the underground parking garage at the Hilton Hotel in Cincinnati. So imagine his surprise when he picked up the car 4 days later and saw it had been driven 85 miles while he was attending the NNTC sessions.

Apparently, one of the garage workers enjoyed a good time cruising around downtown Cincinnati in a new Buick.

2-Minute Old Data

Sometimes NNTC speakers have fresh up-to-date meaningful data to share with attendees for the first time. But when Ellyn Taylor said he was going to deliver up to date information, he meant it.

One year in Des Moines, the farm weather expert from Iowa State University was our luncheon speaker. As I started to introduce Taylor to the luncheon crowd, he walked on stage and was fiddling around with his computer. This was when PowerPoint was still new and the hotel ballroom Internet was spotty, so I had been a little nervous.

I turned the program over to Taylor at 1:17 p.m. His first slide showed the latest Corn Belt weather that was downloaded 2 minutes earlier.

You can't beat that experience for delivering timely information!

Keep on Trucking

Bob Wildermuth has attended most of these mid-winter events. Most years, the Clinton, Wis., no-tiller doesn't drive a car or pickup truck to the event, but instead drives a semi-tractor or an extra-large truck to the different Midwestern locations.

Most importantly, he gets paid for driving to the conference.

To supplement his farm income, Wildermuth often drives for a delivery firm that transports big trucks from southern Wisconsin to cities across the country. Most Januarys, he signs up to drive a truck to a location that will take him through the city where the NNTC is being held.

His wife, Anita, normally accompanies him to the conferences. They tow a car behind the truck, so they have a convenient way to get back to Wisconsin.

Two NNTC Trips in One Year

One year in Indianapolis, a central Illinois no-tiller arrived at the hotel, but couldn't find any information about the event. He didn't find our staff, didn't see any NNTC signs, the hotel staff didn't seem to have a clue about the event and nobody was able to help him. He begin to wonder if he was at the wrong hotel.

As it turned out, he was in the right hotel, but unfortunately had arrived in Indianapolis a week early. He headed home, then made the 3-hour drive again the following week to Indianapolis to attend the conference.

'Red Hot' Visuals

Another year in St. Louis, a No-Till Classroom speaker talked for several minutes before deciding to advance the slides in his Carousel slide tray. When turning on the projector before the session got underway, he'd left a plastic cap on top of the slide tray and the heat from the projector melted the plastic and ruined all 55 of the no-till slides her had planned to show.

He completed his no-till presentation without being able to able to share any of the visuals he had prepared for his talk.

Oh, and by the Way...

Does anyone remember when the wife of a Wisconsin no-tiller brought a parrot to the NNTC one year in Indianapolis? And how the bird squawked and talked with a number of the attendees in the hotel lobby? It wasn't all no-till talk either. ❋

Charting the Course for No-Till

The "Who's Who" of No-Till Development

The page numbers refer to the first mention of these no-till leaders in chapters.

INDEX

"Every session provided at **least one new idea** that I took back to my operation."
— *Tryston Beyrer, Savoy, Ill.*

"I've gained ideas to enhance and improve my current practices, along with insight into how others have adapted their strip-till systems."
— *Mark Richards, Dresden, Ontario*

"There are tremendous opportunities for networking and visiting with a multitude of farmers, educators, suppliers and consultants."
— *Jack Boyer, Reinbeck, Iowa*

"People ask us where we get our information for all the strange things we're implementing and we tell 'em, at the National Strip-Till Conference. We go every year and plan our year around it."
— *Ryan Shaw, Marlette, Mich.*

INDEX

—This chapter was indexed by our granddaughter, Olivia Fitch

"By following crop consultant Phil Needham's suggestions, we've grown no-till wheat yields of over 100 bushels per acre along with better straw quality"
— *Donn Branton, Le Roy, N.Y.*

'IT'S THE SUPER BOWL OF AGRICULTURE'

"Everybody attending the National No-Tillage Conference is always willing to share his or her latest no-till ideas with other attendees," says veteran no-tiller Marion Calmer from Alpha, Ill. **"The latest in no-till innovation is what this annual event is all about."**

I tell everybody that each year's National No-Tillage Conference (NNTC) represents the Super Bowl of ag conferences.

You'll get far more out of life and be a lot happier if you spend your time giving back rather than taking from agriculture. And that's one of the great things that happens at this annual conference.

Everybody attending the NNTC is always willing to share his or her latest no-till ideas. We're willing to talk with growers from next door or across the country and share our latest and greatest no-till secrets on how to grow a better crop. It's this exchange of information that makes the American no-till farmer second to none in the world.

Our ability to move and increase production with ideas that come from the NNTC's farm-er-to-farmer learning opportunity is amazing and the mainstay of this event. It's the best platform anybody could ask for when it comes to moving agriculture in the right direction.

A special thanks to all of the *No-Till Farmer* staff for producing this amazing conference over the years for innovative no-tillers. We've learned many things over the past 26 years about no-till that we've been able to teach to the younger generation of farmers.

The younger generation is starting to talk about cover crops, soil biology and all of the innovation with the electronics in the cab. That's a whole new group of no-tillers that will be there for the next 25 years, representing another group of innovative farmers that'll help drive farming to new levels of efficiency and production.

I tell everybody that you'll hear about the newest no-till innovations at this conference. Then a few years later, you'll see many of these new cropping ideas becoming common practices throughout the country.

No-till innovation is what this event is all about.

— *Marion Calmer, Calmer Farms, Alpha, Ill., and a no-tiller who has attended 24 National No-Tillage Conferences*